Trans Fats in Foods

Trans Fats in Foods

Editors

Gary R. List
USDA, NCAUR
Peoria, Illinois

David Kritchevsky, PhD
The Wistar Institute
Philadelphia, Pennsylvania

W.M. Nimal Ratnayake, PhD
Health Canada
Ottawa, Ontario

AOCS PRESS
Urbana, Illinois

AOCS Mission Statement

To be a global forum to promote the exchange of ideas, information, and experience, to enhance personal excellence, and to provide high standards of quality among those with a professional interest in the science and technology of fats, oils, surfactants, and related materials.

AOCS Press, Urbana, IL 61802

ISBN 978-1-893997-96-7

Library of Congress Cataloging-in-Publication Data

Trans fats in foods / editors, Gary R. List, David Kritchevsky, W.M. Nimal Ratnayake.
p. ; cm.
Includes bibliographical references and index.
ISBN-13: 978-1-893997-96-7 (pbk. : alk. paper)
1. Trans fatty acids. 2. Lipids in human nutrition. I. List, Gary R. II. Kritchevsky, David, 1920-2006. III. Ratnayake, W. M. Nimal.
[DNLM: 1. Trans Fatty Acids--adverse effects. 2. Trans Fatty Acids--analysis. 3. Trans Fatty Acids--metabolism. QU 90 T772 2007]
QP752.T63.T73 2007
612.3'97--dc22

2007001123

Printed in the United States of America.
11 10 09 08 07 5 4 3 2 1

The paper used in this book is acid-free and falls within the guidelines established to ensure permanence and durability.

Printed with soy-based ink.

Dedication

On November 20, 2006, Dr. David Kritchevsky passed away at the tender age of 87 years. This book is dedicated to the memory of this remarkable scholar, mentor, teacher, and leader. His search for truth coupled with his wit and compassion for his fellow man have been an inspiration to all who have been fortunate enough to have known him.

Contents

Preface

In recent years there has been a remarkable growth of interest in the possible adverse health effects associated with consumption of *trans* fatty acids. Studies conducted since the 1990s have consistently shown that consumption of *trans* fats formed during commercial partial hydrogenation increase the risk of heart disease and are more harmful than any other fat. Metabolic studies for instance show that *trans* fats increase blood levels of LDL/HDL-cholesterol to a greater extent than saturated fats. Most of the observational epidemiologic data also point to a greater risk of coronary heart disease from increases in dietary *trans* fat than increases in dietary saturated fats. Although some researchers differed somewhat in their interpretation of the scientific evidence on negative health effects of *trans* fats, various professional health organizations, government health agencies and health conscious consumers have taken notice of the negative health effects of *trans* fats. The American Heart Association and the Panel on Macronutrients of the U.S. National Academies of Science, Institute of Medicine, recommended that *trans* fat consumption be as low as possible. World Health Organization in 2003 recommended that *trans* fat intake be limited to 1% of overall energy intake. In 2004, the Danish government prohibited the use in foods of oils containing more than 2% of industrially produced *trans* fat. In 2005, Canada became the first country to regulate the mandatory labeling of *trans* fat and in 2006, the United States introduced the mandatory declaration of *trans* fats in foods containing 0.5 grams or more per serving.

Although volumes of data have been accumulated on a wide range of issues related to industrially produced *trans* fatty acids, still there are several gaps in our knowledge on *trans* fats, in particularly their role in health and disease. This book captures status of current knowledge on the biochemistry, health effects, analysis, consumption, healthier alternatives to *trans* fats in food formulations, and food labeling legislature of industrial *trans* fats. Over the last 15 years remarkable diversity of biological properties and beneficial effects of conjugated linoleic acid have also been reported. These findings are also critically examined in this book. Currently there is only a limited information on the biological properties and health effects of naturally occurring *trans* fatty acids of dairy products and ruminant meat. This book has devoted a chapter highlighting the current knowledge on the chemistry and health effects of natural *trans* fats.

We hope that the subjects covered in this book will clarify some of the controversial issues on *trans* fat and would benefit consumers, health professionals, health administrators, researchers, students, and the food industry. As the editors of this book, we would like to express our profound gratitude to all the contributors for taking their valuable time and up-to-date contributions and to the AOCS staff for their valuable support in publishing this book.

Gary R. List
David Kritchevsky
W.M. Nimal Ratnayake

Introduction

Trans fatty acids continue to interest the edible oil industry, food processors, health professionals, and the general public. This book has attempted to bring together the latest information on the effect of *trans* fatty acids, health and nutrition, labeling, analysis, and food products. The book begins with a review of the effects of *trans* fatty acids on cardiovascular disease followed by the role of and effects of *trans* fatty acids in cancer. A chapter on the safety of conjugated linoleic acid is included. The literature on the metabolism of *trans* fatty acids compared to non-isomeric fatty acids is reviewed; chapters then follow on the consumption of *trans* fats in North America and the consumption and nutritional characteristics of ruminant *trans* fats. Literature on the reformulation and characteristics of low/zero *trans* fat food products is reviewed. Labeling of *trans* fatty acid is covered. The book concludes with a chapter on the use silver-ion HPLC for the analysis and cauterization of *trans* fats. The editors and contributing authors feel confident that the material assembled here will be useful to those working in the field of fats and oil, food technologists, dieticians, and the medical community.

Gary R. List

Chapter 1

Trans Fatty Acid Effects on Cardiovascular Disease: Animal and Human Studies

Shirley C. Chen[1], David Kritchevsky[2], David J. Baer[1]

[1] Beltsville Human Nutrition Research Center, USDA, Beltsville, MD; [2] Wistar Institute, Philadelphia, PA

Acknowledgements

Supported, in part, by a Research Career Award (HL00734) from the National Institute of Health (USA) and the Commonwealth Universal Research Enhancement Program, Pennsylvania Department of Health (DK)

Introduction

Trans fatty acids (TFA) occur naturally in plants and animals (Sommerfeld, 1983). Dietary TFA arise naturally as the result of biohydrogenation of polyunsaturated *cis*-fatty acids by ruminant bacteria with *trans*-$C_{18:1}$(n-7) (vaccenic acid) being the major product. Other 18:1 positional isomers, such as *trans*-$C_{18:1}$(n-9) (elaidic acid) and small amounts of *cis*-9,*trans*-11 conjugated linoleic acid (CLA) are also found in *trans* fats of ruminant origin. However, commercially manufactured *trans* fats make up the bulk of the dietary *trans* fat in modern diets. Partial hydrogenation of vegetable oils produces numerous geometrical and positional isomers of $C_{18:1}$ and $C_{18:2}$ fatty acids; most of them are $C_{18:1}$ monoenes having the double bonds between n-6 to n-16 with a center around n-7, n-8, and n-10 (Emken, 1995). The same TFA isomers are found in ruminant and industrial fats, but in different profiles. Vaccenic acid is the major component of the TFA in ruminant fat.

Although the name elaidic acid has been used commonly as a generic term to describe $C_{18:1}$ TFA, this *trans*-monoene counterpart of the most prevalent $C_{18:1}$ *cis* monoene, oleic acid [*cis*-$C_{18:1}$(n-9)], is in fact not the most prevalent *trans* monoene in commercially available partially hydrogenated vegetable oils in the U.S. market. Study of the cardiovascular effects of a single *trans* fatty acid species has not been studied in humans. In cell culture it has been shown that the effects on lipoprotein secretion of *trans* and *cis* fatty acids having 16 and 18 carbons are heterogeneous and are determined by the carbon chain length, position of the double bond, and the stereochemical configuration (van Greevenbroek et al., 1998). Thus it is important for readers to bear in mind that unless otherwise specified the generalized terms "elaidic acid," "dietary *trans* fatty acids," or "dietary *trans* fats" used in experimental, clinical, and epidemiological studies collectively represent a mixture of *trans* fats of undefined composition. Although a wide variety of TFA occur in nature, and in our diet, virtually all the data discussed in this review are derived from studies conducted using hydrogenated vegetable or fish oil. Effects unique to CLA are to be covered in a separate chapter.

This review is organized in three sections. First the atherogenic and cholesterolemic effects of TFA in experimental animals are reviewed briefly. Next, epidemiological studies published since 1995 are reviewed individually with emphasis placed on the strength and validity of the calculated association. Finally, clinical and metabolic studies investigating specific cardiovascular disease risk markers consequent to designed dietary changes are reviewed. These studies encompass the effects on blood lipids and lipoproteins; effects on physiological events including immune function and inflammatory responses, hemostasis and endothelial function, glycemic and insulin responses, and blood pressure.

Animal Studies

Cholesterolemic and Atherogenic Effects

Feeding *trans* fats to animals affords researchers the opportunity of observing the impact of *trans* fats directly on specific cardiovascular disease end points, whereas a similar approach cannot be applied to human clinical and epidemiological studies.

Kritchevsky (1983, 1990) reviewed effects of elaidic acid (versus oleic acid), elaidinized olive oil (versus olive oil), linoelaidic acid (versus linoleic acid) and hydrogenated soybean oil (versus soybean oil) in rabbits, swine, and vervet monkeys fed with or without cholesterol for various lengths of time up to 12 months. The levels of *trans* fat in the diets ranged from 0.448–0.84% (*w/w*) to 6g/d elaidic acid for rabbits, 0–8.16% (*w/w*) for swine and 0.048–0.84% (*w/w*) for monkeys. The laboratory of Kummerow (Jackson et al., 1977; Royce et al., 1984; Toda et al., 1985) studied the effects of *trans* fats derived from partially hydrogenated soybean oil (versus lard, beef tallow, high oleic safflower oil) in swine at 1.69–7.76% (*w/w*) of the diet for 4–6 months. In all these studies, serum levels of cholesterol in *trans* and animal fat fed animals were usually higher than those of the control animals. However, *trans* fat-fed and controls were comparable in their atherogenic scores and were less atherogenic than beef tallow and lard.

Mortensen et al. (1992) reported no differences in plasma lipoprotein levels, and atherosclerotic changes in the proximal and distal coronary arteries, abdominal aorta and renal arteries of rabbits fed standard diet and diets with 10% olive oil or 10% margarine added. Their observations were confirmed by Nicolosi et al. (1998) in hamsters fed diets containing uniform triglycerides at 10% (*w/w*). They reported that the plasma lipids of *trans*-$C_{18:1}$ and *cis*-$C_{18:1}$ fed hamsters were comparable and lower than those of the $C_{14:0}$ fed while the *trans*-$C_{18:1}$ fed had the least and $C_{14:0}$ fed had the highest levels of early atherosclerosis. They also reported that the antioxidant properties, as measured by the rate and lag phase of conjugated diene formation and α-tocopherol content of LDL, were similar between the *trans*- and *cis*-18:1 fed animals.

Using a mouse model having LDL as the predominant lipoprotein and thus more akin to humans, Rudel et al. (1998) compared the atherosclerotic outcomes of different fat sources. The group fed the 4% (*w/w*) of a particular *trans* blend (which was also high in saturated fat) had the highest plasma TC, VLDL, and LDL cholesterol levels, and aortic free and esterified cholesterol content than those fed either 4% (*w/w*) palm oil or 4% (*w/w*) oleic acid-rich safflower oil. The authors concluded that *trans* fats were more atherogenic and hypercholesterolemic than either the saturated fat or *cis*-mono fats which were not different. A closer look at the diet composition indicates that although the test diets had the same fat content, the fatty acid composition of the specific pairs of diets compared was unbalanced and skewed against the "*trans* Mono" diet. For example, the total cholesterol-raising fat content of the "*trans* Mono" diet was higher than those of the "*cis* Mono" and "Sat" diets, i.e. the "*trans* Mono" diet contained 2.8 en% TFA and 2.7 en% of saturated fatty acids, whereas the "*cis* Mono" diet had a negligible amount of TFA and 1.2 en% saturated fatty acids, and the "Sat" diet had no TFA and 4.8 en% of saturated fatty acids. Thus, a quantitative conclusion of the cholesterolemic and atherogenic effects of different fatty acid types can not be made conclusively. That *cis*-mono fat is more atherogenic and cholesterolemic than the polyunsaturated fats is the most interesting finding of this study.

Lipoprotein Metabolism

The plasma concentration of cholesterol is determined by a number of metabolic factors, among them the activities of acyl-coenzyme A cholesterol acyltransferase (ACAT), cholesterol ester transfer protein (CETP), and LDL receptor activity. The effects of TFA on these factors have not been examined thoroughly in animal systems.

Sgoutas reported in 1970 that *cis*-$C_{18:1}$(n-9) followed by *cis*-$C_{18:1}$(n-7) are preferred substrates for rat liver microsomal ACAT whereas *trans*-$C_{18:1}$(n-9), *trans*-$C_{18:1}$(n-7), *cc*-$C_{18:2}$, and *tt*-$C_{18:2}$ are not (Sgoutas, 1970). That TFA are less likely to be shifted to the storage pool may be an explanation for the *trans* fat consumption-related elevation of plasma cholesterol. The role of hepatic cholesterol esterification in regulating plasma cholesterol level was reaffirmed by Woollett et al. (1994) a quarter century later. While the latter researchers demonstrated in hamsters that LDL concentration is inversely proportional to the hepatic LDL receptor activity they also showed that hepatic LDL receptor activity is up regulated by *cis*-$C_{18:1}$(n-9), and down regulated by saturated fat ($C_{14:0}$), but not by *trans*-$C_{18:1}$(n-9). The data of Woollett et al. suggest that *trans* fats behave similarly to, but are less hypercholesterolemic than, saturated fat in hamsters.

A study by Dictenberg et al. (1995) found higher plasma lipid levels and lowered liver *cis*-$C_{18:1}$ CE levels and *cis*-$C_{18:1}$CE/*cis*-$C_{18:2}$CE ratios in $C_{12:0}$+$C_{14:0}$, $C_{16:0}$, and *trans*-$C_{18:1}$ fed than in *cis*-$C_{18:1}$ fed young gerbils fed predetermined amounts of diets for growth promotion. In mature gerbils fed weight maintaining *cis*- or *trans*-$C_{18:1}$ diets, plasma lipids were not different after 8 weeks, but the *trans* fat fed gerbils had lowered plasma and hepatic *cis*-$C_{18:1}$/*cis*-$C_{18:2}$ CE ratios. The data are consistent with the suggestion that *trans* and saturated fats behave similarly in regulating the *cis*-$C_{18:1}$ CE pool.

Khosla et al. (1997) reported an increased CETP activity and a lower HDL in cebus monkeys fed a diet enriched with elaidic acid as compared with those fed a similar diet enriched in palmitic acid. The LDL kinetics of the two groups of monkeys were similar suggesting that saturated and *trans* fats exert similar effects on LDL metabolism. The authors interpreted the discrepancy between animal data and human data on LDL response to *trans* fat to the inherent difference in LDL receptor activity and LDL pool size among species.

A study which directly compared elaidic, vaccenic, oleic, and palmitic acid at 10 en% on lipid metabolism was conducted in hamsters by Meijer et al. (2001). Similar to results reported by Nicolosi et al. (1998), comparable LDL cholesterol levels were found in the two TFA and oleic acid fed animals, and higher LDL cholesterol levels in the palmitic acid fed group. Interestingly, the LDL/HDL ratios of the vaccenic and palmitic acid fed hamsters were similar and higher than elaidic acid fed hamsters suggesting a greater cardiovascular disadvantage associated with saturated fat and vaccenic acid than with elaidic acid. On the other hand, CETP activity was similar among all treatment groups. Total liver cholesterol concentration was highest in elaidic acid fed hamsters and lowest in the palmitic acid fed hamsters, although the FC/CE ratio was unaffected. LCAT activity was higher in hamsters fed palmitic or vaccenic acid and lowest in those fed elaidic acid. Altogether, the findings suggest that *trans* and saturated fats exert similar metabolic effects on lipoproteins in laboratory animals, and that elaidic acid is not more harmful to the cardiovascular system as suggested by epidemiological observations (Willett et al., 1993).

Gatto et al. (2002) investigated the effect of *trans* fat on reverse cholesterol transport in rats, a species having predominant lipoproteins in HDL and no cholesterol ester transferase activity, by feeding them diets containing 10 en% saturated, *cis*, or *trans* fat. Compared with *cis*-fat fed rats, *trans* fed rats had lower plasma TC and non-HDL-C, but comparable HDLC. Their observations suggest a role of TFA in regulating apoB-containing lipoproteins.

In summary, the number of animal studies is small and their results suggest a hypolipidemic effect but no exacerbation of atherogenesis. It should be noted that animal experiments are designed to study dietary effects on only one or two markers of CHD and not the array of indicators in human studies. However, the advantage of animal studies is the possibility of performing a necropsy and observing the effects of the experimental modalities on actual coronary disease.

Epidemiological Studies

Studies Prior to 1995

Up to the early 1990s, there were a number of case-control studies examining the relationship of tissue concentration of TFA or *trans* fat intake and clinical risk factors or CAD incidence (Heckers et al., 1997; Thomas et al., 1983, 1987a, 1987b; Hudgins et al., 1991; Triosi et al., 1992; Siguel et al., 1993; Tzonou et al., 1993). Not all identified potential confounders were controlled in these studies and inconsistent or statistically insignificant results were found. These studies have been reviewed by Allison (1995).

Epidemiological studies examining the relationship of intakes of *trans* fats or TFA and CHD incidence, outcome, and risk factors proliferated after Willett and Ascherio concluded that TFA intake is causally related to risk of coronary disease, and advocated publicly for food labeling of *trans* fat content and a reduction of the use of partially hydrogenated vegetable fats (Willett and Ascherio, 1994). These researchers reached the conclusion after examining available human metabolic study data and their own findings of a positive correlation between the intake of foods that are major sources of *trans* isomers and incidence of CHD in a prospective study involving a large cohort of women (Nurses' Health Study) (Willett et al., 1993), and a case-control study in which some confounders were controlled (Ascherio et al., 1994). Intriguingly, a saturated fat effect was not found in these two epidemiological studies (and therefore not adjusted for) and the relative risks for CHD which did not correspond linearly to *trans* fat intake, and overlapped over four of the lower five quintiles (Aro et al., 1995), or all quintiles (Roberts et al., 1995).

Studies Since 1995

Retrospective Studies

Studies Employing Biomarkers

Adipose Tissue TFA Content

Adipose tissue content of TFA is considered a good indicator of long-term dietary intake status. Based on dietary versus adipose tissue *trans*-$C_{18:1}$ levels in rats and mice, Enig et al. (Enig et al., 1990) derived the equation of y (% adipose tissue fat as *trans*-$C_{18:1}$) = 0.97 + 0.44 x (dietary intake of *trans*-$C_{18:1}$ as % of fat) with a correlation coefficient at 0.898. A recent study reported that in humans the best indicators of total TFA intake were *ct*- and *tc* $C_{18:2}$(n-6) (r=0.58 for each), followed by total *trans*-$C_{18:1}$ (r=0.45) and *trans*-$C_{16:1}$ (r=0.16) (Baylin et al., 2002).

EURAMIC Study

In a multicenter study involving eight European countries and Israel (EURAMIC), Aro et al. (1995) investigated the effect of dietary $C_{18:1}$ TFA as reflected in the adipose tissue TFA on risk of acute myocardial infarction by analyzing the content of $C_{18:1}$ TFA in adipose tissue aspiration samples from 671 men with acute myocardial infarction and 717 control men. The odds ratios were determined after controlling for age, BMI, smoking, and socioeconomic status. The adipose *trans*-$C_{18:1}$ content was not different between cases and controls. Reflecting the differences in the habitual fat sources, subjects in the two Spanish centers had significantly lower percentages of adipose *trans*-$C_{18:1}$ and higher oleic acid than subjects in other centers. After subjects from Spain were excluded there was a tendency towards increased risk of acute myocardial infarction in the upper quartiles of *trans*-$C_{18:1}$, but the trend of the multivariate adjusted quartile odds ratios was not significant. The authors concluded that *trans*-$C_{18:1}$, at the intake range exhibited by subjects in the study, did not appear to exert a major effect on risk of acute myocardial infarction. However, they acknowledged that the possibility of a significant impact on risk of acute myocardial infarction in populations with high intake cannot be excluded.

Southampton Study

The adipose tissue content of *trans* isomers of oleic and linoleic acid in 66 cases of sudden cardiac death was compared to that of 286 age and sex matched controls by Roberts et al. (1995). The mean percentage of total TFA ($C_{18:1}$ plus $C_{18:2}$) and individual TFA, expressed as a proportion of all fatty acids, were significantly lower in cases than in healthy controls ($p<0.05$). The estimated univariate relative risk for sudden cardiac death of *trans*-$C_{18:1}$ and *trans*-$C_{18:2}$ fatty acids individually or in combination did not differ significantly from 1.0 in relation to the distribution of these *trans* isomers by quintile in the control population. Despite additional multivariate adjustments (for age, reported smoking, history of hypertension and diabetes, *cis*-oleic and *cis*-linoleic acids) an association of risk with individual TFA or the sum of *trans*-$C_{18:1}$ and *trans*-$C_{18:2}$ could not be established.

Because the conclusions of the two studies cited, where adipose tissue *trans* fat content was used as a marker of dietary intake, are in stark contrast to those made by Willett et al. (1993) and Ascherio et al. (1996) where *trans* fat intakes were estimated by semi-quantitative questionnaires, it is understandable that a number of relevant methodological issues inherent to nutritional epidemiology were put to question, such as potential biases associated with semi-quantitative questionnaires and the capability of epidemiological studies in ascertaining the association of a single dietary component and the interested biological effect(s) when other known and unknown dietary and non-dietary factors of free living populations consuming habitual diets are not or cannot be controlled. It is also another affirmation of the hypothesis generating and not causal nature of epidemiology. That issues identified and debated after the publication of these four papers were taken seriously by epidemiologists is evident, as they have been more vigilantly attended to in subsequently published studies.

Other Studies

Three case-control studies conducted by researchers in three continents reported a positive association between adipose tissue content of TFA and first acute myocardial infarction (MI). In a study of 100 patients and 98 controls, both men and postmenopausal women in Norway, Pedersen et al. (2000) found a significant positive association of adipose content of total TFA with risk of MI. The odds ratio remains significant after adjustment for age, sex, waist-to-hip ratio, smoking, family history of CHD. A similar inverse relation was found individually for adipose content of α-linoleic and α-linolenic acid. The individual correlation between each of these three fatty acids and MI risk disappeared when each was adjusted for the two others. This is because the adipose tissue levels of these three fatty acids were strongly intercorrelated. At the same time, adipose content of very long-chain n-3 fatty acid (20:5, 22:5, and 22:6) was inversely related to MI risk and not affected by adjustment for TFA.

Dietary intake was assessed using a food frequency questionnaire developed for Norwegians, official Norwegian food composition table and the TRANSFAIR fatty acid composition table. There were significant individual correlations between dietary intake and adipose tissue content of very long-chain n-3 fatty acids ($r=0.573$), linoleic acid ($r=0.492$), α-linolenic acid ($r=0.437$), and TFA ($r=0.295$). The observations of an intercorrelation for adipose tissue content of *trans*, of α-linoleic and α-linolenic acid, and an inverse relation between adipose tissue levels of these fatty acids and MI led the authors to suggest that the increased risk associated with these three fatty acids may be connected to TFA or to some other factor associated with margarine consumption.

In Costa Rica, Baylin et al. (2003) assessed the relationship between adipose tissue TFA and the risk of MI in 482 cases and equal number of controls of both genders. The positive association between TFA and MI was attributed mainly to *trans*-$C_{18:2}$, and secondarily to *trans*-$C_{16:1}$. No effect of *trans*-$C_{18:1}$ was found. The authors suggested that the reason for the lack of association between *trans* fat intake and MI in some prior studies may be accounted for by the low intake of *trans*-$C_{18:2}$ in the diets of the populations studied. This is an interesting observation and requires further research. A significant implication to public health not to mention the economic implication to the oil refining industry and consumers can be expected if *trans*-$C_{18:2}$ and *trans*-$C_{16:1}$, not *trans*-$C_{18:1}$, are to be avoided, because partially hydrogenated vegetable oils are the main source of *trans*-$C_{18:1}$ and *trans*-$C_{18:2}$ and the amounts of *trans*-$C_{18:1}$ and *trans*-$C_{18:2}$ may be process-controlled industrially, whereas animal fat is the main source of *trans*-$C_{16:1}$ and its content can only be reduced through reducing the intake of animal products.

The study conducted by Clifton et al. (2004) in Australia examined, in 209 cases collected from 1995–1997 and 179 controls, the association of MI and adipose tissue content and dietary intake of TFA. Cases biopsied before mid-1996, when TFA were eliminated from margarines sold in Australia, had greater levels of *trans*-$C_{18:1}$(n-9) (elaidic acid) and *trans*-$C_{18:1}$(n-7) (vaccenic acid) than controls. TFA intake was estimated based on a 300-item FFQ, and found to be well correlated with TFA intake from margarine before mid-1996. After mid-1996, adipose tissue content of all TFA was not different between cases and controls. In a logistic regression model relating adipose tissue fatty acids to MI, only *trans*-$C_{18:1}$(n-7) was identified as an independent predictor. It was also a predictor of elevated blood pressure, total cholesterol, and triglyceride. It may be reiterated that *trans*-$C_{18:1}$(n-7) is present in both animal and partially hydrogenated vegetable fats, whereas *trans*-$C_{18:1}$(n-9) comes largely from partially hydrogenated vegetable fats. Cases consumed higher quantities of energy, saturated fat and dairy *trans* fat, and comparable amounts of TFA from margarine. Total TFA intake, as % of energy, did not differ between cases and controls. The risk of MI in the highest quintile of TFA intake was more than two times that of the lowest quintile. However, the association between TFA intake and MI was abolished after adjustment for saturated fat intake. Neither total TFA consumed nor the percentage energy from TFA was an independent risk factor for first MI using logistic regression. On the other hand, the percentage of energy from saturates and linolenic acid, respectively, and total energy intake were independent predictors.

Similar to reports made by researchers in Norway and Costa Rica, the findings from Australia support an association between adipose tissue level of TFA and MI. The Australian researchers did not report dietary and adipose *trans*-$C_{18:2}$ levels, possibly because the levels were too low to be detected. However, using the same approach, the different research groups came to different conclusions on the relative effects of *trans* fats of vegetable versus animal origin. Until prospective adipose tissue fatty acid data are available, data generated from retrospective studies, which are prone to bias, should be used with caution.

TFA Content in RBC and Platelet Membranes, and Plasma Phospholipids

It has been observed that there is a direct relationship between high circulating free fatty acid concentration and sudden cardiac death in humans, and arrhythmia in humans and animals. The mechanism of this association has been proposed to be altered membrane function consequent to an exchange between circulating and membrane fatty acids (Katz, 2002). Saturated fatty acid has been associated with arrhythmias in primates. Because physical properties of TFA resemble those of saturated fatty acids, Lemaitre et al. (2002) investigated the association of TFA intake (assessed through its content in red blood cell (RBC) membrane) with the risk of primary cardiac arrest (assessed through the comparison of cases ($n=179$)

identified, over an 11-year period from 1988–1999, from emergency service incident reports of Seattle and suburban King County, Washington, and randomly selected age and sex controls (n=285) from the same community). Cases were free of previous clinically diagnosed heart disease, and a blood sample was obtained at the time of cardiac arrest. No differences between cases and control were found in their RBC content of total TFA, *trans*-$C_{18:1}$ or *trans*-$C_{16:1}$ fatty acid. Mean RBC levels of *trans*-$C_{18:2}$ were higher and very long-chain n-3 fatty acids (DHA+EPA) were lower in cases than controls. Other traditional risk factors for primary cardiac arrest, such as present smoking, diabetes, hypertension, and family history of myocardial infarction or sudden cardiac death were more prevalent in cases. After adjusting for traditional primary cardiac arrest risk factors, RBC membrane levels of total TFA were associated with a modest increase in risk of primary cardiac arrest. However, the magnitude of association was reduced after adjustment for RBC membrane levels of very long chain n-3 fatty acids. Most importantly, consequent to the n-3 fatty acid adjustment, *trans*-$C_{18:1}$, mainly *trans*-$C_{18:1}$(n-9) and *trans*-$C_{18:1}$(n-8), were not associated with risk (odds ratio for interquintile range, 0.8, CI, 1.0–1.2), whereas *trans*-$C_{18:2}$ continued to be highly associated with risk (odds ratio 3.1, CI, 1.7–5.4). No information is available relating to the effect of *trans*-$C_{18:1}$ relative to *trans*-$C_{18:2}$, and the authors did not offer a potential mechanism for the observed risk of *trans*-$C_{18:2}$ and the observed no-risk of *trans*-$C_{18:1}$. In addition, it is unknown if the observed effects of total, and *trans*-$C_{18:1}$ and *trans*-$C_{18:2}$ TFA in this study would be valid after adjustment for other dietary components.

To derive at TFA intake which was not reported in the publication, the authors administered, *post hoc*, a 17-processed food frequency questionnaire to spouses, and linked the intake with the USDA Special Purpose table (USDA, 1993), which reports *trans* fat content of foods collected and analyzed from 1989–1993. It is fair to question the reliability of the intake data, and consequently the validity of RBC membrane as a biomarker of intake because the small number of food items in the food questionnaire would probably have underestimated intake, and recall biases are expected from the spouse recalls of past intake frequencies. Finally, for cases collected after 1993, intakes would have been overestimated as *trans* fat content of margarine, a major source of TFA prior to 1993, has been reduced significantly since. Thus, it is reasonable to conclude the presence of an association between RBC membrane *trans*-$C_{18:2}$ level and cardiac arrest, and not prudent, as these authors did, to conclude the presence of an association of *trans*-$C_{18:2}$ intake with cardiac arrest.

Mensink and Hornstra (1995) noted a good correlation between the *trans*-$C_{18:1}$ content in platelet phospholipids correlated with that in the diet based on a short-term feeding study in men. They suggested its use as an indicator for short-term intake. Subsequently, Hodgson et al. (Hodgson et al., 1996) reported a correlation between platelet content of *trans*-$C_{18:1}$(n-9) (elaidic) as well as *trans*-$C_{18:1}$(n-8), but not *trans*-$C_{16:1}$(n-7), *trans*-$C_{18:1}$(n-6), *trans*-$C_{18:1}$(n-7) (vaccenic), *tt*-$C_{18:2}$(n-6), and total TFA with coronary atherosclerosis assessed by angiography. However, the significance of their finding is unclear because no adjustment was made for any potential confounders. On the other hand, van de Vijver et al. (1996) found no difference in TFA concentration of plasma PL between cases ($n=83$) who had major angiographically documented severe CHD (>80% stenosis) and matched controls ($n=78$) who had minor stenosis on the coronary angiography (<50% stenosis) and no correlations between % TFA in plasma PL and plasma LDL or HDL cholesterol levels. Their data were derived by measuring platelet PL. They concluded that TFA intake is not associated with the risk for CHD.

Dietary Intake Studies

Seven Countries Study

The Seven Countries Study (Kromhout et al., 1995) analyzed the association between intake of individual fatty acids and dietary cholesterol, and serum cholesterol and 25-year mortality from coronary heart disease in 16 cohorts spread among seven countries to establish the determinants of the long-term inter-cohort differences in coronary heart disease mortality. Although food consumption patterns of the 16 cohorts over time were available, the detailed dietary fatty acid composition information was available only in 13 cohorts. Thus, the researchers decided to reconstruct in 1987 the cohort nutrient intakes at baseline (1958–1964) by analyzing in a central laboratory locally collected food composites representing the average baseline food intake of each cohort. The fatty acid composition of retrospectively collected foods was considered to be closely representative because good correlation exists between the two determinations for the 13 cohorts. The authors report a significant population association between *t*18:1 intake and serum cholesterol ($r=0.70$) as well as *trans*-$C_{18:1}$ intake and long-term coronary heart disease mortality ($r=0.78$). The corresponding correlations were, respectively 0.84 and 0.84 for lauric acid, 0.81 and 0.86 for myristic acid, 0.62 and 0.81 for palmitic acid, 0.61 and 0.84 for stearic acid and 0.71 and 0.88 for total saturated fatty acids. These findings are consistent with known effects of saturated fatty acids on serum cholesterol, and suggest similar magnitudes of impact from *trans* and saturated fatty acids. However, because the average intakes of individual saturated fatty acids, TFA, and dietary cholesterol of the different cohorts were highly correlated, the independent effect of the individual fatty acids and dietary cholesterol and coronary heart disease mortality could not be analyzed in multivariate models.

Framingham Study

In 1966–1969, one single 24 h dietary recall was obtained from 832 men who were free of cardiovascular disease and participating in the Framingham Study. The cumulative incidence rates of CHD for these men over categories of margarine intake over 21 years were estimated by Gillman et al. (1997) to test the hypothesis that margarine intake increases the risk of CHD. Margarine intake was categorized into 0, 1–4 tsp/day and ≥ 5 tsp/day. Among the three margarine intake groups, no difference was found in age, blood total and HDL cholesterol levels, BMI, systolic blood pressure, physical activity index, or energy intake. Those who consumed no margarine consumed more saturated fat but less total fat. Within each intake level, incidence rates were not different among users of each of the three types of margarine identified (P/S <1, P/S>1, and a reduced fat margarine made by aeration of regular margarine), thus the data from all types of margarine were combined. CHD incidence was adjusted only for age. An age-adjusted increase in cumulative incidence rate with margarine intake but not with butter intake was found. Further, there was no apparent association with margarine intake in the first 10 years but a strong direct relation in the second 11 years.

There are a number of obvious methodological issues with this research. Most critically is the combining of data from all types of margarine within each margarine intake category. The fatty acid composition of the three types of margarine was not reported but was most likely significantly different in that the saturated fat content was high and *trans* fat content was low in the high P/S margarines, and the converse was true for the low P/S margarines. This is because both saturated and *trans* fats can provide the needed hardness for stick margarine which predominated in the marketplace in the 1960s, and it is apparent that saturated fat provided the hardness for margarine having low P/S ratios, and *trans* fat provided the hardness for the margarine having high P/S ratios. By lumping different types of margarines together, the fatty acid composition of margarine becomes uncertain and is likely inconsistent among intake groups. There is no ground to support the authors' attribution to TFA from margarine for an increased risk of developing CHD as it is entirely possible that the high margarine users had consumed margarines that were low in *trans* fat and high in saturated fat. The other issue pertains to failure in recognizing and accounting for the known changes in fat level and composition of margarine, particularly in the second 10 years, and potential changes in other lifestyle and medical factors over the 21-year period. For these reasons, results reported are not helpful in providing insight into the CHD effect of *trans* fats.

TRANSFAIR Study

TRANSFAIR Study, a multicenter cross-sectional study, examined the association of TFA intake and CVD risk factors (van de Vijver et al., 2000). 327 men and 299 women aged 50–65 years from eight European countries (Finland, France, Greece, Iceland, Portugal, Spain, Sweden, and the Netherlands) participated in the study. The researchers, using a market basket approach, collected, from each of 14 predetermined European countries, up to 100 food products representative of fat intake and contributing to 95% of total fat intake, and analyzed their fatty acid composition, including seven *trans* isomers, in a central laboratory. Dietician-administered and cross-checked meal pattern-based dietary history was taken from participants. The energy and nutrient intakes of participants were then assessed using the TRANSFAIR database and national operational databases. All subjects provided a 12 h fasting blood sample, and a sub-sample from each participating center (n=322) provided subcutaneous adipose tissue. All blood lipids and adipose tissue were analyzed at central laboratories. The dietary intake of TFA was validated by adipose tissue fatty content; Spearman correlation coefficients were 0.67 for total TFA, 0.36 for *trans*-$C_{16:1}$, and 0.63 for *trans*-$C_{18:1}$. In all, the data collection was meticulously planned and executed.

Mean (± SD) intake of TFA was 2.40 ± 1.53 g/day (0.87 ± 0.48 en%) for men and 1.98 ± 1.49 g/day (0.95 ± 0.0.55 en%) for women. Iceland had the highest intake at 1.47 and 1.65 en% for men and women, respectively. Intakes by men and women in the other northern European countries (Finland, Sweden, and the Netherlands) were about 1 en%, and intakes for men and women in the southern European countries (Greece, Portugal, and Spain) ranged from 0.52–0.76 en%. These values are significantly lower than those reported for Americans (Allison et al., 1999; Satchithanandam et al., 2004) and most importantly lower than those previously reported for Europeans, reflecting the outcome of the TFA reduction effort of the European food industry in the latter part of the 1990s (Hulshof, et al., 1999).

The researchers found no associations between total TFA intake and total, LDL, HDL, or LDL/HDL ratio after adjustment for sex, age, smoking status, number of cigarettes, country, energy intake, family history of CVD, angina pectoris, BMI, waist/hip ratio, and alcohol intake. Interestingly, after adjusting for other fatty acid clusters, an inverse association (p=0.03) was found for total TFA intake and total cholesterol level. When the same statistical approach was applied to individual TFA, the following was found: LDL was positively associated with *trans*-$C_{14:1}$(n-5), and inversely associated with *trans*-$C_{18:1}$ and *trans*-$C_{18:2}$; HDL was inversely associated with *trans*-$C_{16:1}$(n-7); LDL/HDL ratio was positively associated with *trans*-$C_{14:1}$(n-5), *trans*-$C_{16:1}$(n-7) and *trans*-$C_{22:1}$; total cholesterol was positively associated with *trans*-$C_{14:1}$(n-5) and *trans*-$C_{22:1}$, and inversely associated with total TFA, *trans*-$C_{18:1}$, and *trans*-$C_{18:2}$. Because within-country analysis of the data revealed no association between TFA intake and total LDL or HDL cholesterol level for all countries, except a small negative association of TFA intake and total cholesterol (p<0.05) in Iceland (which had the highest TFA intake), the pooled data for individual TFA were reassessed after exclusion of Iceland; positive associations for *trans*-$C_{14:1}$(n-5) with LDL and total cholesterol, *trans*-$C_{16:1}$(n-7) and *trans*-$C_{22:1}$ with LDL, LDL/HDL ratio and total cholesterol were identified. Hydrogenated fish oils are the main source of *trans*-$C_{16:1}$(n-7) and *trans*-$C_{22:1}$, the use of which was higher in Iceland and France than the other countries in the study. No association for *trans*-$C_{18:1}$ could be found. The authors thus concluded that they are not associated with an unfavorable serum lipid profile at the current European intake levels of TFA.

Prospective Studies

STARS

St. Thomas' Atherosclerosis Regression Study (STARS) was designed to examine the effect of an intervention diet versus a usual care group and a diet plus cholestyramine group, on plasma cholesterol concentrations on coronary atherosclerosis as quantified by changes in the imaged luminal diameter of the coronary artery. The intervention diet was similar to the NCEP STEP II diet but had lower cholesterol and higher polyunsaturated fat levels. Nutrient intake was assessed by diet history. In the 50 men receiving the lipid-lowering diet or usual diet, Watts et al. (1994, 1996) found that, over 39 months, the decrease in minimum absolute width of coronary segments (MinAWS) was directly related to dietary energy, absolute intakes of total fat, saturated fat, monounsaturated fat, and cholesterol. In the univariate analyses, MinAWS was also highly correlated with intakes of palmitic, stearic, palmitoleic, and elaidic acids. No significant associations were seen with polyunsaturates, carbohydrate, protein, fiber, alcohol, or with the P/S ratio. Total saturates and *trans* unsaturates explained 20% of variance in coronary atherosclerosis progression. Two methodological issues with this study have been identified by other researchers (Kris-Etherton et al., 1997): one, the fatty acid effects were not independent of energy intake and multi-colinerality exists for saturated and TFA, and two, there were no associations between individual fatty acids and coronary artery atherosclerosis by multiple and bivariate ($C_{18:0}$ versus *trans*-$C_{18:1}$) analyses. The authors acknowledged that the ex *post facto* analysis from the dietary intervention in STARS should be viewed as hypothesis generating rather than hypothesis testing (Watts et al., 1997).

Health Professionals Follow-Up Study

The Health Professionals Study tracks the medical history and lifestyle of a large cohort of male health professionals starting in 1986. They completed a 131 item food frequency questionnaire in 1986, and every 2 years thereafter. In 1992, multivariate adjustments for energy intake, BMI, smoking habits, alcohol consumption, physical activity, history of hypertension, family history of myocardial infarction, and profession enabled, the relative risk by quintiles of intake in g/day, dietary fat, primarily saturated fat, total fat and cholesterol, and secondarily linoleic acid, linolenic acid, and *trans*-unsaturated fatty acid for coronary disease of 43,757 men to be assessed (Ascherio et al., 1996). Saturated fat was not associated with myocardial infarction, and had a relative risk of 2.21 (p=0.0016), which became 1.72 (p=0.09) after adjusting for fiber and for fatal coronary heart disease. Total fat, linolenic acid, and cholesterol intakes were associated with neither myocardial infarction nor fatal coronary heart disease.

On the other hand, linolenic acid was found to be mildly associated only with myocardial infarction and only after further adjustment for fiber (p=0.07). Significant associations of *trans*-fat intake with both disease outcomes were observed (relative risks equal to 1.40 and 1.78, respectively) after multivariate analyses (*p* values at 0.01 and 0.04, respectively), but the associations became null (relative risks of 1.21 and 1.41 with *p* values of 0.20 and 0.42, respectively) after the adjustment for fiber. Thus, relative to saturated fats, *trans* fats appear to be somewhat less harmful.

Analyses using intakes of total fat, saturated fat, and TFA as percentage of energy yielded similar results. The same results were also observed at every step after the exclusion of hypercholesterolemic subjects, those subjects who knew their cholesterol count at baseline, those who reported reducing intake of butter, meat, whole milk, or eggs during the last 10 years, and those who had events in the first 4 years of follow-up. It is noteworthy that based on the fatty acid composition of the diets at baseline, the authors expected but could not demonstrate a saturated-fat-intake associated increase in risk.

Overall, the lack of effect of total fat was consistent with findings observed by the same research group in other population cohorts. This is not the only epidemiological study which identified the role of fiber on CHD risk reduction; however, the magnitude of the fiber effect is the most accentuated in this study. The no-effects found for saturated fat and linoleic acid are not consistent with outcomes expected based on established knowledge, again highlighting the methodological problems associated with diet data collection in epidemiological studies.

ATBC Study

The relation of intakes of specific fatty acids and the risk of coronary heart disease was examined, *post-hoc*, in a cohort of 21,930 smoking men aged 50–69 years who were initially free of diagnosed cardiovascular disease and participants in the Finnish Alpha-Tocopherol, Beta-Carotene Cancer (ATBC) Prevention Study (Pietinen et al., 1997). At baseline in 1985–1988 diet intake was assessed using a validated self-administered, modified food frequency questionnaire method, and serum samples were collected. After 6.1 years of follow-up, a significant positive association between the intake of TFA and the risk of coronary death in the supplement group was observed after controlling for age, supplementation assignment, smoking, BMI, blood pressure, intakes of energy, alcohol, and fiber, education and physical activity. The multivariate relative risk of coronary death was 1.39 for men at the top quintile of TFA intake (median = 6.2 g/day) versus men at the lowest quintile of intake (median = 1.3 g/day). Similar outcomes were found for "elaidic acid" (RR= 1.37, median intakes for top and lowest quintiles = 5.6 and 1.3 g/day, respectively) and vegetable TFA (RR= 1.23, median intakes for top and lowest quintiles = 5.1 and 0.1 g/day, respectively), but not for animal TFA (RR= 0.83, median intakes for top and lowest quintiles = 2.5 and 0.6 g/day, respectively). The authors stated that the main reason for high TFA intake in the study was due primarily to heavy use of margarine. However, they failed to discuss the observations that the median intakes of total TFA were accounted for mainly by "elaidic acid," and not by "vegetable *trans* fatty acids" across the first four quintiles of total *trans* fat intakes (e.g., median elaidic acid intakes were 1.3, 1.7, 2.0, 2.7 g/day and "vegetable *trans*-fatty acids" were 0.1, 0.4, 0.8, 1.6 g/day, respectively, for the first four quintiles of total TFA intakes at 1.8, 2.4, 2.9, and 3.5 g/day). Because elaidic acid is a component of animal TFA, these observations could also have been interpreted to suggest that the source of elaidic acid was animal TFA (intakes for the first four quintiles were 0.6, 1.1, 1.5, and 1.9 g/day).

In the multivariate models there was, interestingly, an inverse association between coronary death and saturated fatty acid intake and a positive association for polyunsaturated fatty acids and linoleic acid. Further adjusting simultaneously for serum total- and HDL-cholesterol and other fatty acids resulted in a positive association for omega-3 fatty acid intake from fish, no association for saturated fatty acids and linoleic acid, and an inverse association for linolenic acid with coronary death. It is unclear why the simultaneous adjustment exercise was not performed for TFA and no explanation was provided for the omission.

When correlations between TFA intake and selected coronary heart disease risk factors and other dietary factors were examined individually, no influences of age, smoking, BMI, serum total- and HDL-cholesterol levels, blood pressure, education, or physical activity, intakes of dietary fiber, β-carotene, vitamin C, red meat, or fish across quintiles of TFA intake could be found. Intakes of saturated fatty acids and cholesterol were lowest in the lowest and highest quintile of TFA intake, respectively, and the intake of polyunsaturated fatty acids and vitamin E increased with the intake of TFA.

The 40% increase in relative risk of coronary death for total TFA at the highest quintile of intake versus the lower quintile is in the range of that reported in the early Nurses' Health Study (Aro et al., 1995). Similarly, the multivariate RR for the first four quintiles, and at times all five quintiles, of TFA intakes overlapped depending on the dependent variable examined. Nonetheless, the trend for increasing risk with higher intake of TFA was statistically significant.

Nurses' Health Follow-Up Study

The Nurses' Health Study, started in 1976, and is similar to the Health Professional Study in design. In 1993 Willett and Ascherio (1994) put forth the hypothesis that intake of TFA is causally related to coronary disease risk largely based on results of the 8-year follow-up of the 85,095 women who provided the dietary data in 1980. Out of these women, 80,082 were followed up for an additional 6 years, and the relation of dietary intake of *trans* unsaturated fat and the risk of coronary disease was examined by Hu et al. (1997). With a longer follow-up period, the use of a food questionnaire with an expanded number of food items (from 61 – 116), and better variable adjusting, the results of the Follow-Up Study are more clean-cut than the earlier report (Aro et al., 1995). Relative risks of total, animal, vegetable, saturated, monounsaturated, polyunsaturated, *trans* unsaturated fats, cholesterol and Keys score were estimated after adjusting for age only, as well as for age, time period, BMI, cigarette smoking, menopausal status, parental history of myocardial infarction, multivitamin use, vitamin E supplement use, alcohol consumption, history of hypertension, aspirin use, exercise, cholesterol intake, percentage of energy from protein, and total energy intake. In addition, relative risks of animal fat and vegetable fat were adjusted for each other, respectively, as well as *trans* unsaturated fats. Unlike the ATBC Study, relative risks for saturated, monounsaturated, polyunsaturated, and *trans* unsaturated fats were each adjusted for the other three fat types in this study.

There was no effect associating total fat intake with risk of CHD. The intakes of animal fat, saturated fat, monounsaturated fat, and *trans* fat and Keys score were each associated with an increased risk of disease in age-adjusted analyses. After multivariate adjustment, all associations except for *trans* unsaturated fats and Keys score became insignificant. On the other hand, after multivariate adjustment, intakes of vegetable fat and polyunsaturated fat remained negatively associated with relative risk of coronary heart disease. Because partially hydrogenated vegetable oils are a source of both polyunsaturated and *trans* fats, the authors investigated the relative risk of combinations of the two fats. They found that the risk was the lowest for those who had the lowest intake of *trans* unsaturated fat and the highest intake of polyunsaturated fat, and the risk was the highest for those who had the highest intake of *trans* fat and the lowest intake of polyunsaturated fat. This finding highlights the interactive nature of *trans* fat and polyunsaturated fat which was overlooked in most if not all the prior epidemiological studies. It should be pointed out that researchers had already observed in the 1970s and 1980s (Gottenbos, 1997; Zevenbergen et al., 1989, 1998) that TFA do not exhibit adverse biochemical and physiological effects, including atherogenicity, in experimental animals provided sufficient linoleic acid is present in the diet.

Taking the percentages of total energy from specific types of fat as a continuous variable while adjusting for the intakes of other types of fat, the researchers next analyzed the relative risks of coronary heart disease associated with increases in the percentage of energy from specific types of fat using the diet information collected in 1980 and in subsequent years. The outcomes of the two analyses are comparable, albeit it was weaker with the baseline diet. In both cases, there was a highly significant positive association for the incidence of coronary heart disease and *trans*-unsaturated fat, and a significant inverse association for polyunsaturated fat. There was a positive association for monounsaturated fat, and the association for saturated fat, although positive, was not significant. Applying a multivariate model and adjusting for intakes of *trans* fat and each other, vegetable fat was found to be associated with a reduced risk, while animal fat had no significant association with CHD. By adjusting for confounding effects due to intakes of other fats when the effects of *trans* monounsaturated fat was examined, Hu et al. (1997) were able to quantitatively estimate the change in risk with the replacement of one type of fat by another. For example, replacing 5% energy from saturated fat with energy from unsaturated fats was associated with a 42% lower risk, and replacing 2% of energy from *trans* unsaturated fat with energy from unhydrogenated, unsaturated fats was associated with a 53% lower risk.

Overall, the findings of this study are consistent with those reported by Ascherio et al. (1996). However, with the adjustment of effects of other fats, the association of *trans* monounsaturated fat with CHD risk became stronger and more convincing. The results also lend credible confirmation to the evidence-based public health recommendation of substituting saturated and *trans* fats with unsaturated fats in the diet to lower the risk of CHD.

Zutphen Elderly Study

The association of TFA intake and 10-year risk of CHD was examined by Oomen et al. (2001) in 667 men aged 64–84 from Zutphen, the Netherlands who were free of CHD at baseline. The researchers surveyed their dietary intake in 1985, 1990, and 1995 using the cross-check dietary history method, and calculated TFA intakes by using a time-specific food composition table which was compiled using national data collected in 1985 – 1990, the TRANSFAIR data collected in 1995, and the literature. By collecting dietary intake data at regular intervals and using a time-specific TFA that included a food composition database, the researchers were able to discern changes occurring between 1985 – 1995 in the intake of total TFA and partitioned them by origin into manufactured $C_{18:1}$, manufactured other, and ruminant TFA.

During the 10-year period from 1985 – 1995, average TFA intake decreased from 4.3 – 1.9 of energy. After adjustment for age, BMI, smoking and dietary covariates (energy, alcohol intake, vitamin supplement usage, intakes of saturated, monounsaturated, and polyunsaturated fatty acids, cholesterol, fiber), TFA intake at baseline was positively associated with the 10-year risk of CHD, a finding which is similar in direction to those observed in the Health Professionals Follow-Up Study (van de Vijver et al., 1996), the ATBC Study (Kromhort et al., 1995), and the Nurses' Health Study (Gillman et al., 1997). The relative risk for a difference of 2% energy in TFA intake at baseline was 1.28, with a tight and greater than 1 confidence interval. The Zutphen researchers are to be commended for avoiding many of the methodological pitfalls of the other studies.

The mean intakes of ruminant, manufactured $C_{18:1}$ and other manufactured TFA were 0.7, 2.1, and 1.6 % of energy, respectively. Despite the difference in the amount consumed, the three sources of TFA were found to be comparable in their relative risks for CHD when the relative risks were estimated for each unit of 0.5% energy increment above the baseline. The finding is contrary to the finding of the early Nurses' Health Study (Aro et al., 1995) which reported a nonsignificant inverse relation of *trans* fats of ruminant origin and CHD risk. The findings of the Zutphen Study suggest that it is the amount and not the origin of dietary *trans* fats that should be of public health concern.

Studies of Markers of Inflammation and Endothelial Function

Systemic inflammatory activation is an emerging risk factor for a number of components of metabolic syndrome, including insulin resistance, diabetes, and dyslipidemia. Among the inflammation markers the best studied is c-reactive protein (CRP), which is a relatively moderate predictor of CHD according to a recent meta-analysis of 23 studies (Danesh et al., 2004). The American Heart Association and the Center for Disease Control and Prevention consider CRP a reasonable adjunct to the major risk factors but have not recommended it as a population screening tool (Pearson et al., 2003).

The relation between TFA intake and systemic inflammation was examined by Mozaffarian et al. (2004a) in healthy women in the Nurses' Health Study, and by Mozaffarian et al. in ambulatory heart failure patients (Mozaffarian et al., 2004b). In the Nurses' Health Study, TFA intake was assessed from the average of two semi-quantitative food frequency questionnaires. In 86 heart failure patients, TFA intake was assessed from the RBC membrane TFA level, which as discussed earlier, is not a strong indicator of dietary *trans* fat intake. The inflammatory markers examined included sTNF-R1, sTNF-R2, IL-6, and CRP in the first study, and 11 additional markers in the second study. Lopez-Garcia et al. (2005) analyzed the association of *trans* intake and plasma biomarkers of inflammation and endothelial dysfunction in the same female subcohort (apparently) analyzed by Mozaffarian et al., but adjusted the multiple regression analysis data less rigorously. Mozaffarian found no association of TFA intake with CRP and IL-6, whereas Lopez-Garcia declared the existence of an association for CRP but not for IL-6. Mozaffarian et al. found that TFA intake was positively associated with sTNF-R1 and sTNF-R2 in age, BMI, smoking and physical activity-adjusted analyses, as well as after further adjustment for those confounders operational in more recent analyses of data from the Nurses' Health Study (Watts et al., 1997), namely, aspirin and nonsteroidal anti-inflammatory drug use, alcohol consumption, and intakes of saturated fat, protein, n-6 and n-3 fatty acids, fiber and total energy. Because metabolic studies show high intakes of TFA affect levels of serum lipoproteins, the researchers adjusted for LDL- and HDL-cholesterol, triacylglycerol, and lipoprotein(a), and reported a continued positive association, but reduced in magnitude, of TFA intake and the two TNF-receptors. Thus, on the one hand the data suggest that the TFA effect on inflammatory markers may be partially mediated by serum lipids, and, on the other hand, may partially explain why estimated risks of TFA with CVD in some prior studies are greater than would be predicted by effects on serum lipoproteins alone (Ascherio et al., 1999). In the analysis by Lopez-Garcia et al. sTNF R2, E-Selectin, sICAM-1 and sVCAM-1 were found to be positively correlated with TFA intake.

In patients with heart failure, RBC membrane concentrations of TFA were positively associated with IL-1 receptor antagonist, TNF receptors 1 and 2, monocyte chemoattractant protein 1, and brain natriuretic peptide. These associations remain significant after additional adjustments for BMI, smoking, ejection fraction, NYHA class, ischemic etiology, statin use, and serum glucose. The positive associations with IL-1ß, IL-6, and TNF-a were noted after the additional adjustment. The finding of a negative association of RBC membrane TFA concentration and CRP is inconsistent with the previous finding in healthy women. Unlike the results reported by Lemaitre et al. (2002), the RBC membrane levels of both *trans*-$C_{18:1}$ and *trans*-$C_{18:2}$ were associated with higher inflammatory marker concentration but *trans*-$C_{16:1}$ was not. No association was found for RBC TFA and serum amyloid A, IL-6 receptor, IL-8, endothelin-1, or transforming growth factor ß.

Other than CRP, the absolute and relative biological relevance in dietary induced changes in levels of the other inflammatory and endothelial function markers are yet to be ascertained. Thus, the adverse role assigned to TFA on inflammation and endothelial function deserves to be investigated further.

In summary, results of the cited epidemiological studies, considered in total, show an association of dietary *trans* fat intake with risk of coronary heart disease. The early studies had limitations, mainly with dietary TFA assessment tools and confounder adjustment. Refinement in methodology produced a more biologically plausible association. TFA intake assessment through its content in adipose tissue is more reliable than through its content on RBC membrane. Further research on the effects of TFA intake on the physical and biochemical properties of membrane is indicated as an approach to understand the underlying mechanism of TFA effect. Expansion of these studies may offer new insights into TFA effects on CHD. Since atherosclerosis is described as an inflammatory disease, this is a logical extension of the research effort.

Clinical and Metabolic Studies

There are a large number of clinical studies, which have compared directly and indirectly the effects of *trans* fat versus other fatty acid species on plasma lipoprotein and lipids, and markers of other aspects of heart disease. These subjects, particularly the aspect of TFA effects on lipoproteins, have been extensively reviewed and judged by experts individually and on behalf of authoritative professional and regulatory organizations (Kris-Etherton, 1995; British Nutrition Foundation, 1995; Katan et al., 1995; Khosla and Hayes, 1996; Sebedio and Christie, 1998; Lichtenstein, 1997, 2000; Kritchevsky, 1999; FDA, 1999; Hunter, 2000; NCEP, 2001; Food and Nutrition Board, 2002; WHO, 2003; Stender and Dyerberg, 2004; European Food Safety Authority, 2004; 2005 Dietary Guidelines Advisery Committee, 2005) in the last 10 years. The major consensus of these reports is that LDL-cholesterol level elevates, and HDL-cholesterol level decreases as a consequence of the consumption of TFA. A few of these publications provided detailed summary tables to facilitate study comparison.

Effects on Plasma Lipids and Lipoproteins

The effects of TFA on serum lipids were studied as early as in the mid-1960s. In an elaborate study, De Iongh et al. (1965) fed 72 healthy men a sequence of 17 diets, each containing 33 en% of test fat and fed for 4–6 weeks, over 2 years. The test fats included partially hydrogenated soybean, whale, and fish oils and mixtures of these with sunflower oil and monoglycerides, and butter and margarine. Between the test periods, a diet containing a mixture of sunflower and monoglyceride or triglyceride (90/10) was fed for a total of six times. Serum levels of total lipids and fatty acid composition, cholesterol esters, triglycerides, free fatty acids, free cholesterol, mono/diglycerides, and phospholipids were determined. The detailed feeding regimens were not described but the body weights of subjects were reported not to have changed over the two year period.

One important finding was that the content of TFA in serum CE, TG, and PL is proportional to the *trans* content in diet; this suggests that they are digested, absorbed, transported, and metabolized. Similarly, predicted serum cholesterol levels using the Keys equation (Keys et al., 1959) with the *trans* monoenoic term were consistently higher, and those without the *trans* monoenoic term were more comparable to the observed serum cholesterol levels. These unadjusted observations led the authors to suggest a strong cholesterol-lowering effect of linolenic acid, a cholesterol-raising effect of saturated fatty acids less than suggested by Keys, and a weak cholesterol raising effect of TFA. All clinical studies conducted in later years were a reflection of this study but were more sophisticated in design and statistical analysis which allowed the determination of the effects of TFA independent of many other dietary variables.

The landmark controlled-feeding study by Mensink and Katan (1990) led a renewed interest in the investigation of, and a revision of the view on the effects of TFA on lipoproteins. They fed 61 adults using a randomized, crossover design diets of identical nutrient composition, except 10 en% was either oleic, *trans*-$C_{18:1}$, or saturated fatty acids for 3 weeks each. The authors found that the effect of TFA on the serum lipoprotein profile is at least as unfavorable as that of the cholesterol-raising saturated fatty acids, because they not only raise LDL cholesterol levels but also lower HDL cholesterol levels. In a study of similar design, Zock and Katan (1992) found 7.7 en% *trans* monoenoic acid significantly lowered HDL cholesterol and raised LDL cholesterol relative to linoleic acid. Similar LDL cholesterol responses were reported by Mensink et al. (1992) when *trans* monoenoic acids were compared with oleic acid at 10 en%, and linoleic acid at 8 en%. In both cases, Lp(a) level was also elevated consequent to *trans* fat feeding.

Together with the Dutch researchers, three other research groups were instrumental in advancing our understanding the effect of TFA on serum lipoproteins and lipids, relative to other dietary components. Judd et al. (1994, 1998, 2002), employed a Latin square design or a cross-over to investigate the effects of TFA (fed at 3.8 and 6.6 en% , 2.4–3.9 en%, and 4.2 and 8.3 en%, respectively, and presented as part of an oil blend, or margarine), as compared with those of oleic acid, saturates, stearic acid, carbohydrates, butter and *trans* free margarine in three separate studies. The moderate levels of *trans* fats were used because levels used in the Dutch study were considerably higher than the average intake in most developed countries. They found total TFA and saturated fatty acids, with oleic acid as reference, were comparable in their LDL-cholesterol elevation effects and observed no HDL-cholesterol decrease when *trans* fat intake was moderate (3.8 en%), and a slightly HDL-cholesterol depressing effect only when *trans* fat was consumed at a high level (6.6 en%). With carbohydrate used as a reference TFA and saturates again were comparable in their impact on LDL-cholesterol; however, while TFA (8.3 en%) did not effect HDL-cholesterol, saturates elevated HDL-cholesterol. The changes in apolipoproteins B and AI were found to correspond with changes in the lipoprotein cholesterol. When TFA were considered to be hypercholesterolemic and added to saturated fat, the predicted LDL-cholesterol response, using the Hegsted equation (Hegsted et al., 1993) was close to the observed changes. In addition, the research group (Clevidence et al., 1997) reported a higher *trans* diet induced a slight increase in Lp(a) levels relative to the oleic and moderate *trans* diets, whereas the amounts of saturated fatty acids commonly found in the typical U.S. diet decreased Lp(a) concentration.

Lichtenstein et al. (1993, 1999) conducted an investigation in elderly men and women using a consecutive feeding or a Latin square design. Unlike similar studies conducted by the Dutch and the USDA-Beltsville researchers in which test fats were made by blending hydrogenated and non-hydrogenated vegetable oils, test fats in these studies were in the form of commercially available cooking/baking fats (i.e., corn oil versus corn oil margarine; and soybean oil versus semi-liquid margarine versus soft margarine versus shortening, versus stick margarine, versus butter). They reported a less favorable lipid profile with an increase in the degree of hydrogenation. Relative to the butter fed group, Lp(a) was elevated only in the stick margarine fed group.

The effects of similarly moderate level of *trans* intake were also investigated by Nestel et al. (1992) at the CSIRO Division of Human Nutrition in Australia. Three test diets, each for 3 week duration, were given to mildly hypercholesterolemic men followed a sequential and parallel combined design, and their lipoprotein responses were investigated. The authors reported similar LDL cholesterol levels with diets enriched in elaidic acid and palmitic acid were higher than a diet enriched in oleic acid. Palmitic acid feeding led to a higher HDL-cholesterol level than that as a result of elaidic and oleic acid feedings which were not different. Lp(a) was significantly elevated compared with the other two diets. In another study by this group (Noakes and Clifton, 1998), in which the investigation of TFA effects was a secondary objective. They compared canola oil blend with TFA, TFA-free canola oil blend, polyunsaturated oil blend with TFA, and TFA-free polyunsaturated oil blend. They reported an outcome similar to that of Judd et al. (Judd et al., 1998). Compared with butter, TFA-free margarines were equal to or more effective than margarine containing TFA in lowering LDL-cholesterol.

Thus the findings from these four research groups are consistent, that is, TFA elevate LDL-cholesterol similar to saturates, and at a higher level of intake TFA depress while saturates elevate HDL-cholesterol. After reviewing 20 studies in which butter exchanged with margarine products having different physical forms (i.e., soft, stick) and different amounts of *trans* fat Zock and Katan (1997) came to a conclusion similar to that of Lichtenstein et al. (2003), that replacement of 30 g of butter per day by soft tub margarine would theoretically predict a reduction in coronary heart disease risk of 10%, while replacement of butter by hard, high-*trans* margarines would have no effect, in subjects having constant intakes of energy and fat. The risk parameter used was total/HDL cholesterol ratio.

Other studies which utilized either a randomized, crossover, or a parallel design and which had as the major objective to study the effects of TFA on serum lipoprotein levels are described briefly here. Almendingen et al. (Almendingen et al., 1995) compared the effects of partially hydrogenated fish oil (PHFO, with *trans* fat at 8.0 en%) versus partially hydrogenated soybean oil (PHSO, with *trans* fat at 8.5 en%) versus butter in young men. They found that only PHFO elevated LDL-cholesterol while both PHFO and PHSO lowered HDL-cholesterol compared with butter. Compared with PHSO, PHFO caused a greater reduction in HDL-cholesterol and resulted in a higher LDL/HDL ratio. Chisholm et al. (1996) compared butter versus margarine, which contributed the same amount of fat and as part of a low-fat diet fed to adult hypercholesterolemic men and women. Subjects consuming the margarine diets had HDL-cholesterol and Lp(a) levels similar to, and had 10% lowered LDL-cholesterol and apoB levels than subjects consuming the butter diet. Aro et al. (1997) reported that *trans* fat at 8.7en% increased LDL/HDL ratio and Lp(a) whereas stearic acid at 9.3 en% had no effect on LDL/HDL ratio and raised Lp(a) less than *trans* fats relative to a dairy fat based diet, in healthy men and women.

In young women, Muller et al. (1998a) found that replacing saturated fatty acids in a margarine formulated with palm oil with saturated and TFA from PHSO led to decreases in HDL-cholesterol and apoA-1 without significant impact on LDL/HDL and Lp(a) ratio. Diets high in saturates and *trans* produced similar levels of TC, LDL-C, apoB, TG, and Lp(a). Saturated fat feeding resulted in higher levels in HDLC and ApoA but the ratio of TC/HDLC was similar to that seen in *trans* fat feeding. In another study, they (Muller et al., 1998b) found that replacing TFA in a margarine formulated with PHFO and containing 7.7 en% *trans* by a mixture of vegetable oils containing 1.1 en% *trans* led to significant decreases in LDL-cholesterol, LDL/HDL ratio and ApoB/ApoA-I ratio, with no changes in total- and HDL-cholesterol, triacylglycerols, apoB, and Lp(a). Sundram et al. (1997) compared *trans*-$C_{18:1}$ at 5.5 en% with equal amounts of *cis*-$C_{18:1}$, $C_{16:0}$, or $C_{12:0}$ + $C_{14:0}$ in young Malaysian men and women fed a moderate fat and cholesterol free diet. The *trans*-rich fat significantly elevated total cholesterol and LDL cholesterol relative to the $C_{16:0}$- and *cis*-18:1 rich fats, and depressed HDL cholesterol and elevated Lp(s) relative to all dietary treatments. More recently, de Roos et al. (2001) compared 9.2 en% TFA in a 37.4 en% fat diet (en% of saturates, *cis* monoenes, *trans* monoenes, and polyunsaturates were 12.9, 8.4, 9.2, and 4.7, respectively) with 6.8 en% lauric acid in a 41 en% fat diet (en% of saturates, *cis* monoenes, *trans* monoenes, and polyunsaturates were 22.9, 7.9, 0.3, and 6.9 respectively) in 29 healthy men and women. They observed lower serum total- and HDL-cholesterol levels, and an elevated LDL/HDL ratio, with no change of LDL-cholesterol and triglycerides at the end of the *trans*-rich diet period. Based on these findings they suggest substituting lauric acid for *trans* fats in manufacturing foods in which hard fats are required for the needed functionality.

During the deodorization process, up to 40% of α-linolenic in rapeseed oil may be converted to *trans* α-linolenic acid. Pedersen (Pedersen, 2001) reported that as much as 50% ingested α-linolenic acid may be *trans* isoforms. The TransLinE Study (Sebedio et al., 2000; Vermunt et al., 2001) was designed to investigate the effects of *trans* α-linolenic acid on plasma lipids and platelet fatty acid composition in a randomized, double blind parallel study involving 88 healthy men and conducted in three European research centers. Through controlled deodorization of rapeseed oil, two oils containing the same amount of *trans*-$C_{18:1}$ with one free of and the other rich in *trans* α-linolenic acid rich oil were produced and incorporated individually into an otherwise *trans* free diet which was fed for 6 weeks. Subjects in the "low" *trans* diet group consumed an average of 269 mg/d of *trans*-$C_{18:1}$; subjects in the "high" *trans* diet group consumed 366 mg/d (0.6 en%) of *trans*-$C_{18:1}$, and 1410 g/d of *trans*-$C_{18:3}$. The researchers observed the incorporation of *trans*-$C_{18:3}$ and *trans*-$C_{20:5}$ into plasma triglyceride, phospholipids, and cholesterol-esters of subjects in the "high" *trans* group, but no treatment differences in plasma levels of TC, LDLC, HDLC, TG, ApoB, ApoA1, and ApoB/ApoA1 ratio. However, the TC/HDL ratio was found to be higher in subjects that consumed the "high" *trans* diet.

Thus, the totality of evidence from clinical trial data, generated from men and women within broad spectra of demographical and physiological status, has established that *trans* fats qualitatively exert a more unfavorable impact on lipoprotein risk factors than saturated fats, particularly on TC/HDLC or LDLC/HDLC which are considered by some but not all researchers in the field as better predicators of CHD risk than individual cholesterol levels (Natarajan et al., 2003). The quantitative effects of *trans* fats relative to saturated fats are less certain. Adopting a meta-analytical approach, Mensink et al. (2003) estimated the effects of *trans*, relative to that of saturated, mono-, and polyunsaturated fats and carbohydrates, individually and in combination, on serum cholesterol levels and TC/HDL ratio based on a database that consisted of 60 controlled trials.

They concluded that *trans* MUFA were the most harmful macronutrient in terms of the ratio of total to HDL cholesterol. At the same time, they expressed some reservation on the stand-alone utility of the risk marker of TC/HDLC, particularly because based on the use of this marker, lauric acid must be considered a healthy fat.

Based on the linear regression coefficients for *trans* and saturated fats, they determined that it takes 1.8 en% saturated fats to induce a magnitude of change in TC/HDLC similar to that obtained with 1 en% of *trans* MUFA when they were individually substituted, isoenergetically, for a 1:1:1 mixture of carbohydrates, *cis* MUFA, and *cis* PUFA, meaning *trans* MUFA post 1.8 times greater risk than saturates. However, when carbohydrates alone are used as the reference, *trans* MUFA post 7.3 times greater risk than saturates. The reference-dependent difference in the *trans* MUFA effects relative to saturates on TC/HDLC is a reminder of the diverse impacts on TC and HDLC of saturated, unsaturated, and TFA and carbohydrates. It should be pointed out that the equivalency of *trans* MUFA to saturates relative to carbohydrates derived from the meta-analysis of controlled feeding trials is significantly smaller than reported by epidemiological studies using disease end point in risk calculation. For example, the Nurses' Health Follow-Up Study (Hu et al., 1997) reported that every 5 en% intake from saturated fat was associated with a 17% increase in the risk of CHD, while every 2 en% intake from *trans* fat was associated with a 93% increase of risk compared with equivalent energy intake from carbohydrates. This translates into an estimated 14 times greater risk associated with *trans* MUFA than saturates. The doubling in magnitude of the relative risk of CHD disease estimated based on disease end point versus plasma markers of risk thus suggests that TFA may also affect CHD risk factors other than plasma cholesterol levels more than saturated fats do.

It is important to note that most of the dietary studies cited above used TFA levels considerably greater than those seen in the average diet. In addition, a linear effect at intake levels closer to realistic ones has been assumed but not demonstrated.

Effects on Physiological Events

Immune Function and Inflammatory Responses

Atherosclerosis has been recognized as a chronic inflammatory disease (Ross, 1999). Both quantity and quality of dietary fats have been shown to affect immune and inflammatory responses (Hwang, 2000). TFA can conceivably affect immune and inflammatory response through their impact on essential fatty acid requirement and metabolism, and on cell membrane fluidity. However, only after their effects on serum lipids and lipoproteins were established did researchers start to consider the effects of TFA on immune and inflammatory functions, and thus literature in these areas is limited. Two well executed randomized, crossover controlled feeding studies which reported immune function and inflammatory responses to TFA are reviewed below.

Han et al. (2002) investigated the effects of diets containing hydrogenated fat in 11 moderately hypercholesterolemic men and women over 50 years old who were fed diets containing 30 en% as fat, of which 2/3 were provided as soybean oil, soybean oil-based stick margarine, or butter. Each diet was given for 32 days. The effects of different dietary fats on serum lipid profiles were as expected. The production of the inflammatory cytokines TNF-α and IL-6 following the consumption of the diet containing stick margarine was higher than the diet containing soybean oil but not different from the diet containing butter. No effects attributable to TFA on cellular immune responses as determined by IL-1ß, delayed type hypersensitivity response, lymphocyte proliferation, IL-2, PGE2 production were found. A positive correlation of TC/HDL ratio and TNF-α was observed. Lichtenstein et al. (2003) reported no effect of dietary fat type on C-reactive protein in subjects fed diets in which the major source of fat was provided by different commercially available fat products (soybean oil, semi-liquid margarine, soft margarine, shortening, traditional stick margarine and butter in their study subjects).

Baer et al. (2004) investigated the effects of several fatty acids on C-reactive protein and several other inflammatory markers, namely fibrinogen, IL-6, and E-selectin. Fifty healthy men consumed six controlled diets, each for 5 weeks. The diets were identical in all aspects except 8% of fat or fatty acids was replaced in all diets with carbohydrates (CHO), oleic acid, TFA, stearic acid, TFA plus stearic acid at 1:1 ratio, and $C_{12:0}+C_{14:0}+C_{16:0}$ (LMP). The level of C-reactive protein after the consumption of the 8 en% TFA diet was not different from those of the 8 en% stearic acid diet and the 8 en% LMP diet, but higher than those of the CHO, oleic acid, and the TFA+stearic acid combined diets. They observed a stearic acid associated elevation in fibrinogen concentrations over all other diets. The effect of TFA on IL-6 was comparable to those of both stearic acid and LMP, statistically not different from CHO, and higher than oleic acid. The level of E-selectin, a marker for leukocyte adhesion to endothelial sites of inflammation, was highest after TFA, and lowest after oleic acid.

Results of these studies suggest a causative role of TFA in modulating the short-term response in the levels of markers of the inflammation process. Whether this finding applies on a longer-term basis and can be extended to population studies to explain the difference found in the relative risks of TFA observed from epidemiological versus clinical studies remain to be investigated.

Hemostasis and Endothelial Function

From the same randomized, crossover controlled feeding study discussed earlier in this review in which Almendingen et al. (1995) found an HDL-cholesterol lowering effect of PHFO relative to PHSO, the observation of higher levels of plasminogen activator inhibitor type 1 antigen and activity from PHSO intake led the Norwegian researchers to conclude that PHSO is antifibrinolytic relative to PHFO and butter (Almendingen et al., 1996). In the same study, levels of factor VII, fibrinopeptide A, D-dimer, tissue plasminogen activator or ß-thromboglobulin were not affected by the type of test fats (PHFO versus PHSO versus butter). Hemostatic factors were also determined for subjects that participated in the lipoprotein study by Aro et al. (1997) where the Finnish researchers compared PHVO with stearic acid and with the dairy fat used in the run-in diet. Their findings (Mutanen et al., 1997) were similar to those reported by Almendingen et al., that is, plasma fibrin degradation products and D-dimer, factor VII coagulant activity, tissue type plasminogen activity and plasminogen activator inhibitor activity were not affected by the experimental diets. Effects of stearic acid and TFA on platelet activation and endothelial PGI2 production in vivo and in vitro were also found to be similar (Turpeinen et al., 1998). The level of *trans* fat used in this study was comparable to that used by Baer et al. (2004) and much higher than either the average intake in the US population (Allison et al., 1999) or the level consumed by subjects in the highest quintile of intake in epidemiological studies such as those by Willett et al. (1993) and Ascherio et al. (1994).

No dietary fat effects could be detected in fasting blood levels of fibrinogen, PAI-1 activity, PAI-1 antigen, and *t*-PA activity in subjects in the PHFO versus VO study by Muller et al. (1998b), and *t*-PA antigen, PAI-1 activity, PAI-1 antigen, factor VII, and fibrinogen in subjects in the PHSO versus saturates study by the same group of researchers (Muller et al., 1998a). When the same hemostatic variables were measured in 1.5 h postprandial blood samples collected at different times of the day from subjects in the PHSO versus saturates study, *t*-PA activity was found to be the only one that is affected by dietary fat; it was significantly higher after subjects consumed the high saturated diet than either the high *trans* diet or the high PUFA diet. Because postprandial lipemia has been reported to affect hemostatic status (Byrne et al., 1998) and because some hemostatic variables have been reported to show a diurnal pattern (Marckmann et al., 1993), it is difficult to determine if the observed elevation in *t*-PA activity reflects a normal diurnal rhythm or a postprandial phenomenon.

The TRANSLinE researchers (Armstrong et al., 2000) also investigated the effect of *trans* isomers of α-linolenic acid on a number of platelet aggregation and haemostatic factors. No effect was found on platelet aggregation, platelet thromboxane production, fibrinogen levels, activities of factor VII, activated factor VIIa, and plasminogen activator inhibitor activity. In a randomized, cross over controlled feeding study, Sanders et al. (1998) fed 29 men each of diets containing *trans*-$C_{18:1}$ versus oleate at 10en% for two weeks. When consuming the *trans*-$C_{18:1}$ containing diet, subjects had higher TC/HDLC and ApoB/ApoA1 ratios without changes in plasma concentrations of fasting fibrinogen, D-dimer concentration, factor VII coagulant, plasminogen activator inhibitor type 1, and tissue plasminogen activator.

The effect of dietary fatty acid composition and postprandial lipid concentration and hemostatic factors was recently studied by Tholstrup et al. (2003). After a 12-h overnight fast, healthy young men were given one of six test meals in random order, and FVII activation in blood samples collected 2, 4, 6, and 8 h afterward. The meals were prepared so that the fat content was 50.6 en% and its intake of each meal was 1 g for each kg of body weight. All fats (stearic acid, palmitic acid, palmitic + myristic, oleic, *trans* $C_{18:1}$, linoleic) increased FVII activation. However, the effects of TFA on postprandial activation of blood coagulation factor VII and fibrinolysis actually resembled those of the unsaturated fatty acids and were less than saturated fatty acid.

Endothelial function as assessed by flow-mediated vasodilation (FMD) of the brachial artery correlates well with risk factors and markers of CHD (Cohon et al., 2000). de Roos et al. determined the FMD of subjects participated in a study reviewed earlier (de Roos et al., 2001) in which effects on lipoproteins of *trans* fat and saturated fat (rich in lauric acid) were compared. Subjects experienced an average decrease in FMD by 29% when they switched from the saturated rich diet (FMD±SD = 6.2 ± 3.0%) to the *trans* rich diet (FMD±SD = 4.4 ± 2.3%) (de Roos et al., 2001). This is the first study which examined the short-term effects of TFA on FMD.

Studies of hemostatic factors conducted in plasma or serum showed no effect of TFA. When endothelial function was assessed by FMD, a TFA effect was observed. This suggests a direct effect on tissue which is not expressed in markers analyzed in vitro, and that more in situ studies are necessary.

Glycemic and Insulin Responses

There are a limited number of studies of TFA effects on insulin sensitivity, all of which are negative. Berlin et al. (1994) reported no difference in effects of *trans* versus saturated fat on erythrocyte membrane fluidity and insulin binding in subjects participating in the first *trans* fat study conducted by Judd et al. (1994) reviewed earlier. In a randomized crossover study, Louheranta et al. (1999) fed 14 subjects diets containing 36.6–37.9 en% fat with test fat, either TFA or oleic acid, 5.1 and 5.2 en%, respectively. Lipoprotein responses to TFA versus MUFA were similar to those reported in other studies, but glucose effectiveness, insulin sensitivity index, and acute insulin response were not different between the diets. Christiansen et al. (1997) failed to detect any difference in postprandial glycemia in 16 obese patients with NIDDM who were given, 30 en% fat diets containing 20 en% of saturated fat *cis*-MUFA or *trans* MUFA in a randomized crossover study. Postprandial insulinemia was similar between the SAFA and *t*MUFA and lower in *c*MUFA. In a randomized, double-blind, crossover study in which 25 adult subjects consumed. Three diets containing 9 en% of palmitic acid, oleic acid, or *trans*-$C_{18:1}$ for 4 weeks each, Lovejoy et al. (2002) reported no diet effect on insulin secretion. Subjects oxidized the least fat over 8 h on the M diet and the most on the T diet. Fasting insulin and glucose levels of subjects were not different in the study by Lichtenstein et al. (2003) in which different commercial soybean-derived fat products were compared.

Blood Pressure

Available information on blood pressure response to TFA ingestion is scarce and indicates the nonexistence of an effect. In a randomized, crossover controlled dietary study, Zock et al. (1993) compared the effect of 8 en% stearic acid, TFA, or linoleic acid, on blood pressure in 55 normotensive young women and men and found no treatment difference. Lichtenstein et al. (2003) measured a number of physiological and biochemical responses to the consumption of soybean oil, semi-liquid margarine, soft margarine, shortening, traditional stick margarine and butter in their study subjects. The lipid, lipoprotein, and CRP responses were reported earlier in this review (Lichtenstein et al., 1999; Han et al., 2002). They found no effect of dietary fat type on blood pressure. In a study having parallel design with 24 men in each arm, Dyerberg et al. (2004) compared 20 g TFA from PHSO versus 4 g n-3 PUFA from fish oil in diets containing comparable en% from total fat, saturates, cismonounsaturates, and polyunsaturates. Subjects consuming the TFA-enriched diet had lower HDL-cholesterol levels. No difference between the two groups in heart rate variability, arterial dilatory capacity, compliance, and distensibility was detected. Systolic, diastolic, and mean arterial pressures did not change in response to TFA feeding.

Conclusion

The majority of epidemiological and clinical studies in humans suggest a negative effect of TFA ingestion on CHD risk. At the current U.S. average intake level, TFA effects resemble those seen with dietary saturated fats in many ways. However, TFA are more harmful than saturated fats at higher intake levels. The epidemiological studies vary in methods of data collection and analysis and differ principally in the derived levels of risk. Their results indicate a negative effect of TFA on CHD risk factors and an increase in fatal and non-fatal endpoints. Clinical studies of levels of plasma lipid and lipoprotein CHD risk markers such as TC, HDLC, LDLC, etc., however, are generally uniform. Analysis of TFA effects, however, do not indicate the same level of risk as that indicated by epidemiological studies. Dietary studies have generally used elevated levels of TFA not usually seen in customary diets and more studies using realistic levels of TFA are indicated. Assessment of plasma indicators of inflammation are just beginning and should be expanded. Animal studies have shown that TFA raise serum lipids and lipoprotein levels but have virtually no effect on atherosclerosis. When non-invasive methods of measuring actual atherosclerotic involvement become standardized and generally accepted, it will be possible to assess unequivocally the various dietary effects. Research is further complicated by the number of TFA, natural and man-made, in the average diet. While the absence of an involvement of a specific TFA does not give the intellectual satisfaction of having identified one specific culprit (the smoking gun), the overall data have shown that ingestion of excess TFA has deleterious effects on CHD risk. Total deletion of TFA in the diet is impossible, but moderation in their use is attainable and desirable. The advisors to the 2005 edition of Dietary Guidelines for Americans recommend an intake of 1 en% (2005 Dietary Advisory Committee, 2004).

References

2005 Dietary Guidelines Advisory Committee. Section 4. Fats in Science Base. 2005 Dietary Guidelines Advisory Committee Report. Issued on line in 2004 at www/health.gov/dietaryguidelines/dga2005/report/.

Nutrient Data Fat and Fatty Acid Content of Selected Foods Containing Trans-Fatty Acids Special Purpose Table No. 1. Agriculture Research Service, U.S. Department of Agriculture. http://www.nal.usda.gov/fnic/foodcomp, 1993.

Allison D.B. Epidemiology. In trans Fatty Acids and Coronary Heart Disease Risk. Report of the Expert Panel on trans Fatty Acids and Coronary Heart Disease. *Am J Clin Nutr.* **1995,** *62,* 670S–8S.

Allison D.B.; S.K. Egan; L.M. Barraj; C. Caughman; M. Infante; J.T. Heimbach. Estimated Intakes of trans Fatty Acid and Other Fatty Acids in the US Population. *J Am Diet Assoc.* **1999,** *99*, 166–74.

Almendingen K.; O. Jordal; P. Keirulf; B. Sandstad; J.I. Pedersen. Effects of Partially Hydrogenated Fish Oil, Partially Hydrogenated Soybean Oil, and Butter on Serum Lipoproteins and Lp(a) in Men. *J Lipid Res.* **1995,** *36*, 1370–84.

Almendingen K.; I. Seljeflot; B. Sandstad; J.I. Pedersen. Effects of Partially Hydrogenated Fish Oil, Partially Hydrogenated Soybean Oil, and Butter on Hemostatic Variables in Men. *Arterioscler Thromb Vasc Biol.* **1996,** *16*, 375–80.

Armstrong R.A.; J.M. Chardigny; B. Beaufrere; L. Bretillon; S.H.F. Vermunt; et al. No Effect of Dietary trans Isomers of a-linolenic Acid on Platelet Aggregation and Haemostatic Factors in European Healthy Men: the TransLinE Study. *Thromb Res.* **2000,** *100*, 133–41.

Aro A.; M. Jauhiainen; R. Partanen; I. Salminen; M. Mutanen. Stearic Acid, trans Fatty Acids, and Dairy Fat: Effects on Serum and Lipoprotein Lipids, Apolipoproteins, Lipoprotein(a), and Lipid Transfer Proteins in Healthy Subjects. *Am J Clin Nutr.* **1997,** *65*, 1419–26.

Aro A., A.F.M. Kardinaal; I. Salminen; J.D. Kark; R.A. Riemersma; et al. Adipose Tissue Isomeric trans Fatty Acids and Risk of Myocardial Infarction in Nine Countries: the EURAMIC study. *Lancet.* **1995,** *345*, 273–8.

Ascherio A.; C.H. Hennekens; J.E. Burring; C. Master; M.J. Stampfer; W.C. Willett. Trans-fatty Acids Intake and Risk of Myocardial Infarction. *Circulation.* **1994,** *89*, 94–101.

Ascherio A.; E.B. Rimm; E.L. Giovannucci; D. Spiegelman; M. Stampfer; W.C. Willett. Dietary Fat and Risk of Coronary Heart Disease in Men: Cohort Follow up Study in the United States. *BMJ.* **1996,** *313*, 84–90.

Ascherio A.; M.B. Katan; P.L. Zock; M.C. Stampfer; W.C. Willett. Trans Fatty Acids and Coronary Heart Disease. *N Engl J Med.* **1999,** *340*,1994–8.

Baer D.J.; J.T. Judd; B.A. Clevidence; R.P. Tracy. Dietary Fatty Acids Affect Plasma Markers of Inflammation in Healthy Men Fed Controlled Diets: a Randomized Crossover Study. *Am J Clin Nutr.* **2004,** *79*, 969–73.

Baylin A.; E.K. Kabagambe; A. Ascherio; D. Spiegelman; H. Campos. High 18:2 trans-fatty Acids in Adipose Tissue are Associated with Increased Risk of Nonfatal Acute Myocardial Infarction in Costa Rican Adults. *J Nutr*. **2003,** *133*, 1186–91.

Baylin A.; E.K. Kabagambe; X. Siles; H. Campos. Adipose Tissue Biomarkers of Fatty Acid Intake. *Am J Clin Nutr*. **2002,** *76*, 750–7.

Berlin E.; S.J. Bhathena; J.T. Judd; B.A. Clevidence; R.C. Peters. Human Erythrocyte Membrane Fluidity and Insulin Binding are Independent of Dietary trans Fatty Acids. *J Nutr Biochem*. **1994,** *5*, 591–8.

Trans fatty acids. Report of the British Nutrition Foundation Task Force. British Nutrition Foundation, London, 1995.

Byrne C.D.; N.J. Wareham; N.D. Martensz; S.E. Humphries; J.C. Metcalfe; D.J. Grainger. Increased PAI Activity and PAI-1 Antigen Occurring with an Oral Fat Load: Associations with PAI-1 Genotype and Plasma Active TGF-levels. *Atherosclerosis*. **1998,** *140*, 45–53.

Chisholm A.; J. Mann; W. Sutherland; A. Duncan; M. Skeaff; C. Frampton. Effect on Lipoprotein Profile of Replacing Butter with Margarine in a Low Fat Diet: Randomized Crossover Study with Hypercholesterolaemic Subjects. *BMJ*. **1996,** *312*, 931–4.

Christiansen E.; S. Schnider; B. Palmvig; E. Tauber-Lassen; O. Pedersen. Intake of a Diet High in trans Monounsaturated Fatty Acids or Saturated Fatty Acids. Effects on Postprandial Insulinemia and Glycemia in Obese Patients with NIDDM. *Diabetes Care*. **1997,** *20*, 881–7.

Clevidence B.A.; J.T. Judd; R.J. Schaefer; J.L. Jenner; A.H. Lichtenstein; et al. Plasma Lipoprotein (a) Levels in Men and Women Consuming Diets Enriched in Saturated, cis-, or trans-monounsaturated Fatty Acids. *Artereoscler Throm Vasc Biol*. **1997,** *17*, 1657–61.

Clifton P.M.; J.B. Keogh; M. Noakes. Trans Fatty Acids in Adipose Tissue and the Food Supply are Associated with Myocardial Infarction. *J Nutr*. **2004,** *134*, 874–9.

Cohon J.D.; J.H. Drury; J. Ostdiek; J. Finn; B.R. Babu; et al. Benefits of Lipid Lowering on Vascular Reactivity in Patients with Coronary Artery Disease and Average Cholesterol Levels: a Mechanism for Reducing Clinical Events? *Am Heart J*. **2000,** *139*, 734–8.

Danesh J.; J.G. Wheeler; G.M. Hirshfield; S. Eda; G. Eiriksdottir; et al. C-reactive Protein and Other Circulating Markers of Inflammation in the Predication of Coronary Heart Ddisease. *N Engl J Med*. **2004,** *350*, 1387–97.

de Iongh H.; R.K. Beerthuis; C. den Hartog; L.M. Dalderup; P.A.F. van der Spek. The Influence of Some Dietary Fats on Serum Lipids in Man. *Bibl Nutr Dieta*. **1965,** *7*, 137–52.

de Roos N.M.; M.L. Bots; M.B. Katan. Replacement of Dietary Saturated Fatty Acids by *trans* Fatty Acids Lowers Serum HDL Cholesterol and Impairs Endothelial Function in Healthy Men and Women. *Arterioscler Thromb Vasc Biol*. **2001,** *21*, 1233–7.

de Roos N.M.; E.G. Schouten; M.B. Katan. Consumption of a Solid Fat Rich in Lauric Acid Results in a More Favorable Serum Lipid Profile in Healthy Men and Women than Consumption of a Solid Fat Rich in *trans*-fatty Acids. *J Nutr*. **2001,** *131*, 242–5.

Dictenberg J.B.; A. Pronczuk; K.C. Hayes. Hyperlipidemic Effects of trans Fatty Acids are Accentuated by Dietary Cholesterol in Gerbils. *J Nutr Biochem*. **1995,** *6*, 353–61.

Dyerberg J.; D.C. Eskesen; P.W. Andersen; A. Astrup; B. Buemann; et al. Effects of trans- and n-3. Unsaturated Fatty Acids on Cardiovascular Risk Markers in Healthy Males. An 8 Weeks Dietary Intervention Study. *Clin Nutr*. **2004,** *58*, 1062–70.

Emken E.A. Physicochemical properties, intake, and metabolism. In Trans Fatty Acids and Coronary Heart Disease Risk. Report of the Expert Panel on Trans Fatty Acids and Coronary Heart Disease *Am J Clin Nutr*. **1995,** *62*, 659S–69S.

Enig M.G.; S. Atlas; M. Keeney; J. Sampugna. Isomeric trans Fatty Acids in the U.S. Diet. *J Am Coll Nutr*. **1990,** *5*, 471–6.

European Food Safety Authority. Opinion of the Scientific Panel on Dietetic Products, Nutrition and Allergies on a request from the Commission related to the presence of trans fatty acids in foods and the effect on human health of the consumption of trans fatty acids. *EFSA J*. **2004,** *81*, 1–49.

Food and Drug Administration. Food Labeling: Trans Fatty Acids in Nutrition Labeling, Nutrient Content Claims, and Health Claims. *Federal Register* **1999,** *64*, 62746–62825.

Food and Nutrition Board, Institute of Medicine. *Letter report on dietary reference intakes for trans fatty acids*. Drawn from the Report on Dietary Reference Intakes for Energy, Carbohydrate, Fiber, Fat, Fatty Acids, Cholesterol, Protein, and Amino Acids. 2002. available online at www.iom.edu/fnb.

Gatto L.M.; M.A. Lyons; A.J. Brown; S. Samman. Trans Fatty Acids Affect Lipoprotein Metabolism in Rats. *J. Nutr.* **2002,** *132*, 1242–8.

Gillman M.W.; L.A. Cupples; D. Gagnon; B.E. Millen; R.C. Ellison; W.P. Castelli. Margarine Intake and Subsequent Coronary Heart Disease in Men. *Epidemiology* **1997,** *8*, 144–9.

Gottenbos J.J. Biological effects of trans fatty acids. In *Dietary Fats and Health*; E.G. Perkins; W.J. Visek, Eds.; American Oil Chemist's Society: Champaign, IL, 1997; pp. 375–390. .

Han S.N.; L.S. Leka; A.H. Lichtenstein; L.M. Ausman; E.J. Schaefe; S.N. Meydani. Effect of hydrogenated and saturated, relative to polyunsaturated, fat on immune and inflammatory responses of adults with moderate hypercholesterolemia. *J. Lipid Res*. **2002,** *43*, 445–52.

Heckers H.; M. Korner; T.W.L. Tuschen; F.W. Melcher. Occurrence of Individual trans-Isomeric Fatty Acids in Human Myocardium, Jejunum and Aorta in Relation to Different Degrees of Atherosclerosis. *Atherosclerosis* **1997,** *28*,189–98.

Hegsted D.M.; L.M. Ausman; J.A. Johnson; G.E. Dallal. Dietary Fat and Serum Lipids: an Evaluation of the Experimental Data. *Am. J. Clin. Nutr.* **1993,** *57*, 875–83.

Hodgson J.M.; M.L. Wahlqvist; J.A. Boxall; N.D. Balazs. Platelet trans Fatty Acids in Relation to Angiographically Assess Coronary Artery Disease. *Atherosclerosis* **1996,** *120*, 147–54.

Hu F.B.; M.J. Stampfer; J.E. Manson; E. Rimm; G.A. Colditz; et al. Dietary Fat Intake the Risk of Coronary Heart Disease in Women. *New Engl J Med* **1997,** *337*, 1491–9.

Hudgins L.C.; J. Hirsch; E.A. Emken. Correlation of Isomeric Fatty Acids in Human Adipose Tissue with Clinical Risk Factors for Cardiovascular Disease. *Am J Clin Nutr* **1991,** *53*, 474–82.

Hulshof K.F.; M.A. van Erp-Baart; M. Anttolainen; W. Becker; S.M. Church; et al. Intake of Fatty Acids in Western Europe with Emphasis on trans Fatty Acids: the TRANSFAIR Study. *Eur J Clin Nutr* **1999,** *53*, 143–57.

Hunter J.E. Safety and Health Effects of Isomeric Fatty Acids. In: *Fatty Acid in Foods and Their Health Implications*. C.K. Chow, Ed.; Marcel Dekker: New York, 2000; pp. 667–686.

Hwang D. Fatty Acids and Immune Responses-a New Perspective in Searching for Clues to Mechanism. *Annu Rev Nutr* **2000,** *20*, 431–56.

Jackson R.L.; J.D. Morrisett; H.J. Penal; A.M. Grotto; A. Cameo; et al. Influence of Dietary trans-fatty Acids Onswine Lipoprotein Composition and Structure. *J Lipid Res* **1977,** *18*, 182–90.

Judd J.T.; B.A. Clevidence; R.A. Muesing; J. Wittes; M.E. Sunkin; J.J. Podczasy. Dietary trans fatty acids: effects on plasma lipids and lipoproteins of healthy men and women. *Am J Clin Nutr* **1994,** *59*, 861–8.

Judd J.T.; D.J. Baer; B.A. Clevidence; R.A. Muesing; S.C. Chen; et al. Effects of Margarine Compared with Those of Butter on Blood Lipid Profiles Related to Cardiovascular Disease Risk Factors in Normolipemic Adults Fed Controlled Diets. *Am J Clin Nutr* **1998,** *68*, 768–77.

Judd J.T.; D.J. Baer; B.A. Clevidence; P. Kris-Etherton; R.A. Muesing; M. Iwane. Dietary cis and trans Monounsaturated and Saturated FA and Plasma Lipids and Lipoproteins in Men. *Lipids* **2002,** *37*, 123–31.

Katan M.B.; P.L. Zock; R.P. Mensink. Trans Fatty Acids and Their Effects on Lipoproteins in Humans. *Ann Rev Nutr.* **1995,** *15*, 473–93.

Katz A.M. Trans-fatty Acids and Sudden Cardiac Death. *Circulation* **2002,** *105*, 669–71.

Keys A.; J.T. Anderson; F. Grande. Serum Cholesterol in Man: Diet Fat and Intrinsic Responsiveness. *Circulation* **1959,** *19*, 210–4.

Khosla P.; K.C. Hayes. Dietary trans-monounsaturated Fatty Acids Negatively Impact Plasma Lipids in Humans: Critical Review of the Evidence. *J Am Coll Nutr* **1996,** *15*, 325–39.

Khosla P.; T. Hajri; A. Pronczuk; K.C. Hayes. Replacing Dietary Palmitic Acid with Elaidic Acid (t-C18:1? 9) Depresses HDL and Iincreases CETP Activity in Cebus Monkeys. *J Nutr* **1997,** *127*, 531S–6S.

Kris-Etherton P.M. Trans Fatty Acids and Coronary Heart Disease Risk. Ed. Report of the Expert Panel on Trans Fatty Acids and Coronary Heart Disease. *Am J Clin Nutr* **1995,** *62*, 655S–708S.

Kris-Etherton P.M.; S. Yu; T.D. Etherton; R. Morgan; K. Moriarty; D. Shaffer. Letter to the Editor. *Am. J. Clin. Nutr.* **1997,** *65*, 1088–90.

Kritchevsky D. Influence of trans Unsaturated Fat on Experimental Atherosclerosis. In *Dietary Fats and Health*. Perkins E.G.; W.J. Visek, Eds. American Oil Chemists' Society, Champaign, IL: 1983; pp. 403–13.

Kritchevsky D. Trans unsaturated fat in nutrition and health. In *Edible Fats and Oils Processing: Basic Principles and Modern Practices*. Erickson D.R., Ed. American Oil Chemists' Society, Champaign, IL: 1990; pp. 158–65.

Kromhout D.; A. Menotti; B. Bloemberg; C. Aravanis; H. Blackburn; et al. Dietary Saturated and trans Fatty Acids and Cholesterol and 25-year Mortality From Coronary Heart Disease: the Seven Countries Study. *Prev. Med.* **1995,** *24*, 308–15.

Kritchevsky D. Trans Unsaturated Fat in Health and Disease. In *Lipids in Health and*

Nutrition. J.H.P. Tyman, Ed. The Royal Society of Chemistry, Cambridge, UK: 1999; pp.32–47.

Lemaitre R.N.; I.B. King; T.E. Raghunathan; R.M. Pearce; S. Weinmann; et al. Cell Membrane trans-fatty Acids and the Risk of Primary Cardiac Arrest. *Circulation* **2002,** *105*, 697–701.

Lichtenstein A.H.; L.M. Ausman; W. Carrasco; J.L. Jennet; J.M. Ordovas; E.J. Schaefer. Hydrogenation Impairs the Hypolipidemic Effect of Corn Oil in Humans. Hydrogenation, trans Fatty Acids, and Plasma Lipids. *Arterioscler Thromb.* **1993,** *13*, 154–61.

Lichtenstein A.H. for the Nutrition Committee. Trans fatty Acids, Plasma Lipid Levels, and Risk of Developing Cardiovascular Disease. *Circulation* **1997,** *95*, 2588–90.

Lichtenstein A.H.; L.M. Ausman; S.M. Jalbert; E.J. Schaefer. Effects of Different Forms of Dietary Hydrogenated Fats on Serum Lipoprotein Cholesterol Levels. *N. Engl. J. Med.* **1999,** *340*, 1933–40.

Lichtenstein A.H. Trans Fatty Acids and Cardiovascular Disease Risk. *Curr Opin Lipidol* **2000,** *11*, 37–42

Lichtenstein A.H.; A.T. Erkkila; B. Lamarche; U.S. Schwab; S.M. Jalbert; L.M. Ausman. Influence of Hydrogenated Fat and Butter on CVD Risk Factors: Remnant-Like Particles, Glucose and Insulin, Blood Pressure and C-reactive Protein. *Atherosclerosis* **2003,** *171*, 97-107.

Lopez-Garcia E.; M.B. Schulze; J.B. Meigs; J.E. Manson; N. Rifai; et al. Consumption of trans Fatty Acids is Related to Plasma Biomarkers of Inflammation and Endothelial Dysfunction. *J Nutr* **2005,** *135*, 562–6.

Louheranta A.M.; A.K. Turpeinen; H.M. Vidgren; U.S. Schwab; M.I.J. Uusitupa. A High-trans Fatty Acid Diet and Insulin Sensitivity in Young Healthy Women. *Metabolism* **1999,** *48*, 870–5.

Lovejoy J.C.; S.R. Smith; C.M. Champagne; M.M. Most; M. Lefevre; et al. Effects of Diets Enriched in Saturated (palmitic), Monounsaturated (oleic), or trans (elaidic) Fatty Acids on Insulin Sensitivity and Substrate Oxidation in Healthy Adults. *Diabetes Care* **2002,** *25*, 1283–8.

Marckmann P.; B. Sandstrom; J. Jespersen. Dietary Effects on Circadian Fluctuation in Human Blood Coagulation Factor VII and Fibrinolysis. *Atherosclerosis* **1993,** *101*, 225–34.

Meijer G.M.; A. van Tol; T.J.C. van Berkel; J.A. Weststrat. Effect of Dietary Elaidic Versus Vaccenic Acid on Blood and Liver Lipids in the Hamster. *Atherosclerosis* **2001,** *157*, 31–40.

Mensink R.P.; M.B. Katan. Effect of Dietary trans Fatty Acids on High-density and Low-density Lipoprotein Cholesterol Levels in Healthy Subjects. *N Engl J Med* **1990,** *323*, 439–45.

Mensink R.P.; P.L. Zock; M.B. Katan; G. Hornstra. Effect of Dietary cis and trans Fatty Acids on Serum Lipoprotein[a] Levels in Humans. *J Lipid Res* **1992,** *33*, 1493–501.

Mensink R.P.; G. Hornstra. The Proportion of trans Monounsaturated Fatty Acids in Serum Triacylglycerols or Platelet Phospholipids as an Objective Indicator of Their Short-term Intake in Healthy Men. *Br J Nutr* **1995,** *73*, 605–12.

Mensink R.P.; P.L. Zock; A.D.M. Kester; M.B. Katan. Effects of Dietary Fatty Acids and Carbohydrates on the Ratio of Serum Total to HDL Cholesterol and on Serum

Lipids and Apolipoproteins: a Meta-Analysis of 60 Controlled Trials. *Am J Clin Nutr* **2003,** *77*, 1146–55.

Mortensen A.; P.L. Espensen; B.F. Hansen; P. Ibsen. The influence of Dietary Olive Oil and Margarine on Aortic Cholesterol Accumulation in Cholesterol-fed Rabbits Maintained at Similar Plasma Cholesterol Level. *Atherosclerosis* **1992,** *96*, 159–70.

Mozaffarian D.; T. Pischon; S.E. Hankinson; N. Rifai; K. Joshipura; W.C. Willett; E.B. Rimm. Dietary Intake of trans Fatty Acids and Systemic Inflammation in Women. *Am J Clin Nutr* **2004a,** *79*, 606–12.

Mozaffarian D.; E.B. Rimm; I.B. King; R.L. Lawler; G.B. McDonald; W.C. Levy. trans Fatty Acids and Systemic Inflammation in Heart Failure. *Am J Clin Nutr* **2004b,** *80*, 1521–5.

Muller H.; O. Jodal; P. Kierulf; B. Kirkhus; J.I. Pedersen. Replacement of Partially Hydrogenated Soybean Oil by Palm Oil in Margarine Without Unfavorable Effect on Serum Lipoproteins. *Lipids* **1998a,** *33*, 879–87.

Muller H.; O. Jordal; I. Seljeflot; P. Kierulf; B. Kirkhus; O. Ledsaak; J.I. Pedersen. Effect on Plasma Lipids and Lipoproteins of Replacing Partially Hydrogenated Fish Oil With Vegetable Fat in Margarine. *Br J Nutr* **1998b,** *80*, 243–51.

Mutanen M.; A. Aro. Coagulation and Fibrinolysis Factors in Healthy Subjects Consuming High Stearic or trans Fatty Acid Diets. *Throm Haemost* **1997,** *77*, 99–104.

Natarajan S.; H. Glick; M. Criqui; D. Horowitz; S.R. Lipsitz; B. Knosian. Cholesterol Measures to Identify and Treat Individuals at Risk for Coronary Heart Disease. *Am J Prev Med.* **2003** Jul, **25**, 50–7.

National Cholesterol Education Program (NCEP) Executive Summary of the Third Report of the National Cholesterol Education Program Expert Panel on Detection, Evaluation, and Treatment of High Blood Cholesterol in Adults (Adult Treatment Panel III). *JAMA* **2001,** *102*, 2284–99.

Nestel P.J.; M. Noakes; B. Belling; R. McArthur; P. Clifton; E. Janus; M. Abbey. Plasma Lipoprotein Lipid and Lp(a) Changes With Substitution of Elaidic Acid for Oleic Acid in the Diet. *J Lipid Res* **1992,** *33*, 1029–36.

Nicolosi R.J.; T.A. Wilson; E.J. Rogers; and D. Kritchevsky. Effects of Specific Fatty Acid (8:0, 14:0, cis-18:1, trans-18:1) on Plasma Lipoproteins, Early Atherogenic Potential, and LDL Oxidative Properties in the Hamster. *J Lipid Res* **1998,** *39*, 1972–80.

Noakes M.; P.M. Clifton. Oil Blends Containing Partially Hydrogenated or Interesterified Fats: Differential Effects on Plasma Lipids. *Am J Clin Nutr* **1998,** *68*, 242–7.

Oomen C.M.; M.C. Ocke; E.J.M. Feskens; M.J. van Erp-Baart; F.J. Kok; D. Kromhout. Association Between trans Fatty Acid Intake and 10-year Risk of Coronary Heart Disease in the Zutphen Elderly Study: a Prospective Population-based Study. *Lancet* **2001,** *357*, 746–51.

Pearson T.A.; G.A. Mensah; R.W. Alexander; J.L. Anderson; R.O. Cannon; et al. Markers of Inflammation and Cardiovascular Disease. Application to Clinical and Public Health Practice. A Statement for Healthcare Professionals From the Centers for Disease Control and Prevention and the American Heart Association. *Circulation* **2003,** *107*, 499–511.

Pedersen J.I.; J. Ringstad; K. Almendingen; T.S. Haugen; I. Stensvold; D.S. Thelle. Adipose Tissue Fatty Acids and Risk of Myocardial Infarction – a Case-control Study. *Eur J Clin Nutr* **2000,** *54*, 618–25.

Pedersen J.I. More on trans Fatty Acids. *Br J Nutr* **2001,** *85*, 249–50.

Pietinen P.; A. Ascherio; P. Korhonen; A.M. Hartman; W.C. Willett; D. Albanes; J. Virtamo. Intake of Fatty Acids and Risk of Coronary Heart Disease in a Cohort of Finnish Men. *Am J Epidemiol* **1997,** *145*, 876–87.

Roberts T.L.; D.A. Wood; R.A. Riemersma; P.J. Gallagher; F.C. Lampe. Trans Isomers of Oleic and Linoleic Acids in Adipose Tissue and Sudden Cardiac Death. *Lancet* **1995,** *345*, 278–82.

Ross R. Atherosclerosis: an Inflammatory Disease. *N Engl J Med* **1999,** *340*, 115–26.

Royce S.M.; R.P. Holmes; T. Taiga; F.A. Kummerow. The Influence of Dietary Isomeric and Saturated Fatty Acids on Atherosclerosis and Eicosanoid Synthesis in Swine. *Am J Clin Nutr* **1984,** *39*, 215–22.

Rudel L.L.; K. Kelley; J.K. Sawyer; R. Shah; M.D. Wilson. Dietary Monounsaturated Fatty Acids Promote Aortic Atherosclerosis in LDL Receptor-null, Human apoB100-overexpressing Transgenic Mice. *Arteriosclero Thromb Vasc Biol* **1998,** *18*, 1818–27.

Sanders T.A.; F.R. Oakley; D. Crook; J.A. Cooper; G.J. Miller. High Intakes of trans Monounsaturated Fatty Acids Taken for 2 Weeks do not Influence Procoagulant and Fibrinolytic Risk Markers for CHD in Young Healthy Men. *Br J Nutr* **1998,** *89*, 767–76.

Satchithanandam S.; C.J. Oles; C.J. Spease; M.M. Brandt; M.P. Yurawecs; J.L. Rader. Trans, Saturated, and Unsaturated Fat in Foods in the United States Prior to Mandatory trans-fat Labeling. *Lipids* **2004,** *39*, 11–8.

Sebedio J.L.; W.W. Christie. *Trans Fatty Acid in Human Nutrition*. The Oily Press: Dundee, Scotland, UK, 1998.

Sebedio J.L.; S.H.F. Vermunt; J.M. Chardigny; B. Beaufrere; R.P. Mensink; et al. The Effect of Dietary trans a-linolenic Acid on Plasma Lipids and Platelet Fatty Acid Compositing: the TransLinE Study. *Euro J Clin Nutr* **2000,** *54*, 104–13.

Sgoutas D.S. Effect of Geometry and Position of Ethylenic Bond Upon Acylcoenzyme A-cholesterol-o-acyltransferase. *Biochemistry* **1970,** *9*, 1826–33.

Siguel E.N.; R.H. Lerman. Trans-fatty Acid Patterns in Patients with Angiographically Documented Coronary Artery Disease. *Am J Cardiol* **1993,** *71*, 916–20.

Sommerfeld M. Trans Unsaturated Fatty Acids in Natural Products and Processed Foods. *Prog Lipid Res* **1983,** *22*, 221–33.

Stender S.; J. Dyerberg. Influence of trans Fatty Acids on Health. *Ann Nutr Metab* **2004,** *48*, 61–6.

Sundram K.; A. Ismail; K.C. Hayes; R. Jeyamalar; R. Pathmanathan. Trans (elaidic) Fatty Acids Adversely Affect the Lipoprotein Profile Relative to Specific Saturated Fatty Acids in Humans. *J. Nutr* **1997,** *127*, 514S–20S.

Tholstrup T.; G.J. Miller; A. Bysted; B. Sandstorm. Effect of Individual Dietary Fatty Acids on Postprandial Activation of Blood Coagulation Factor VII and Fibrinolysis in Healthy Young Men. *Am J Clin Nutr* **2003,** *77*, 1125–32.

Thomas L.H.; J.A. Winter; R.G. Scott. Concentration of 18:1 and 16:1 trans Unsaturated Fatty Acids in the Adipose Body Tissue of Decedents Dying of Ischaemic Heart Disease Compared with Controls: Analysis by Gas Liquid Chromatograph. *J Epidemiol Community Health* **1983,** *37*, 16–21.

Thomas L.H.; S.O. Olpin; R.G. Scott; M.P. Wilkins. Coronary Heart Disease and the Composition of Adipose Tissue Taken at Biopsy. *Hum Nutr Food Sci Nutr* **1987a,**

41F, 167-72.

Thomas L.H.; J.A. Winter. Ischaemic Heart Disease and Consumption of Hydrogenated Marine Oils. *Hum Nutr Food Sci Nutr* **1987b,** *41F*, 153–65.

Toda T.; Y. Toda; V.K. Yamamoto; F.A. Kummerow. Comparative Study of the Athergenicity of Dietary trans, Saturated and Unsaturated Fatty Acids on Swine Coronary Arteries. *J Nutr Sci Vitaminol* **1985,** *31*, 233–41.

Triosi R.; W.C. Willett; S.T. Weiss. Trans-fatty Acid Intake in Relation to Serum Lipid Concentration in Adult Men. *Am J Clin Nutr* **1992,** *56*, 1019–24.

Turpeinen A.M.; J. Wubert; A. Aro; R. Lorenz; M. Mutanen. Similar Effects of Diets Rich in Stearic Acid or trans-fatty Acids on Platelet Function and Endothelial Prostacyclin Production in Humans. *Arterioscler Thromb Vasc Biol* **1998,** *18*, 316–22.

Tzonou A.; A. Kalandidi; A. Trichopoulou. Diet and Coronary Heart Disease: a Case-Control Study in Athens, Greece. *Epidemiology* **1993,** *4*, 511–6.

van de Vijver L.P.L.; G. van Poppel; A. van Houwelingen; D.A.C.M. Kruyssen; G. Hornstra. Trans Unsaturated Fatty Acids in Plasma Phospholipids and Coronary Heart Disease: a Case-control Study. *Atherosclerosis* **1996,** *126*, 155–61.

van de Vijver L.P.; A.F. Kardinaal; C. Couet; A. Aro; A. Kafatos; et al. Association Between trans Fatty Acid Intake and Cardiovascular Risk Factors in Europe: the TRANSFAIR Study. *Eur J Clin Nutr* **2000,** *54*, 126–35.

van Greevenbroek M.M.J.; M.G. Robertus-Teunissen; D.W. Erkelens; T.V.A. de Bruin. Lipoprotein Secretion by Intestinal Caco-2 Cells is Affected Differently by trans and cis Unsaturated Fatty Acids: Effect of Carbon Chain Length and Position of the Double Bond. *Am J Clin Nutr* **1998,** *68*, 561–7.

Vermunt S.H.F.; B. Beaufrere; R.A. Riemersma; J.-L. Sebedio; J.-M. Chardigny; R.P. Mensink. Dietary trans a-linolenic Acid From Deodorized Rapeseed Oil and Plasma Lipids and Lipoproteins in Healthy Men: the TransLinE Study. *Br J Nutr* **2001,** *85*, 387-92.

Watts G.F.; P. Jackson; S. Mandalia; J.N. Brunt; E.S. Lesis; D.J. Coltart; B. Lewis. Nutrient Intake and Progression of Coronary Artery Disease. *Am J Cardiol* **1994,** *73*, 328–32.

Watts G.F.; P. Jackson; W. Burke; B. Lewis. Dietary Fatty Acids and Progression of Coronary Artery Disease in Men. *Am J Clin Nutr* **1996,** *64*, 202–9.

Watts G.F.; V. Burke. Letter to the Editor. *Am J Clin Nutr* **1997,** *65*, 1090.

Willett W.C.; M.J. Stampfer; J.E. Manson; G.A. Colditz; F.E. Speizer; et al. Intake of trans Fatty Acids and Risk of Coronary Heart Disease Among Women. *Lancet* **1993,** *341*, 581–5.

Willett W.C.; A. Ascherio. Trans Fatty Acids: are the Effects Only Marginal? *Am J Public Health* **1994,** *84*, 722–4.

Woollett L.A.; C.M. Daumerie; J.M. Dietschy. Trans-9-octadecenoic Acid is Biologically Neutral and Does Not Regulate the Low Density Lipoprotein Receptor as the cis Isomer Does in the Hamster. *J Lipid Res* **1994,** *35*, 1661–73.

World Health Organization. Diet, Nutrition and the Prevention of Chronic Diseases. Chapter 5.4 Recommendations for Preventing Cardiovascular Disease. In *WHO Technical Report Series 916, Report of a joint WHO/FAO Expert Consultation*: 2003; pp 60–8.

Zevenbergen J.L.; E. Haddeman. Lack of Effects of trans Fatty Acids on Eicosanoid Biosynthesis with Adequate Intake of Linoleic Acid. *Lipids* **1989,** *24*, 555–63.

Zevenbergen J.L.; U.M. Houtsmuller; J.J. Gottenbos. Linoleic Acid Requirement of Rats Fed trans Fatty Acids. *Lipids* **1998,** *23*, 178 – 86.

Zock P.L.; M.B. Katan. Hydrogenation Alternatives: Effects of trans Fatty Acids and Stearic Acid Versus Linoleic Acid on Serum Lipids and Lipoproteins in Humans. *J Lipid Res* **1992,** *33*, 399–410.

Zock P.L.; R.A.M.T. Blijlevens; J.H.M. de Vries; M.B. Katan. Effects of Stearic Acid and trans Fatty Acids Versus Linoleic Acid on Blood Pressure in Normotensive Women and Men. *Euro J Clin Nutr* **1993,** *47*, 437–44.

Zock P.L.; M.B. Katan. Butter, Margarine and Serum Lipoproteins. *Atherosclerosis* **1997,** *131*, 7–16.

Chapter 2

Trans Fats and Cancer

David Kritchevsky, PhD

The Wistar Institute, 3601 Spruce Street, Philadelphia, PA 19104

There are multiple dietary sources of *trans*-unsaturated fatty acids (TFA). Sommerfield (1983) summarized data regarding plant sources of TFA, ranging from palmitoleic (16:1*t*) to parimaric [18:4 (9*c*,11*t*,13*t*,15*c*)] acid. Ackmann (1998) has cataloged all the TFA which occur in nature. TFA are also present in meat of ruminant animals and in dairy products. However, the major source of TFA in the Western diet is partially hydrogenated vegetable or fish oils.

Interest in the biological effects of TFA dates to the 1950s (Melnick, 1954) but their possible relation to coronary risk has heightened interest and expanded research in this area. Virtually all publications relating to biological effects of TFA discuss them as if they had consistently similar structures. However, both naturally occurring and man-made TFA-containing fats contain a spectrum of unsaturated double bonds, both *cis* and *trans*, distributed between carbon atoms 4 and 16. Wolff et al. (1998) have shown milk fats containing *trans* double bonds between carbon atoms C_4 and C_{16} with most of the unsaturation being found at C_{11}. Margarines and shortenings contain *trans* double bonds between C_6 and C_{16} with the peak occurrence at C_9 but with appreciable levels at C_6–C_8 and C_{10}. Dutton (1979) found a similar distribution. There is virtually no information regarding biological effects of *cis* or *trans* double bonds at specific sites and we do not know if any particular *trans* fatty acid is responsible for the observed effects. The presence of *trans* double bonds alters the stereochemistry of fats; this difference may also contribute to the particular properties of fatty acids with *cis* or *trans* double bonds. For example, stearic acid (18:0) is a linear molecule with a width of 25 nm. Linoleic acid [18:2 (9*c*,12*c*)] is almost L-shaped and its width is 13 nm; however *trans,trans* linoleic acid [18:2 (9*t*,12*t*)] is linear and its width is 32 nm.

In the United States from 1989–1991 the mean intake of TFA was 4.7 g/d, representing about 2.6% of total energy. About one-quarter of the TFA consumed may come from ruminant sources (Allison et al., 1999). It has been shown that virtually no dietary fat is totally free of *trans* unsaturated fatty acids, (Ackmann et al., 1974; Sébédio et al., 1988). The varied nature of dietary TFA must be kept in mind when discussing their biologic effects although the general view seems to be of an 18 carbon monoene.

The bulk of publications regarding TFA and disease are centered on coronary

TABLE 2.1
Influence of Dietary Fats on DMBA-induced Mammary Tumors in Rats[a]

Fat (20%)	Iodine value	Rats with Tumors (%)	Tumors/Tumor-bearing Rat	Tumors/30 Rats
Sunflower seed oil	148	96.6	4.8 ± 0.31	130
Corn oil	130	90.0	4.0 ± 0.62	110
Cottonseed oil	123	93.3	4.5 ± 0.75	127
Soybean oil	139	100.0	3.4 ± 0.46	103
Olive oil	84	86.6	4.5 ± 0.42	117
Lard	59	93.3	3.4 ± 0.47	97
Tallow	39	80.0	3.0 ± 0.49	72
Butter	34	86.6	3.3 ± 0.41	88
Coconut oil	6	96.6	2.5 ± 0.36	73

[a]After Carroll and Khor, 1971.

heart disease and the relatively few epidemiological or experimental studies on TFA and cancer speak to virtually no effect.

Effects of TFA on carcinogenesis may be related, in part, to their resemblance stereochemically and biologically to saturated fatty acids. Carroll and Khor (1971) compared the effects of various dietary fats on dimethylbenz(a)anthracene (DMBA)-induced mammary tumors in rats. (Table 2.1). Both incidence (% rats with tumors) and multiplicity (tumors/tumor-bearing rat) were higher in the rats fed unsaturated fats. The results obtained with olive oil (principally monoene) were similar to those obtained with the saturated fats. Ip et al. (1985) demonstrated the essentiality of linoleic acid for growth of DMBA-induced tumors which may explain, in part, Carroll and Khor's data.

Selenskas et al. (1984) compared the effects of different levels (5 or 20%) of corn oil, a "*cis*" fat (blend containing 54.7% 18:1*c*) and a "*trans*" fat (blend containing 38.3% 18:1*t* and 19.2% 18:1*c*) on DMBA-induced mammary tumors in female Sprague-Dawley rats. At both levels of dietary fat intake the rats fed the *trans* fat exhibited the lowest tumor incidence (Table 2.2). Erikson et al. (1984) examined the effects of fats containing TFA on latency and tumor growth of transplanted line 168 mammary tumor cells in mice. They observed no difference in latency or rate of tumor growth between mice fed *cis* or *trans* fats. Watanabe et al. (1985) studied the effects of PHC (partially hydrogenated corn oil) (42%

TABLE 2.2
Influence of Different Levels of *trans*-fat, *cis*-fat, and Corn Oil on DMBA-induced Tumors in Rats (25 rats/group)[a]

	Fat					
	trans		*cis*		Corn Oil	
Level	5 %	20%	5 %	20%	5 %	20%
Tumor incidence (%)	9	18	16	27	28	72
Total tumors	16	32	24	40	44	80
Tumors/tumor-bearing rat	2.3 ± 0.5	2.3 + 0.4	2.7 + 0.9	2.7 ± 0.4	2.5 ± 0.3	3.6 ± 0.5

[a]After Selenskas et al., 1984.

TABLE 2.3
Effect of Different Dietary Levels of *trans* Fat or Corn Oil on AOM-induced Colon Cancer in F344 Rats (30 rats/group)[a]

Diet	% Fat	% Rats with Tumors	Tumors/Rat ± SD
Trans			
Low	5.9	63	1.10 ± 0.14
Intermediate	11.8	67	1.33 ± 0.16
High	17.6	57	1.10 ± 0.12
Corn oil			
Low	5.0	67	1.07 ± 0.13
Intermediate	13.6	63	1.17 ± 0.12
High	23.5	93	1.77 ± 0.16

[a] After Reddy et al., 1985.

trans fatty acid) or olive oil on 1,2-dimethylhydrazine (DMH)-induced colon tumors in Fischer 344 rats. There were no differences in incidence or multiplicity of colon tumors. Reddy et al. (1985) reported on effects of dietary *trans* fat or corn oil on azoxymethane (AOM)-induced colon tumors in female F344 rats. They fed three levels of *trans* fat (1.8, 3.5, or 5.3%) and compared them with three levels of corn oil—low fat (2.8% 18:2), intermediate fat (7.6% 18:2). and high fat (13.2% 18:2). Their results are summarized in Table 2.3. Dietary TFA did not affect tumor incidence in the low and intermediate fat groups and reduced it by 39% in the high fat group.

Sugano et al. (1989) studied the effects of dietary *cis* and *trans* fats on DMH-induced colon tumors in the Wistar-Furth-Osaka strain of rats, which are prone to development of colon cancer. The *trans* diet contained 39% 18:1*t* and 23.2% 18:1*c* whereas the *cis* diet contained 62% 18:1*c*. The diets were fed to both male and female rats and elicited no differences in weight gain or food intake. Table 2.4 shows that between the fats there were no differences in tumorigenicity. Effects of the various fats on production of eicosanoids are summarized in Table 2.5. Earlier, Watanabe and Sugano (1986) had shown that *trans* fat did not enhance growth of DMBA-induced mammary tumors in rats. Comparison of diets containing native or hydrogenated (71% *trans*) soybean oil for effects on DMBA-induced mammary carcinogenesis in rats showed no difference (Kritchevsky et al., 1992).

In a review of *trans* fatty acid effects on carcinogenesis Ip (1997) concluded, "In summary, six studies showed that *trans* monoene fatty acids behave similarly to *cis* monoene fatty acids with respect to modulation of tumor development. Increase the intake of *trans* fat (at the expense of *cis* fat) does not produce an adverse outcome as far as cancer risk is concerned." This view was expressed 8 years ago and nothing to alter the conclusion has appeared in the intervening period. The results are consistent despite the use of different strains and genders of rats and different chemical carcinogens.

The data relating to TFA effects in experimental carcinogenesis are sparse and there are even fewer data (all epidemiological) on TFA effects on cancer incidence in humans. The epidemiological data are based on correlation of tissue

TABLE 2.4
Incidence of Intestinal Tumors in DMH-treated Rats Fed *cis* or *trans* Fat (± SEM)[a]

Group	Small Intestine (%)	Colon (%)	Tumors/tumor-bearing rat
Cis fat			
Male	60.0	70.0	1.0 ± 0.3
Female	71.4	7.14	0.7 ± 0.2
Trans fat			
Male	66.7	88.9	1.4 ± 0.3
Female	57.1	57.1	0.9 + 0.2

[a]After Sugano et al., 1989.

TABLE 2.5
Production of Eicosanoid in DMH-treated Rats Fed *cis* or *trans* Fat[a]

Diet	Aortic PGI Concentration (ng/mg)	Plasma TXAs Production (pg/mL)
Cis Fat		
Male	5.91 ± 0.51	229 ± 46
Female	6.80 ± 0.51	480 ± 73
Trans Fat		
Male	5.12 ± 0.37	195 ± 34
Female	7.89 ± 1.00	548 ± 37

[a]After Sugano et al., 1989

or dietary TFA with incidence of cancer.

Kohlmeier (1997) reviewed data on biomarkers of fatty acids and breast cancer. Her own data, based on the EURAMIC study (Kohlmeier et al., 1997) showed a significant positive correlation between tissue levels of TFA and breast cancer. Of four other studies cited in the review one showed a positive correlation, one a negative correlation, and two showed no correlation. Among her conclusions, "An increased risk from breast cancer from *trans* fatty acid intake cannot be ruled out on the basis of these studies." One might add that perhaps it shouldn't be ruled in, either.

A study of the association of fat intake with the risk of breast cancer (Holmes et al., 1999) revealed no association with total fat, type of fat, or type of fatty acid. In a cohort of 88,795 women, studied between 1980 and 1994, the relative risk for *trans* unsaturated fat was 0.92.

McKelvey et al. (1999) carried out a case-control study of colorectal adenomatous polyps and consumption of foods containing partially hydrogenated oils. They could find no correlation between the risk of colorectal adenomas and consumption of foods containing TFA. After adjustment, the odds ratios for TFA intake between <2 g/d and 6+ g/d ranged between 0.86 and 1.00.

Slattery et al. (2001) reviewed the data on TFA and colon cancer and found a weak association in women but not in men. Those at greatest risk were estrogen-negative, post-menopausal women. The general conclusion was that it seemed prudent to avoid foods containing TFA.

Both experimental and epidemiological data suggest very little, if any, correlation between the intake of foods containing *trans* unsaturated fatty acids and cancer risk, but all recommend caution.

The data discussed previously have been related, in most part, to foods containing *trans* monoenes. Conjugated linoleic acid (CLA) is an inclusive term for a family of octadecadienoic fatty acids. The major naturally occurring member of the family is the 9*c*,11*t* isomer. Synthetic CLA is a mixture of roughly equal levels of the 9*c*,11*t*, and 10*t*,12*c* isomers (about 40–45% of each). The individual isomers are now available for research purposes. CLA has been shown to inhibit chemically induced tumors in rats (Ip et al., 1990), transplanted tumors in mice (Cesano et al., 1998), as well as growth of tumor cells in vitro (Schultz et al., 1992). CLA is discussed in depth elsewhere in this volume. Reviews of its anti-tumorigenic activity may be found in the recent literature (Kritchevsky, 2000; Kritchevsky and Pariza, 2004).

References

Ackmann, R.G.; S.N. Hooper; D.L. Hooper. Linoleic Acid Artifacts From the Deodorization of Oils. *J. Am. Oil Chem. Soc.* **1974,** *51*, 42–49.

Ackmann, R. Natural *Trans* Fatty Acids in: *Trans Fatty Acids in Human Nutrition.* Sebedio J.L.; W.W. Christie, Eds. The Oily Press: United Kingdom, 1998; pp. 303–311.

Allison, D.B.; B.K. Egan; L.M. Barraj; et al. Estimated Intakes of *trans* Fatty and Other Fatty Acids in the US Population. *J. Am. Diet. Assoc.* **1999,** *99*, 166–174.

Carroll, K.K.; H.T. Khor. Effects of Level and Type of Dietary Fat on Incidence of Mammary Tumors Induced in Female Sprague-Dawley Rats by 7,12-dimethylbenzanthracene. *Lipids* **1971,** *6*, 415–420.

Cesano, A.; S. Visonneau; J.A. Scimeca; et al. Opposite Effects of Linoleic Acid and Conjugated Linoleic Acid on Human Prostatic Cancer in SCID Mice. *Anticancer Res.* **1998,** *18*, 833–838.

Dutton, H.J. Hydrogenation of fats and its significance. in: *Geometrical and Positional Fatty Acid Isomers*. E.A. Emken; H.J. Dutton, Eds. AOCS Press: Champaign, IL, 1979; pp 1–16.

Erikson, K.L.; D.S. Schlanger; D.A. Adams; et al. Influence of Dietary Faty Acid Concentration and Geometric Configuration on Murine Mammary Tumorigenesis and Experimental Metastasis. *J. Nutr.* **1984,** *114*, 1834–1842.

Holmes, M.D.; D.J. Hunter; G.A. Colditz; et al. Association of Dietary Intake of Fat and Fatty Acids with Risk of Breast Cancer. *JAMA.* **1999,** *281*. 914–920.

Ip, C.; C.A. Carter; M. Ip. Requirement of Essential Fatty Acid for Mammary Tumorigenesis in the Rat. *Cancer Res.* **1985,** *45*, 1997–2001.

Ip, C.; S.F. Chin; J.A. Scimeca; et al.. Mammary Cancer Prevention by Conjugated Dienoic Derivatives of Linoleic Acid. *Cancer Res.* **1990,** *50*, 6118–6124.

Ip, C. Review of the Effects of *trans* Fatty Acids, Oleic Acids and n-3 Polyunsaturated Fatty Acids and Conjugated Linoleic Acid on Mammary Carcinogenesis in Animals. *Am. J. Clin. Nutr.* **1997,** *66*, 1523S–1529S.

Kohlmeier, L. Biomarkers of Fatty Acid Exposure and Breast Cancer Risk. *Am. J. Clin. Nutr.* **1997,** *66*, 1548S–1556S.

Kohlmeier, L.; N. Simonsen; P. van't Peet; et al. Adipose Tissue *trans* Fatty Acids and

Breast Cancer in the EURAMIC Study. *Cancer Epidemiol. Biomarkers Rev.* **1997,** *6*, 705–710.

Kritchevsky D.; M.M. Weber; M.D. Klurfeld. Influence of Different Fats (soybean oil, palm olein or hydrogenated soybean oil) on Chemically-induced Mammary Tumors in Rats. *Nutr. Res.* **1992,** *12*, S175–S179.

Kritchevsky, D. Animutagenic and Some Other Effects of Conjugated Linoleic Acid. *Br. J. Nutr.* **2000,** *83*, 459–465.

Kritchevsky, D.; M.W. Pariza. Conjugated linoleic acid as a tumor preventive agent in: *Cancer Chemoprevention Vol 1:Promising Cancer Chemoprevention Agents*. Kelloff G.J.; E.T. Hawk; C.C. Sigman, Eds. Humana Press: Totowa, NJ, 2004; pp. 583–590.

McKelvey, W.; S. Greenland; M.-J. Chen; et al. A Case-control Study of Colorectal Adenomatous Polyps and Consumption of Foods Containing Partially Hydrogenated Oils. *Cancer Epidemiol. Biomarkers Rev.* **1999,** *8*, 519–524.

Melnick, D. Biological Utiliazation of Fatty Acid Isomers. *J. Am. Oil Chem. Soc.* **1954,** *31*, 63–71.

Reddy B.S.; T. Tanaka; B. Simi. Effect of Different Levels of Dietary *trans* Fat or Corn Oil on Azoxymethane-induced Colon Carcinogenesis in F344 Rats. *J. Natl. Cancer Inst.* **1985,** *75*, 791–798.

Schultz, T.D.; B.P. Chew; W.R. Seaman; et al. Inhibitory Effect of Conjugated Dienoic Derivatives of Linoleic Acid and β Carotene on the in vitro Growth of Human Cancer Cells. *Cancer Lett.* **1992,** *63*, 125–133.

Sébédio, J.L.; A. Grandgaard; J. Prevost. Linoleic Acid Isomers in Heat Treated Sunflower Oils. *J. Am. Oil Chem. Soc.* **1988,** *65*, 362–366.

Selenskas, S.L.; M. Ip; C. Ip. Similarity Between *trans* Fat and Saturated Fat in the Modification of Rat Mammary Carcinogenesis. *Cancer Res.* **1984,** *44*, 1321–1326.

Slattery, M.L.; J. Benson; K-N. Nia; et al. *Trans* Fatty Acids and Colon Cancer. *Nutr. Cancer* **2001,** *39*, 170–175.

Sommerfield, M. *Trans* Unsaturated Fatty Acids in Natural Products and Processed Food. *Prog. Lipid Res.* **1983,** *22*, 221–233.

Sugano M.; M. Watanabe; K. Yoshida; et al. Influence of Dietary *cis* and *trans* Fats on DMH-induced Colon Tumors, Steroid Excretion and Eicosanoid Production in Rats Prone to Colon Cancer. *Nutr. Cancer* **1989,** *12*, 177–187.

Watanabe, M.; M. Sugano. Effects of Dietary *cis* and *trans*-monene Fats on 7,12-dimethylbenzanthracene-induced Rat Mammary Tumors. *Nutr. Rep. Intl.* **1986,** *33*, 163–169.

Watanabe, M.; T. Koga; M. Sugano. Influence of Dietary *cis* and *trans* Fat on 1,2-dimethylhydrazine-induced Colon Tumors and Fecal Steroid Excretion in Fischer 344 Rats. *Am. J. Clin. Nutr.* **1985,** *42*, 475–484.

Wolff, R.L.; D. Precht; and J. Molkentrin. Occurrence and distribution profiles of *trans* 18:1 acids in edible fats of natural origin. in: *Trans Fatty Acids in Human Nutrition.* Sebedio J.L.; W.W. Christie, Eds. The Oily Press: Dundee, United Kingdom, 1998; pp. 1–33.

Chapter 3

Safety and Efficacy of Conjugated Linoleic Acid

Michael W. Pariza[a] and Martin P. Yurawecz[b]

[a]Food Research Institute, Department of Food Microbiology and Toxicology, University of Wisconsin–Madison
[b]Consultant, Boca Raton, FL

Supported in part by gift funds administrated through the Food Research Institute at the University of Wisconsin–Madison. M.W. Pariza is an inventor of CLA-use patents that are assigned to the Wisconsin Alumni Research Foundation (WARF).

The first report indicating that fried ground beef contained both bacterial mutagens and antimutagens was published in 1978 (Pariza et al., 1979). The finding of mutagens in grilled beef was confirmatory, but evidence for antimutagens was a new observation. The publication (Pariza et al., 1979) concluded with a speculative prediction, "... it may also be found that the mutagenic inhibitory activity inhibits carcinogenesis." Subsequently it was established that this was indeed the case (Pariza et al., 1985) and the new anticarcinogen was identified as conjugated linoleic acid, for which the acronym "CLA" was coined (Ha et al., 1987).

Further investigation established that CLA inhibited carcinogenesis in several animal models. These findings led many groups to investigate the biochemical mechanism(s) of action of CLA. In the course of this work several additional potential applications were identified, as indicated in Table 3.1 (Pariza et al., 2001; Henriksen et al., 2003; Belury, 2003; Noone et al., 2002; Nagao et al., 2003a; Kritchevsky, 2003; Atkinson and Whigham, 2003; Albers et al., 2003; Bassaganya-Riera et al., 2002, 2003; Whigham et al., 2002, 2001; Miller et al., 1994; Cook et al., 1993). While the potential use of CLA to control body fat gain in humans and animals has received the most recent attention in both popular and scientific publications, the use of CLA to enhance immunity while also reducing inflammation and other adverse effects typically associated with immune enhancement may be of even greater importance (Albers et al., 2003; Bassaganya-Riera et al., 2002, 2003; Whigham et al., 2002, 2001; Miller et al., 1994; Cook et al., 1993).

Two Biologically Active CLA Isomers

It must be emphasized that CLA is not a single substance. Rather, the term CLA refers to a class of conjugated dieonic isomers of linoleic acid. Interestingly all of the known physiological effects of CLA are induced by just two isomers: *c*9*t*11

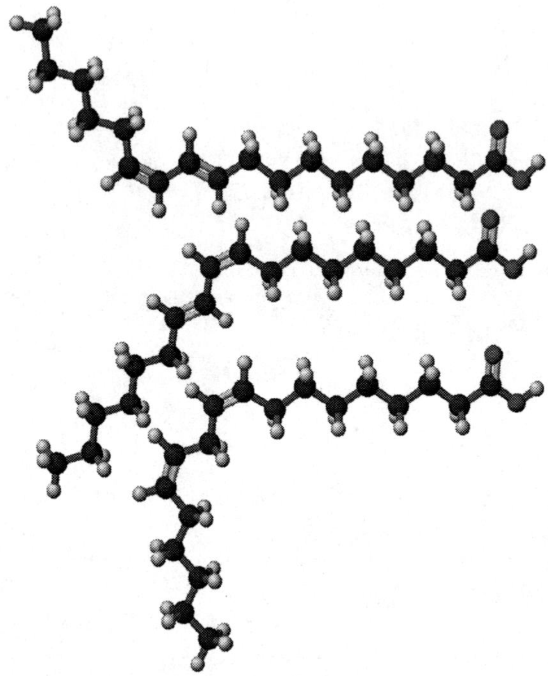

Fig. 3.1. Structures of the biologically active isomers of CLA. Reprinted with permission from the Journal of Chemical Education 1996;73(12):A302–A303. Copyright Division of Chemical Education, Inc.

TABLE 3.1
Some of the Reported Physiological Effects of CLA (3-21)

Inhibits carcinogenesis
Enhances immune function
Reduces inflammation
Reduces the catabolic effects of immune stimulation
Reduces asthma in animal models
Reduces atherosclerosis
Enhances growth of young rodents
Reduces body fat gain
Enhances lean body mass gain
Reduces the negative effects of weight-loss diets
Reduces symptoms of diabetes in some models
Reduces hypertension

CLA and *t*10*c*12 CLA (Fig. 3.1) (Pariza et al., 2001). This is not to say that other CLA isomers are without biological activity, but rather that a biological activity that is not induced by either *c*9*t*11 or *t*10*c*12 CLA has not been identified.

The *c*9*t*11 and *t*10*c*12 CLA isomers act alone or in concert to produce their effects. For example, *t*10*c*12 CLA is solely responsible for the reduction of body fat gain (Park et al., 1999) while *c*9*t*11 CLA enhances growth and feed efficiency in young rodents (Pariza et al., 2001). In other cases the isomers act together to induce an effect: both *c*9*t*11 and *t*10*c*12 CLA appear to be equally effective in inhibiting chemically induced mammary carcinogenesis in rodent models in part by inhibiting angiogenesis (Ip et al., 2002; Masso-Welch et al., 2002) whereas *t*10*c*12 CLA appears to be more effective than *c*9*t*11 CLA in inhibiting the proliferation of MCF-7 breast cancer cells via elicitation of a p53 response (Kemp et al., 2003). The biologically-active CLA isomers inhibit prostate cancer metastasis apparently in part via modulation of the expression of protein kinase C isoforms (Song et al., 2004). There are also instances where the two isomers appear to act in opposition (Roche et al., 2002). Hence, the multiple physiological effects that are reported for CLA (Table 3.1) appear to be the result of multiple interactions of the biologically active CLA isomers with numerous metabolic signaling pathways (Pariza et al., 2001).

Body Fat Reduction by CLA

Figure 3.2 depicts a model that we developed for the proposed mechanism of body fat control by *t*10*c*12 CLA, the isomer responsible for inducing this physiological effect. A major effect of *t*10*c*12 CLA appears to be the inhibition of adipocyte lipoprotein lipase (LPL) activity, which in turn reduces lipid uptake

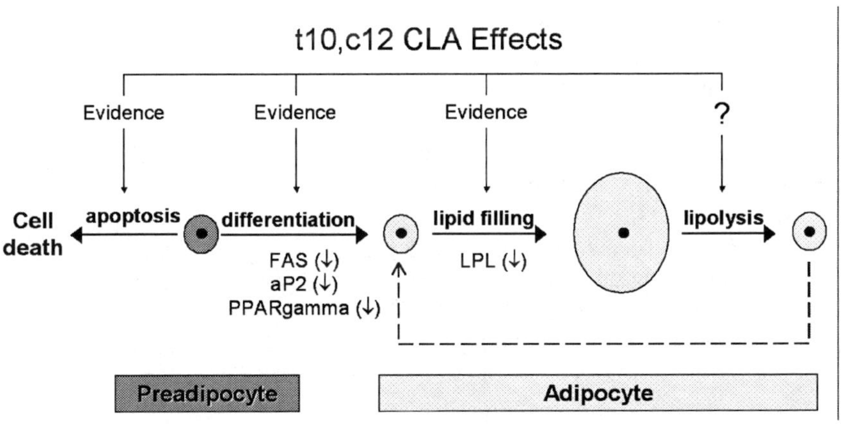

Fig. 3.2. A model for the effects of *trans*-10,*cis*-12 CLA on adipocytes and preadipocytes (Pariza et al., 2001).

into adipocytes (Park et al., 1997, 1999). The *t*10*c*12 CLA isomer also affects preadipocyte differentiation (Kang, et al., 2003), but a possible effect of CLA on lipolysis in adipocytes has not been proved (Xu et al., 2003). There is evidence that CLA enhances fat oxidation in skeletal muscle (Park et al., 1997), which may explain the enhancement of oxygen consumption and energy expenditure reported in CLA-fed OLETF rats (Nagao et al., 2003b).

The mechanisms summarized in Fig. 3.2 are supported mostly by data from studies with mice and cultured mouse 3T3-L1 adiposities. However, *t*10*c*12 CLA reduced triglyceride accumulation in cultured human adipocytes and modified markers of differentiation as well as the uptake and oxidation of glucose and oleate in cultured human preadipocytes (Brown et al., 2003a, 2003b).

We should consider how these findings may affect the design and interpretation of human clinical trials. First, in most animal models the reduction in body fat appears to be due to reductions in body fat accretion, not to reductions in preexisting body fat that had accumulated prior to the initiation of the experiment. In fact the mouse is the only species that has been clearly shown to lose accumulated body fat when fed a diet supplemented with *t*10*c*12 CLA (Pariza et al., 2001). We have speculated that this may occur because mice may depend more on fat combustion for energy than larger species and hence are more sensitive to CLA-induced reductions in fat storage in adipocytes.

Secondly, CLA induces significant reductions in body fat accretion without significantly reducing body weight. For example female and male weanling ICR mice fed a diet supplemented with 0.5% CLA (composed of approximately equal amounts of the *c*9*t*11 and *t*10*c*12 isomers) for 4 weeks exhibited, respectively, 60 and 57% reductions in body fat accretion relative to controls. However, in this study, the CLA-fed animals exhibited no significant differences in body weight relative to controls (Park et al., 1997).

Finally, there are considerable differences among species with regard to reductions in body fat accretion in response to dietary CLA (Kim et al., 2002). Mice are the most sensitive species, followed in order by hamsters and rats. Direct comparisons of pigs versus rodents have not been performed, but it would appear likely that pigs are intermediate, that is, more like hamsters than mice or rats (Ostrowska et al., 2003).

To date the published human clinical trials have almost all been designed to test the hypothesis that CLA ingestion will reduce preexisting body fat in adult humans (Gaullier et al., 2002). In general the protocols involved giving about 3 g/d of encapsulated CLA as a free fatty acid, with the highest intake in one study of 7.2 g/d (Gaullier et al., 2002). These trials have produced mixed results, but in general when the highest quality commercially available grades of CLA were used, there was an indication of significant body fat mass reduction relative to placebo controls (Gaullier et al., 2002). Perhaps not surprisingly there are also indications that CLA may prove to be most effective in reducing fat mass and increasing lean mass when combined with enhanced physical activity. However

CLA intake was not associated with significant reductions in body weight, which is consistent with data from animal studies (Pariza et al., 2001; Park et al., 1997).

Recently Kamphuis et al. (Kamphuis et al., 2003a, 2003b) tested the hypothesis that CLA might reduce the regaining of body fat and body weight in overweight adults who had undergone weight loss. All subjects received a very-low-calorie diet for 3 weeks followed by a 13 week intervention period during which they ate ad libitum but were given CLA (1.8 or 3.6 g) or placebo each day. The subjects consuming CLA (either dose level) exhibited significantly enhanced regaining of fat-free mass relative to controls, and a significantly enhanced resting metabolic rate. Interestingly, measures of appetite (hunger, satiety and fullness) were also favorably, dose-independently affected by CLA ingestion. However, as one might expect from the animal data, the regaining of body weight per se was not reduced by CLA ingestion. Blood lipids, glucose, and insulin were also not affected by CLA ingestion. These findings indicate that CLA may be most effective in controlling body fat accretion in anabolic humans, rather than reducing accumulated body fat per se.

CLA Safety

Under U.S. federal law, a substance that is not deemed to be safe when used as intended cannot be added to food. Animal tests are typically used in preclinical evaluations of the safety of new food ingredients. CLA safety has been evaluated in several well-conducted animal toxicological studies.

Scimeca (1998) conducted a 36-week feeding trial with Fischer 344 rats. The animals were fed either control diet or diet supplemented with 1.5% CLA, a level approximately 30-times greater than humans would ingest at 3 g CLA/day. Food disappearance, body weights, cageside examinations, and hematological and histopathological analyses of 15 major organs were conducted. No adverse effects were observed.

O'Hagan and Menzel (2003) conducted a 90-day rat feeding trial on a commercial preparation of conjugated linoleic acid consisting of equal amounts of the *c*9*t*11 and *t*10*c*`12 CLA isomers in the form of glycerides. They also subjected the CLA preparation to a battery of in vitro genotoxicity tests that are typically utilized in food ingredient safety assessment. They concluded that the No Observed Adverse Effect Level (NOAEL) for males and females, respectively, was 2433 mg/kg bw/day and 2728 mg/kg bw/day.

In addition, Schulte et al. (2003) and Pfeiffer et al. (2003) conducted comprehensive toxicological evaluations of CLA methyl esters in dogs and pigs, using standard toxicological protocols approved by European (OECD) Guidelines. These investigators concluded that CLA methyl esters did not produce adverse effects in the test species when fed at 5% of the diet.

The safety of commercial CLA preparations has also been evaluated in numerous human clinical trials. For human application, CLA quality is a topmost issue. Typically commercial CLA preparations intended for human use consist almost entirely (i.e., >90%) of the two biologically active isomers (Fig. 3.1) in approximately equal amounts (i.e., about 45% each), as reviewed by Gaullier et al. (2002). There is no evidence to indicate that such high-quality CLA, when consumed at 3–6 g/d, will induce adverse effects in healthy humans (Noone et al., 2002; Albers et al., 2003; Kamphuis, 2003a, 2003b; Eyjolfson, 2004; Whigham, 2004).

Despite these conclusions, some researchers have expressed concerns about the potential safety of CLA for humans (Kelley et al., 2003; Larsen et al., 2003; Riserus et al., 2002). Some of these concerns arise from mouse studies, where CLA-supplemented diets have been associated with the inductions of fatty liver, insulin resistance, and lipodystrophy. Others arise from human clinical trials involving specific patient groups, in particular men with metabolic syndrome, where data have been interpreted to indicate that CLA supplementation might enhance lipid peroxidation, inflammation, and unfavorable changes in serum lipids. There is also experimental and human clinical evidence indicating that CLA ingestion by nursing females might reduce milk fat. We should now consider these concerns in light of the overall scientific literature data base on CLA.

It is well documented that fatty liver is induced in mice fed CLA-supplemented diets (Pariza et al., 2001). However it should be emphasized that this finding appears to be limited to mice, and has not been reported for any other species to date. Hamsters fed CLA, and female rats fed a diet supplemented with 15% CLA, also exhibit enlarged livers but this is due to hypertrophy, not fat accumulation (O'Hagan and Menzel, 2003). It is important to note that neither fatty liver nor liver hypertrophy are considered by toxicologists to be toxic effects (Plaa, 1986). Moreover, the liver hypertrophy observed in female rats fed diet supplemented with 15% CLA was reversed when the animals were switched to a diet free of CLA (O'Hagan and Menzel, 2003).

The induction of mild insulin resistance in mice fed CLA-supplemented diets has been studied extensively and appears to be related to experimental conditions. For example, the effect was observed in mice fed a diet supplemented with 1% CLA for 39 days (DeLany et al., 1999) but not in mice fed a diet supplemented with 0.5% CLA for 4 or 49 days (Xu et al., 2003). (It should be noted that diet containing 0.5% CLA is sufficient to maximally induce body fat reduction in mice, and that there is no additional benefit from a diet supplemented with 1% CLA).

Additionally, even feeding mice a diet containing excessive levels of CLA does not appear to induce an irreversible enhancement of insulin resistance. To the contrary, diabetic C57BLKS-Lepr^{db}/lepr^{db}(*db/db*) mice maintained on a diet supplemented with 1.2% CLA for 23 weeks exhibited significant reductions in blood glucose and improved insulin sensitivity even in the face of an oral glucose challenge (Hamura et al., 2001a, 2001b). It is possible that this transient increase in insulin resistance results from CLA-induced changes in adipocyte dynamics, i.e., it occurs as older less responsive cells are replaced by younger more metabolically active adipocytes. It should also be noted that in the Zucker fatty rat, dietary CLA restores sensitivity and reduces symptoms of diabetes (Henriksen et al., 2003; Belury, 2003).

It has been reported that mice fed diets supplemented with CLA develop lipodystrophy, which in turn induces a number of negative metabolic consequences. However, like fatty liver, the induction of lipodystrophy has not been reported to occur in other species fed CLA-supplemented diets and it is possible that this effect is seen in mice because mice are so sensitive to CLA-induced body fat reduction. Increasing the amount of fat in CLA-supplemented diets substantially reduces the lipodystrophic effect (Tsuboyama-Kasaoka et al., 2003).

Finally, it should be noted that despite these seemingly negative consequences in mice, feeding diets supplemented with CLA significantly extended the lifespan of NZB/WF1 mice, which are prone to developing Lupus erythematosus (Yang et al., 2003). This finding is consistent with the conclusion that CLA does not induce toxic effects. It is all the more important because mice appear to be the most sensitive and responsive known species to the effects of CLA on lipid metabolism (Kim et al., 2002).

Riserus et al. (2002) investigated the effects of CLA in men with metabolic syndrome. They compared a typical high-quality CLA preparation consisting of equal amounts of *c*9*t*11 and *t*10*c*12 CLA with a supplement that is not commercially available, which was enriched for *t*10*c*12 CLA but contained very little *c*9*t*11 CLA. Riserus et al. (2002) reported that the patients ingesting CLA displayed elevated VLDL coupled with reduced HDL, enhanced levels of urinary isoprostane (interpreted as indicating increased oxidation), and enhanced C-reactive protein in serum (interpreted as indicating enhanced inflammation). These apparent negative effects were significantly elevated relative to placebo in the patients taking the *t*10*c*12 CLA-enriched supplement, but were generally not significantly affected in patients taking the typical commercially available CLA supplement relative to placebo. Riserus et al. concluded that these findings indicate that CLA may enhance cardiovascular disease risk.

Others have questioned these conclusions. For example, Wahle et al. (2004) noted that Riserus et al. (2002) reported the increased urinary isoprostane induced by *t*10*c*12 CLA stress did not result in reduced blood antioxidants, but rather was actually correlated with increased blood vitamin E. Wahle et al. (2004) concluded that this finding casts doubt on the physiological relevance of the increased urinary isoprostane levels induced by *t*10*c*12 CLA ingestion.

It is also very well documented in many mammalian models that CLA reduces inflammation, contrary to the conclusions of Riserus et al. (2002). For example, pigs fed a commercial mixture of CLA isomers displayed reduced inflammation (Bassaganya-Riera, 2002), and dietary CLA not only reduced atherosclerosis in animal models (Kritchevsky, 2003) but also reduced preestablished atherosclerotic lesions in mice (Toomey et al., 2003) and rabbits (Kritchevsky et al., 2004).

Additionally, the authors (Noone et al., 2002) of a comprehensive clinical study of the effects on serum lipids of a high-quality CLA preparation consisting of *c*9*t*11 and *t*10*c*12 CLA in approximately equal amounts concluded, "The study confirms that some of the cardioprotective effects of CLA that were shown in animal studies are relevant to man."

Riserus et al. (2002) also reported that the men with metabolic syndrome in their study who were given the *t*10*c*12 CLA supplement exhibited enhanced insulin resistance. Again, there was no significant difference in this parameter between the placebo group and the group receiving the typical commercially-available high quality CLA isomer mixture, a finding that is also consistent with a previous report from this group (Riserus et al., 2001). One should also keep in mind the interesting effect of dietary CLA on insulin resistance in C57BLKS-Lepr^{db}/lepr^{db}(*db/db*) mice (Hamura et al., 2001a; 2001b): in the short term (a few weeks of feeding) insulin resistance was enhanced, but by 23 weeks of CLA-feeding the mice exhibited significant reductions in blood glucose and improved insulin sensitivity. Hence, it is clearly possible that treating metabolic syndrome patients with CLA for a longer time period than used by Riserus et al. (2002) might indeed result in improvement.

It is well documented that *t*10*c*12 CLA reduces milk fat. This effect has been most thoroughly studied in cows (Peterson et al., 2003) but is also observed in lactating women who consume CLA supplements (Masters et al., 2002). The authors of the latter study (Masters et al., 2002) concluded that the reduction of milk fat might lead to less energy availability for the nursing infant. However, in a study with rats (Chin et al., 1994) it was found that the pups-nursing dams fed CLA-supplemented diet actually grew to a larger body size. Since very little *t*10*c*12 CLA carries over to milk, whereas the *c*9*t*11 CLA isomer is concentrated in milk, it may be concluded that *c*9*t*11 CLA may be a growth factor for young rats, as it appears to be for young mice (Pariz et al., 2001). Similar findings were obtained in a study with pigs where its authors concluded, "Irrespective of the dietary fat supplied in the starter period, piglets reared on the CLA sows had greater final body and warm carcass weights ($P<0.01$), and greater feed intake ($P=0.02$) than piglets reared on the [control] sows" (Bee, 2000). It seems likely that in these studies (Chin et al., 1994; Bee, 2000) the nursing animals compensated for reduced calories due to milk fat reduction by consuming more milk. In other studies involving pigs (Bontempo et al., 2004), maternal CLA ingestion did not affect nursing piglet growth but did positively affect a number of immunological variables in both the lactating sows and nursing piglets.

CLA and FDAs *Trans* Fat Definition

At the request of the U.S. Food and Drug Administration (FDA) the Institute of Medicine (IOM) prepared a report on *trans* fat that included the definition: "*Trans* fatty acids are unsaturated fatty acids that contain at least one double bond in the *trans* configuration." In the November 1999 proposal, FDA defined *trans* fatty acids as: "… unsaturated fatty acids that contain one or more isolated (i.e., non-conjugated) double bonds in a *trans* configuration" (64 FR 62746 at 62757).

Most of the comments on the definition of *trans* fat supported the proposed definition that excludes fatty acids with conjugated bonds, stating that *trans* fatty acids with conjugated bonds are metabolized differently than those with non-conjugated bonds and that this definition adequately identifies the fatty acids that the FDA intended to be covered by the rule. The FDA definition was consistent with the test article measured as indicated in the references cited by the IOM report.

It should also be noted that some CLA isomers do not contain *trans* double bonds, e.g., the *cis*-9, *cis*-11-$C_{18:2}$ and other *cis,cis* CLA isomers that have been identified in foods and biological extracts. In addition, both highly processed foods (e.g., partially hydrogenated oils) and less highly processed foods (e.g., dairy products) that contain CLA also contain *trans* fat (as defined by the FDA rule), and currently available data indicate that in these foods CLA is a minor component when compared to the content of total *trans* fat (as defined by the FDA rule).

References

Albers R.; R.P.J. van der Wielen; E.J. Brink; H.F.J. Hendriks; V.N. Dorovska-Taran; I.C.M. Mohede. Effects of cis-9, trans-11 and trans-10, cis-12 Conjugated Linoleic Acid (CLA) Isomers on Immune Function in Healthy Men. *European J Clin Nutr* **2003,** *57*, 595–603.

Atkinson R.L.; L.D. Whigham. Effects of CLA on body composition and metabolic variables in humans. 94th AOCS Annual Meeting and Expo, 2004;4 (abstr). Internet: (http://www.aocs.org/archives/am2003/session.asp?strsession=ana1) (accessed 1 August 2003).

Bassaganya-Riera J.; R.M. Pogranichniy; S.C. Jobgen; P.G. Halbur; K.J. Yoon; M. O'Shea; I. Mohede; R. Hontecillas. Conjugated Linoleic Acid Ameliorates Viral Infectivity in a Pig Model of Virally Induced Immunosuppression. *J Nutrit* **2003,** *133*, 3204–14.

Bassaganya-Riera J.; R. Hontecillas; D.C. Beitz. Colonic Anti-inflammatory Mechanisms of Conjugated Linoleic Acid. *Clinical Nutrit* **2002,** *21* (6), 451–459.

Bee G. Dietary Conjugated Linoleic Acid Consumption During Pregnancy and Lactation Influences Growth and Tissue Composition in Weaned Pigs. *J Nutrit* **2000,** *130* (12), 2981–89.

Belury, M.A. Conjugated Linoleic Acids in type 2 diabetes mellitus: Implications and potential mechanisms. In *Advances in conjugated linoleic acid research, Volume 2.* Sebedio J.-L.; Christie W.W.; Adlof R., Eds. Champaign, IL: AOCS Press, 2003; 302–15.

Bontempo V.; D. Sciannimanico; G. Pastorelli; R. Rossi; F. Rosi; C. Corino. Dietary Conjugated Linoleic Acid Positively Affects Immunologic Variables in Lactating Sows and Piglets. *J. Nutrition* **2004,** *134* (4), 817–24.

Brown J.M.; M.S. Boysen; S.S. Jensen; R.F. Morrison; J. Storkson; R. Lea-Currie; M. Pariza; S. Mandrup; M.K. McIntosh. Isomer-specific Regulation of Metabolism and PPAR Signaling by CLA in Human Preadipocytes. *J Lipid Res* **2003a,** *44*, 1287–1300.

Brown J.M.; M.K. McIntosh. Conjugated Linoleic Acid in Humans: Regulation of Adiposity and Insulin Sensitivity. *J. Nutr.* **2003b,** *133*, 3041–6.

Chin S.F.; J.M. Storkson; K.J. Albright; M.E. Cook; M.W. Pariza. Conjugated linoleic Acid is a Growth Factor for Rats as Shown by Enhanced Weight Gain and Improved Feed Efficiency. *J Nutrit* **1994,** *124* (12), 2344–49.

Cook M.E.; C.C. Miller; Y. Park; M. Pariza. Immune Modulation by Altered Nutrient Metabolism - Nutritional Control of Immune-induced Growth Depression. *Poultry Sci* **1993,** *72* (7),1301–5.

Current citations to the published scientific literature on CLA, see http://www.wisc.edu/fri/clarefs.htm.

DeLany J.P.; F. Blohm; A.A. Truett; J.A. Scimeca; D.B. West. Conjugated Linoleic Acid Rapidly Reduces Body Fat Content in Mice Without Affecting Energy Intake. *Am J Physiol* **1999,** *276*, R1172–R1179.

Eyjolfson V.; L.L. Spriet; D.J. Dyck. Conjugated Linoleic Acid Improves Insulin Sensitivity in Young, Sedentary Humans. *Med and Sci in Sports and Exercise* **2004,** *36* (5), 814–20.

Gaullier J.-M.; G. Breven; H. Blankson; O. Gudmondsen. Clinical Trial Results Support a Preference for Using CLA Preparations Enriched with Two Isomers Rather Than Four Isomers in Human Studies. *Lipids* **2002,** *37*, 1019–25.

Ha Y.L.; N.K. Grimm, M.W. Pariza. Anticarcinogens From Fried Ground Beef: Heat Altered Derivatives of Linoleic Acid. *Carcinogenesis* **1987,** *8*, 1881–7.

Hamura M.; H. Yamotoya; S. Kudo. Long-term Administration of Glycerides Rich in Conjugated Linoleic Acid (CLA) to Diabetic C57BLKS-Leprdb/leprdb(db/db) Mice. *J Oleo Sci* **2001a,** *50*, 883–8.

Hamura M.; H. Yamotoya; S. Kudo. Glycerides Rich in Conjugated Linoleic Acid (CLA) Improve Blood Glucose Control in Diabetic C57BLKS-Leprdb/leprdb(db/db) Mice. *J Oleo Sci* **2001b,** *50*, 889–94.

Henriksen E.J.; M.K. Teachey; Z.C. Taylor; S. Jacob; A. Ptock; K. Kramer; O. Hasselwander. Isomer-specific Actions of Conjugated Linoleic Acid on Muscle Glucose Transport in the Obese Zucker Rat. *Amer J. Physiol.* **2003,** *285* (1), E98–E105.

Ip C.; Y. Dong; M.M. Ip; S. Banni; G. Carta; E. Angioni; E. Murru; S. Spada; M.P. Melis; A. Saebo. Conjugated Linoleic Acid Isomers and Mammary Cancer Prevention. *Nutrit. Cancer* **2002,** *43* (1), 52–8.

Kamphuis M.M.J.W.; M.P.G.M. Lejeune; W.H.M. Saris; M.S. Westerterp-Plantenga. The Effect of Conjugated Linoleic Acid Supplementation After Weight Loss on Body Weight Regain, Body Composition, and Resting Metabolic Rate in Overweight Subjects. *Inter. J. Obesity* **2003a,** *27* (7), 840–7.

Kamphuis M.M.J.W.; M.P.G.M. Lejeune; W.H.M. Saris; M.S. Westerterp-Plantenga. Effect of Conjugated Linoleic Acid Supplementation After Weight Loss on Appetite and Food Intake in Overweight Subjects. *European J Clin Nutr* **2003b,** *57* (10), 1268–74.

Kang K.H.; W. Liu; K.J. Albright; Y. Park M.W. Pariza. trans-10,cis-12 CLA Inhibits Differentiation of 3T3-L1 Adipocytes and Decreases PPAR gamma Expression. *Biochem. Biophys. Res. Comm.* **2003,** *303* (3), 795–9.

Kelley D.S.; K.L. Erickson. Modulation of Body Composition and Immune Cell Functions by Conjugated Linoleic Acid in Humans and Animal Models: Benefits vs. Risks. *Lipids* **2003,** *38* (4), 377–86.

Kemp M.Q.; B.D. Jeffy; D.F. Romagnolo. Conjugated Linoleic Acid Inhibits Cell Proliferation Through a p53-dependent Mechanism: Effects on the Expression of G1-Restriction Points in Breast and Colon Cancer Cells. *J Nutrit* **2003,** *133* (11), 3670–7.

Kim M.R.; Y. Park; K.J. Albright; M.W. Pariza. Differential Responses of Hamsters and Rats Fed High-fat or Low-fat Diets Supplemented with Conjugated Linoleic Acid. *Nutrit. Res.* **2002,** *22* (6), 715–22.

Kritchevsky D. Conjugated linoleic acid in experimental atherosclerosis. In *Advances in Conjugated Linoleic Acid Research, Volume 2*. Sebedio J.-L.; Christie W.W.; Adlof R., Eds. AOCS Press: Champaign, IL 2003; 293–301.

Kritchevsky D.; S.A. Tepper; S. Wright; S.K. Czarnecki; T.A. Wilson; R.J. Nicolosi. Conjugated Linoleic Acid Isomer Effects in Atherosclerosis: Growth and Regression of Lesions. *Lipids* **2004,** *37* (7), 611–6.

Larsen T.M.; S. Toubro; A. Astrup. Efficacy and Safety of Dietary Supplements Containing Conjugated Linoleic Acid (CLA) for the Treatment of Obesity: Evidence from Animal and Human Studies. *J. Lipid Res.* **2003,** 44 (12), 2234–41.

Lee K.N.; D. Kritchevsky; M.W. Pariza. Conjugated Linoleic Acid and Atherosclerosis in Rabbits. *Atherosclerosis* **1994,** *108* (1), 19–25.

Masso-Welch P.A.; D. Zangani; C. Ip; M.M. Vaughan; S. Shoemaker; R.A. Ramirez; M.M. Ip. Inhibition of Angiogenesis by the Cancer Chemopreventive Agent Conjugated Linoleic Acid. *Cancer Res.* **2002,** *62* (15), 4383–9.

Masters N.; M.A. McGuire; K.A. Beerman; N. Dasgupta; M.K. McGuire. Maternal Supplementation with CLA Decreases Milk Fat in Humans. *Lipids* **2002,** *37* (2), 133–8.

Miller C.C.; Y. Park; M.W. Pariza; M.E. Cook. Feeding Conjugated Linoleic Acid to Animals Partially Overcomes Catabolic Responses Due to Endotoxin Injection. *Biochem Biophys Res Comm* **1994,** *198* (3), 1107–12.

Nagao K.; N. Inoue; Y.M. Wang; J. Hirata; Y. Shimada; T. Nagao; T. Matsui; T. Yanagita. The 10-trans,12-cis Isomer of Conjugated Linoleic Acid Suppresses the Development of Hypertension in Otsuka Long-Evans Tokushima fatty rats. *Biochem. Biophys. Res. Comm.* **2003a,** *306* (1), 134–8.

Nagao K.; Y.M. Wang; N. Inoue; S.Y. Han; Y. Buang; T. Noda; N. Kouda; H. Okamatsu; T. Yanagita. The 10-trans, 12-cis Isomer of Conjugated Linoleic Acid Promotes Energy Metabolism in OLETF Rats. *Nutrition* **2003b,** *19* (7-8), 652–6.

Noone E.J.; H.M. Roche; A.P. Nugent; M.J. Gibney. The Effect of Dietary Supplementation Using Isomeric Blends of Conjugated Linoleic Acid on Lipid Metabolism in Healthy Human Subjects. *Brit J Nutrit* **2002,** *88*, 243–51.

O'Hagan S.; A. Menzel. A Subchronic 90-day Oral Rat Toxicity Study and in vitro Genotoxicity Studies with a Conjugated Linoleic Acid Product. *Food Chem. Toxicol.* **2003,** *41* (12), 1749–60.

Ostrowska E.; D. Suster; M. Muralitharan; R.F. Cross; B.J. Leury; D.E. Bauman; F.R. Dunshea. Conjugated Linoleic Acid Decreases Fat Accretion in Pigs: Evaluation by Dual-energy X-ray Absorptiometry. *J. Nutrit.* **2003,** *89* (2), 219–29.

Pariza, M.W.; S.H. Ashoor; F.S. Chu; D.B. Lund. Effects of Temperature and Time on Mutagen Formation in Pan-fried Hamburger. *Cancer Lett* **1979,** *7*, 63–9.

Pariza M.W.; W.A. Hargraves. A Beef-derived Mutagenesis Modulator Inhibits Initiation of Mouse Epidermal Tumors by 7,12-dimethylbenz[a]anthracene. *Carcinogenesis* **1985,** *6*, 591–3.

Pariza M.W.; Y. Park; M.E. Cook. The Biologically-active Isomers of Conjugated Linoleic Acid. *Prog Lipid Res* **2001,** *40*, 283–98.

Park Y.; K.J. Albright; W. Liu; J.M. Storkson; M.E. Cook; M.W. Pariza. Effect of Conjugated Linoleic Acid on Body Composition in Mice. *Lipids* **1997,** *32* (8), 853–858.

Park Y.; J.M. Storkson; K.J. Albright; W. Liu; M.W. Pariza. Evidence That the trans-10,cis-12 Isomer of Conjugated Linoleic Acid Induces Body Composition Changes in Mice. *Lipids* **1999,** *34* (3), 235–41.

Peterson D.G.; E.A. Matitashvili; D.E. Bauman. Diet-induced Milk Fat Depression in Dairy Cows Results in Increased trans-10, cis-12 CLA in Milk Fat and Coordinate Suppression of mRNA Abundance for Mammary Enzymes Involved in Milk Fat Synthesis. *J Nutrit* **2003,** *133* (10), 3098–102

Pfeiffer A.M.; W. Kaufmann; J. Braun; B. Kaesler. Target Animal Safety of Conjugated Linoleic Acid (CLA) on Young Growing Pigs. 94th AOCS Annual Meeting and Expo, 2004;4 (abstr). Internet: (http://www.aocs.org/archives/am2003/session.asp?strsession=ana2 (accessed 1 August 2003).

Plaa G.L. Toxic responses of the liver. In *Casarett and Doull's Toxicology, The Basic Science of Poisons, 3rd edition.* Klaassen C.D.; M.O. Amdur; J. Doull, Eds. Macmillan Publishing Company: New York, NY, 1986; 286–309.

Riserus U.; L. Berglund; B. Vessby. Conjugated Linoleic Acid (CLA) Reduced Abdominal Adipose Tissue in Obese Middle-aged Men with Signs of the Metabolic Syndrome: a Randomised Controlled Trial. *Int J Obesity* **2001,** *25* (8), 1129–1135.

Riserus U.; S. Basu; S. Jovinge; G.N. Fredrikson, J. Arnlov; B. Vessby. Supplementation with Conjugated Linoleic Acid Causes Isomer-dependent Oxidative Stress and Elevated C-reactive Protein - A Potential Link to Fatty Acid-induced Insulin Resistance. *Circulation* **2002,** *106* (15), 1925–29.

Roche H.M.; E. Noone; C. Sewter; S. McBennett; D. Savage; M.J. Gibney; O'Rahilly; A.J. Vidal-Puig. Isomer-dependent Metabolic Effects of Conjugated Linoleic Acid: Insights from Molecular Markers Sterol Regulatory Element-binding Protein-1c and LXR. *Diabetes* **2002,** *51*, 2037–44.

Schulte S.; O. Hasselwander; F.-W. Rensmann; B. Kaesler; A.M. Pfeiffer. Safety Assessment of Conjugated Linoleic Acid (CLA) Esters in the Future Application Field as a Feed Additive in Growing Pigs. 94th AOCS Annual Meeting and Expo, 2004;4 (abstr). Internet: (http://www.aocs.org/archives/am2003/session.asp?strsession=ana2 (accessed 1 August 2003).

Scimeca J.A. Toxicological Evaluation of Dietary Conjugated Linoleic Acid in Male Fischer 344 Rats. *Food Chem Toxicol* **1998,** *36*, 391–5.

Song H.J.; A.A. Sneddon; P.A. Barker; C. Bestwick; S.N. Choe; S. McClinton; I. Grant; D. Rotonodo; S.D. Heys; K.W.J. Wahle. Conjugated Linoleum Acid Inhibits Proliferation and Modulates Protein Kinas C Isoforms in Human Prostate Cancer Cells. *Nutrition and Cancer* **2004,** *49* (1), 100–8.

Toomey S.; H. Roche; D. Fitzgerald; O. Belton. Regression of Pre-established Atherosclerosis in the apoE(-/-) Mouse by Conjugated Linoleic Acid. *Biochem Soc Trans* **2003,** *31*, 1075–79.

Tsuboyama-Kasaoka N.; H. Miyazaki; S. Kasaoka; O. Ezaki. Increasing the Amount of

Fat in a Conjugated Linoleic Acid-supplemented Diet Reduces Lipodystrophy in Mice. *J. Nutrit.* **2003,** *133* (6), 1793–9.

Wahle K.W.J.; S.D. Heys; D. Rotondo. Conjugated Linoleic Acids: Are They Beneficial or Detrimental to Health? *Progress in Lipid Res.* **2004,** *43* (6), 553–87.

Whigham L.D.; E.B. Cook; J.L. Stahl; R. Saban; D.E. Bjorling; M.W. Pariza; M.E. Cook. CLA Reduces Antigen-induced Histamine and PGE(2) Release From Sensitized Guinea Pig Tracheae. *Amer J Physiol* **2001,** *280* (3), R908–12.

Whigham L.D.; A. Higbee; D.E. Bjorling; Y.H. Park; M.W. Pariza; M.E. Cook. Decreased Antigen-induced Eicosanoid Release in Conjugated Linoleic Acid-fed Guinea Pigs. *Amer J Physiol* **2002,** *282* (4), R1104–12.

Whigham L.D.; M. O'Shea; I.C.M. Mohede; H.P. Walaski; R.L. Atkinson. Safety Profile of Conjugated Linoleic Acid in a 12-month Trial in Obese Humans. *Food Chemical Toxicol.* **2004,** *42* (10), 1701–9.

Xu X.F.; J. Storkson; S.H. Kim; K. Sugimoto; Y. Park; M.W. Pariza. Short-term Intake of Conjugated Linoleic Acid Inhibits Lipoprotein Lipase and Glucose Metabolism but Does Not Enhance Lipolysis in Mouse Adipose Tissue. *J Nutrit* **2003,** *133* (3), 663–7.

Yang M.D.; M.E. Cook. Dietary CLA Decreased Weight Loss and Extended Survival Following Outset of Kidney Failure in NZB/WF1 Mice. *Lipids* **2003,** *38*, 21–4

Chapter 4

Metabolism of *Trans* and *Cis* Fatty Acid Positional Isomers Compared to Non-Isomeric Fatty Acids

Edward Emken

Midwest Research Consultants, Princeville, IL 61559-8964

The predominant isomers in partially hydrogenated vegetable oil (PHVO) include about 16 *trans* and *cis* monounsaturated C_{18} positional isomers and several polyunsaturated C_{18} isomers that contain at least one *trans* double bond. Diets also contain relatively small amounts of conjugated linoleic acid isomers (CLA) from dairy fat and linolenic acid isomers formed by heat treatment of vegetable oils. The composition and abundance of the fatty acid isomers in these dietary fats have been described in detail (Craig-Schmidt, 1992; Emken, 1995; Sébédio and Christie, 1998; Yurawecz et al., 1999). This review is not intended to be all encompassing. The main focus is to review the metabolism of isomeric fatty acids in dietary fats and provide some comparisons to the metabolism of common dietary fatty acids (16:0, 18:0, 9*c*-18:1, 18:2n-6, and 18:3n-3).

Since the early 1970s, fatty acid isomers have been the source of considerable controversy because of reports that they are associated with health and nutritional issues. This concern prompted many metabolic, nutritional and health related studies and numerous journal reviews, books, and expert committee reports. The following are representative examples of reviews that contain limited to significant metabolic information (Emken and Dutton, 1979; Senti, 1985; Emken, 1991; Chow, 1992; Emken, 1984, 1995, 1997; Sébédio and Christie, 1998; Yurawecz et al., 1999; Sébédio et al., 2003).

A large number of enzymes are required to metabolize dietary fatty acids. Some of the enzymes required for major metabolic processes include lipases (absorption), Acyl CoA:diacylglycerol transferase (triglyceride synthesis), microsomal and peroxisomal β-oxidation enzymes, desaturases, elongases and phospholipid acyltransferases, lecithin:cholesterol acyltransferase, and acyl-CoA:cholesterol acyltransferase (incorporation). A wide variety of in vitro, animal, and human studies have provided considerable insight and information for understanding the metabolism of *trans* and *cis* fatty acid isomers and illustrate the complexity of the mechanisms that regulate dietary fat metabolism. In vivo studies in several animal species have been used to investigate essentially all aspects of isomeric fatty acid metabolism. Cultured cell, single cell organisms, isolated enzymes, and incubated tissue cells have provided additional details related to specific enzymes and mechanisms. Many of these in vivo and in vitro studies have used specific isomers labeled with radioisotopes (carbon-14 or tritium) to provide definitive metabolic information. In a similar manner, specific fatty acid isomers labeled with a stable isotope (carbon-13 and deuterium) have provided much of the metabolic information available for humans.

Absorption

Absorption of dietary triacylglycerol (TAG) requires (1) emulsification of the TAG, (2) hydrolysis by pancreatic lipase to yield a mixture of 1,2- and 2,3-diacylglycerol, 2-acylmonoglycerol and fatty acids, (3) passive transport of the hydrolysis products through the intestinal mucosa cells, and (4) reassembly of the TAG structure by the monoacylglycerol acyltransferase-diacylglycerol transferase pathway and/or the α-glycerolphosphate pathway. The reassembled triacylglycerols are packaged into chylomicron particles and transported via the lymphatic system to the circulation system. However, this process is not nearly as simple as outlined. Dietary fat absorption has been well studied and described in detail (Brindley, 1974; Borgstrom, 1977; Friedman and Nylund, 1980) but many aspects are still not well understood.

The results from animal and human studies provide considerable evidence that over 95% of most isomeric and non-isomeric C_{18} fatty acids (including 18:0) in PHVO are well absorbed when fed as part of a mixed fatty acid diet containing a variety of triacylglycerol molecular structures (Coots, 1964a; Ono and Fredrickson, 1964; Karmen et al., 1983; Emken, 1983, 1986, 1994; Brink et al., 1995; Baer et al., 2003). Studies (generally rat) that report relatively poor absorption for 9*t*-18:1 and 18:0 appear to have employed experimental protocols that compromised absorption (Clifford et al., 1986; Bernard et al., 1987; Mani et al., 1994; Aoyama et al., 1995).

Good absorption of the C_{18} fatty acid isomers in PHVO implies that enzymes involved in fat absorption (pancreatic lipase, diglyceride acyltransferase) and in TAG biosynthesis are not impaired by double bond configuration or position. Exceptions are some positional 18:1 and conjugated linoleic acid (CLA) isomers present in dairy fat but not in PHVO. Double bond position inhibited pancreatic lipase hydrolysis of triacylglycerols that contained *cis* 18:1 isomers with the double bond located at the Δ2 to Δ7 positions (Heimermann et al., 1973). The acyl position of 9*t*-18:1 in TAG structures had no effect on lipase hydrolysis (Jensen et al., 1964). The percent absorption of CLA was about 75–80% in humans (Emken et al., 2002) and is consistent with human jejunum data that showed partial absorption of linolenic acid was via portal transport (Surawicz, et al., 1981). In rats, % absorption of CLA was dependent on triglyceride structure and ranged from 87–95% (Chardigny et al., 2003). Lymphatic recovery of CLA in rats was 62% (Sugano, et al., 1997) and 95% (Martin et al., 2000) when compared to 18:2n-6 recovery and formation of absorbable products was lower for CLA than for 18:2n-6. Also, there is a minor difference between in vitro lipase hydrolysis of 9*c*,11*t*-18:2 and 10*t*,12*c*-18:2 (Martin et al., 2000).

Clearance or turnover of deuterium-labeled 18:1 isomers (Emken, 1984, 1986) and CLA (Emken et al., 2002) isomers from circulating human chylomicron TAG were similar to labeled non-isomeric fatty acids and suggest that lipoprotein lipase is not sensitive to fatty acyl chain structure. However, the *sn* acyl TAG position does appear to influence CLA lipase hydrolysis (Martin et al., 2000) and β-oxidation (Chardigny et al., 2003). The effect of TAG structure on CLA metabolism is interesting since TAG structure has been reported to influence the metabolism of non-isomeric fatty acids also (Kubow, 1996; Renaud et al., 1995; Emken et al., 2004). However, recent evidence shows that rearrangement of chylomicron TAG structures is partially responsible for the predominate location of saturated fatty acids at the *sn*-1 and *sn*-3 TAG position and unsaturated fatty acids at the *sn*-2 position of human plasma TAG (Emken et al., 2004). This TAG restructuring process would diminish metabolic effects associated with dietary TAG structure. No information is available but it is likely that chylomicron TAG rearrangement would redistribute *cis* and *trans* mono- and di-unsaturated fatty acid isomers in a similar preferentially manner between the *sn*-1, *sn*-2, and *sn*-3 acyl positions of chylomicron TAG.

β-Oxidation

Mammalians use two β-oxidation systems, mitochondrial and peroxisomal, for the degradation of fatty acids to CO_2. Different sets of several enzymes are utilized by these two β-oxidation systems, and the enzymes of both systems are sensitive to fatty acid structures. The peroxisomal β-oxidation system compliments the mitochondrial system and serves to oxidize fatty acids longer than 24 carbons plus a variety of other substrates such as branched-chain and dicarboxylic fatty acids (Osmundsen et al., 1991). In contrast to the mitochondrial system, peroxisomal oxidation is reported to chain-shorten fatty acids by only 2–3 β-oxidation cycles. Thus, 14:1 and 12:1 products would be formed from chain-shortened 18:1 isomers and would be available for elongation to 16:1 isomers. Mitochondrial β-oxidation has been estimated to be responsible for about 70% of dietary fatty acid oxidation (Masters and Crane, 1984) but this is a contentious issue (Osmundsen et al., 1991). For example, mitochondrial fatty acid oxidation by the classical isomerase dependent pathway is reported to be responsible for over 85% of C_{18} fatty acid oxidation (Ren, 2003).

In addition to the classical desaturase-dependent pathway, a mitochondrial NADPH reductase-dependent pathway is present that is used for the oxidation of unsaturated enoyl CoA intermediates that are not readily oxidized by the classic isomerase dependent pathway (Cuebas and Schultz, 1982; Tserng and Jin, 1991). For example, a sequence of reactions that involve enoyl CoA reductases are utilized to convert unsaturated intermediates, such as 5*c*-14:1 and 4*c*-14:1 CoA formed from 9*c*-18:1 and 8*c*-18:1 after two β-oxidation cycles, to substrates suitable for further β-oxidation. The mitochondrial and peroxisomal β-oxidation systems have been extensively studied and details related to the specific enzymes, pathways, regulation mechanisms, and interactions between the two oxidation systems have been reviewed (Masters and Crane, 1984; Gurr and James, 1971; Mead et al., 1986; Osmundsen et al., 1991; Reddy and Hashimoto, 2001).

Isomerase-Dependent Pathway

Oxidation of unsaturated fatty acids by the classical mitochondrial isomerase-dependent pathway involves the formation of the fatty acid CoA ester followed by several β-oxidation cycles that remove two carbon segments of the fatty acid acyl chain per cycle until a carbon-carbon double bond is encountered. At this point, the sequence of enzymatic reactions required to continue oxidation depend on the position and configuration of the original double bond. If the double bond was originally in an odd numbered position, a Δ3-Δ2 *trans*-enoyl-CoA isomerase is required to conjugated the double bond with the CoA ester carbonyl group but it is not required for fatty acids with an even numbered double bond position. If the original double bond was in the *cis* configuration, a 3-hydroxy acyl-CoA epimerase is required to convert the D-3-hydroxy-CoA stereoisomer to the L-3-hydroxy isomer. The epimerase isomerization step is not required when the original double bond is in the *trans* configuration. Note that if there were no other factors involved, the predicted oxidation rates based on number of reaction steps would follow the order: even # *trans* > odd # *trans* = even # *cis* > odd # *cis*. This theoretical relative order is supported, to some extent, by both in vitro and in vivo results for oxidation of 18:1 isomers. For example, in vitro oxidation rates for *cis* and *trans* octadecenoate CoA ester positional isomers by rat liver mitochondria and heart mitochondria preparations show an odd-even double bond position effect (Lawson and Holman, 1981). *Cis* 18:1 positional isomers with even-numbered double bonds were oxidized more slowly than the immediately adjacent odd-numbered *cis* isomer. Most *trans* 18:1 isomers with even-numbered double bonds were oxidized more rapidly than adjacent odd-numbered *trans* isomers and the oxidation rates of 8c-18:1 and 9*t*-18:1 were similar.

Reductase-Dependent Pathway

A second mitochondrial β-oxidation pathway, (reductase-dependent pathway) utilizes a 5-enoyl CoA reductase or a Δ3,5-Δ2,4-dienoyl-CoA isomerase and a 2,4-dienoyl-CoA reductase to convert 3*t*,5c-acyl CoA and 2*t*,4*c*-acyl CoA ester intermediates to 2*t*-enoyl acyl CoA esters that can then be further oxidized by the usual sequence of reactions (Cuebas and Schultz, 1982; Luo et al., 1994; Du Plessis, 1979; Tserng and Jin, 1990; Ren and Schulz, 2003). Since this second mitochondrial pathway involves the reduction of conjugated diene β-oxidation intermediates, it is an attractive pathway for oxidation of conjugated fatty acid isomers. In fact, it may be that the reductase-dependent pathway is the reason β-oxidation of 9*c*,11*t*-18:2 and 10*t*,12*c*-18:2 by rat liver slices was about two-fold higher than for 9*c*-18:1 and that oxidation of 9*c*,11*t*-18:2 was 20% higher than 10*t*,12*c*-18:2 (Gruffat et al., 2003). However, inspection of the β-oxidation pathways does not reveal an explanation for the 78% higher amount of oxidation observed for CLA in the 1,3 acyl TAG position compared to CLA in 2-acyl TAG position (Chardigny et al., 2003).

9t- versus 9c-18:1 Oxidation

Several studies with isolated mitochondria and peroxisomes from rat liver have also shown that the β-oxidation of 9*t*-18:1 is greater than for 9*c*-18:1 (Guzman et al., 1999; Ide and Sugano, 1984; Menon et al., 1983; Neat et al., 1981). For example, total oxidation of 9*t*-18:1 was 1.7 times higher than 9*c*-18:1, in rat mitochondria (Guzman et al., 1999) and 50% higher in peroxisomes (Neat et al., 1981) and in vivo human studies have shown that carbon-13 labeled 9*t*-18:1 was more highly oxidized than 9*t*-18:1-^{13}C (Delany et al., 2000). Total fat oxidation was also about 20% higher for human subjects fed high *trans* diets than subjects fed a high 9*c*-18:1 diets (Lovejoy et al., 2002), which suggests *trans* fatty acids upregulate fatty acid β-oxidation. Although the classical mitochondrial isomerase-dependent pathway is responsible for oxidation of over 85% of the non-conjugated dietary *cis* and *trans* unsaturated fatty acids, in vitro rat liver studies show that the peroxisomal pathway contributes to oxidation of C_{18} fatty acids and preferentially oxidized 9*t*-18:1 relative to 9*c*-18:1(Guzman et al., 1999). Calculations based on acid soluble oxidation products in rat liver indicate that oxidation of 9*t*-18:1 by peroxisomes was about 43% higher than for 9*c*-18:1 in vitro. However, there are also studies that show no little difference between the oxidation of 9*t*- and 9*c*-18:1 (Ono and Fredrickson, 1964; Anderson and Coots, 1967; Ide, and Sugano, 1986; Royce and, Holmes, 1984; Ide et al, 1987) or that 9*t*-18:1 is oxidized more slowly than 9*c*-18:1 (Ide et al., 1987; Anderson, 1967; Lanser et al., 1986). An example are results from human heart homogenates, which are nearly devoid of peroxisomes, that show oxidation of 9*t*-18:1 was equal to 9*c*-18:1 but 9*t*-18:1 was more slowly oxidized by rat heart homogenates (Lanser et al., 1986). A recent human study that compared oxidation of carbon-13 labeled 9*t*-18:1, 9*c*-18:1, 18:2n-6 and 18:3n-3 in adult women, found no difference between the 9 h cumulative oxidation of ^{13}C-9*t*-18:1 (19%), ^{13}C-9*c*-18:1 (20%), and ^{13}C-18:3n-3 (19%) but ^{13}C-18:2n-6 (12%) cumulative oxidation was significantly lower (McCloy et al., 2004).

Oxidation of 18:1 Positional Isomers

β-oxidation of octadecenoic acid isomers produces an array of metabolic intermediates, some of which accumulate and escape from the mitochondria. For example, the oxidation of 8, 9, 10, and 11-18:1 *cis* and *trans* isomers during the perfusion of rat heart resulted in the leakage of the 5*t*-14:1 and 5*t*-12:1 metabolites formed from the oxidation of 9*t*- and 11*t*-18:1 (Willebrands and Van der Veen, 1966). The other *cis* and *trans* positional isomers did not produce detectable levels of similar chain-shortened fatty acid metabolites. A comparison of 18:0, 9*c*-18:1 and 9*t*-18:1 β-oxidation by rat liver and heart mitochondria has recently confirmed the formation of the 5*t*-14:1 metabolite of 9*t*-18:1 and that it escapes from the β-oxidation cycle (Liang, et al, 2004). No accumulation of similar metabolites was observed for 18:0 and 9*c*-18:1. However, results from oxidation studies using carboxyl and methyl ^{13}C labeled fatty acids suggest that, except for 9*c*-18:1, most C_{18} fatty acids are not completely catabolized to CO_2 because intermediates escape from mitochondria (Delany et al., 2000).

Oxidation of 18:2 Isomers

A number of studies have investigated the β-oxidation of various mono- and di-*trans* methylene-interrupted polyunsaturated fatty acid isomers (Ono and Fredrickson, 1964; Ide and Sugano, 1984; Fukuda et al., 1993; Coots, 1964b; Hiltunen et al., 1983; Chen et al., 2001; Bretillon et al., 1998b, 2001) and non-methylene interrupted conjugated linoleic and linolenic acid isomers (Gruffat et al., 2003; Sakono et al., 1999; Demizieux et al., 2002; Sergiel et al., 2001). Beta-oxidation rates for 9*t*,12*t*- and 9*c*,12*t*-/9*t*,12*c*-18:2 isomers are not greatly different than the rate for *trans* 18:1 isomers (Ide and Sugano, 1984; Fukuda et al., 1993; Coots, 1964; Anderson and Coots, 1967). The β-oxidation rates for 9*c*,12*t*- + 9*t*,12*c*-18:2 mixtures are typically much higher than 18:2n-6 and intermediate between *t*,*t*-18:2 and 18:2n-6 (Ide and Sugano, 1984; Fukuda et al., 1993; Coots, 1964), although there are exceptions that show no differences (Ono and Fredrickson, 1964; Anderson and Coots, 1967). The above relative β-oxidation rate data are reasonably consistent with % oxidation results obtained by a whole body balance study with rats (Chen et al., 2001). In this study, the % oxidation of a 9*c*,13*t*-18:2 was also reported and found to be similar to the 9*c*,12*t*- and 9*t*,12*c*-18:2 isomers.

However, the 9*t*,12*c*-18:2 and 9*c*,12*t*-18:2 isomers are not equally oxidized. A study in fasting rats, fed individual 9*t*,12*c*-18:2-^{14}C and 9*c*,12*t*-18:2-^{14}C isomers, observed higher $^{14}CO_2$ production from rats administered the 9*c*,12*t*-18:2 isomer and oxidation of the 9*t*,12*c*-18:2 isomer was similar to labeled 18:2n-6 (Bretillon et al., 1998b). The same study also compared the oxidation of 9*c*,12*t*,15*t*-18:3 and 9*t*,12*c*,15*c*-18:3 to 18:3n-3 and 18:2n-6. Total oxidation of the *t*-18:3 isomers was similar to 18:3n-3 but oxidation of the *t*-18:3 fatty acid isomers was about 15% higher than 9*t*,12*c*-18:2 and 18:2n-6. In a related study, the oxidation by male subjects of ^{13}C labeled 9*c*,12*t*-18:2 and 9*c*,12*c*,15*t*-18:3 was compared to 18:2n-6 and 18:3n-3 by measuring $^{13}CO_2$ excretion over 48 h. The results showed that oxidation of 9*c*,12*t*-18:2, 9*c*,12*c*,15*t*-18:3 and 18:3n-3 were similar (23.4–24.7%) and significantly higher than for 18:2n-6 (14.0%) (Bretillon et al, 2001). These results are consistent with results from the rat studies that show *trans* 18:2 isomers are oxidized more rapidly than 18:2n-6. For example, a whole-body balance study with rats showed that apparent oxidation (ca. 74%) of the 9*c*,12*t*-, 9*t*,12*t*- and 9*c*,13*t*-18:2 isomers was similar to and much higher than 18:2n-6 oxidation (ca. 40%) and slightly less than 9*t*-18:1oxidation (80%) (Chen et al., 2001).

Compared to 18:1 and 18:2 isomers, oxidation of conjugated linoleic acid isomers (9*c*,11*t*-18:2 and 10*t*,12*c*-18:2) has not been extensively investigated. There does appear to be a large difference between the oxidation of the two primary CLA isomers present in dietary fats (Gruffat et al., 2003; Sakono et al., 1999; Sergiel et al., 2001). For example, oxidation of 9*c*,11*t*-18:2 is reported to be 20% higher than 10*t*,12*c*-18:2 (Gruffat et al., 2003) and CLA oxidation is about two-fold higher than 9*c*-18:1 (Gruffat et al., 2003, Sergiel et al., 2001). Results with rats report that β-oxidation of ^{14}C-CLA isomers was about 15% greater than 9*c*,12*c*-18:2 (Sergiel et al., 2001). An interesting three-fold increase in peroxisomal β-oxidation activity occurred for rats fed a conjugated 18:3 diet compared to rats fed CLA, 18:2n-6 or 18:3n-3 diets but only a modest increase occurred in mitochondrial β-oxidation activity for conjugated 18:3 fed rats. The result implies that conjugated 18:3 may stimulate peroxisomal β-oxidation because it is more difficult to oxidize by the mitochondrial β-oxidation pathway than CLA (Koba et al., 2002).

The human plasma TAG selectivity values in Fig. 4.1 reflect removal of fatty acids from the TAG pool. The TAG selectivity values for the fatty acid isomers are all negative relative to 9*c*-18:1 and are consistent with a higher β-oxidation

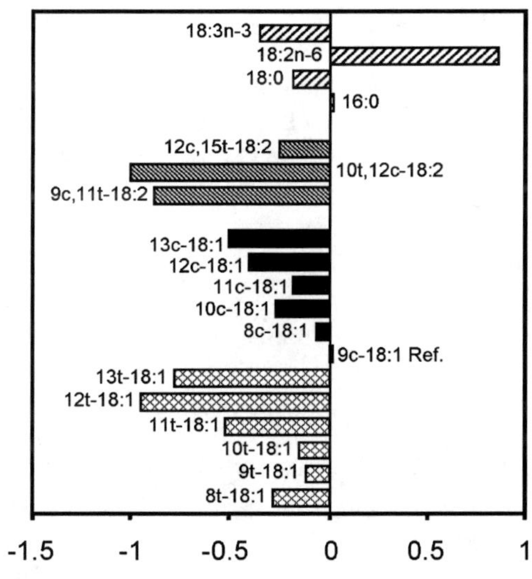

Fig. 4.1. Selectivity values for human plasma triacylglycerol.
Positive selectivity values indicate preferential incorporation and negative values indicate discrimination relative to 9*c*-18:1. Selectivity values are relative values and are equivalent to % difference divided by 100.

rate reported for 9*t*-18:1 by most in vitro and in vivo studies. The selectivity values for 9*t*-18:1, 18:2n-6 and 18:3n-3 are consistent with results from a recent carbon-13 labeled fatty acid study that reported a correlation between tracer levels in plasma TAG and β-oxidation (McCloy et al., 2004). This relationship is not surprising since the majority of ingested fatty acids are oxidized rather than transferred to tissue phospholipids and cholesterol esters. The negative TAG selectivity values provide indirect support for observations that oxidation rates for isomeric fatty acids are generally higher than for 9*c*-18:1 and for the relative differences between the oxidation rates for individual isomers. The approximately 50% lower than expected *trans* fatty acid content of animal and human adipose tissue and the relative distribution pattern of positional 18:1 isomers in adipose tissue based on intake data also provides indirect evidence that most *trans* and *cis* 18:1 isomers are preferentially oxidized relative to 9*c*-18:1 and that individual isomers are oxidized at different rates (Ohlrogge et al., 1981; Hudgins et al., 1991; Emken 1995).

There are inconsistencies between results from the various oxidation studies which may be due to several factors such as the amount and proportions of fatty acids in diet (Neat et al., 1981; Becker, 1984; Malewiak et al., 1988), age (Wisneski et al., 1987), adaptation to high *trans* intake (Royce and Holmes, 1984; Hsu and Kummerow, 1977; Giudetti et al., 2003; Flatmark et al., 1988), experimental protocol (fasted versus non-fasted, oral versus injection, in vitro versus in vivo), isotope label position in fatty acid substrates (Delany et al., 2000), gender (Menon and Dhopeshwarka, 1983) and genetics (Ide and Sugano, 1986; Ide et al., 1987). Based on the combined available data, an approximate overall pattern for the relative β-oxidation rates is 9*t*,12*t*-18:2 = 9*t*-18:1 > 9*c*-18:1 = 9*c*,12*t*-18:2 = mono-*trans* 18:3 = 18:3n-3 > 9*t*,12*c*-18:2 = 18:2n-6.

Desaturation

Mammalian systems contain three oxygen-dependent desaturases Δ9, Δ6, and Δ5 that have varying degrees of substrate specificity and provide a capability for introducing *cis* double bonds into essentially all isomeric and non-isomeric fatty acids present in PHVO. Nearly all of the fatty acids and their desaturated products can also be chain-elongated by the addition of acyl units to the carboxyl end of the fatty acid chain. The general desaturation-elongation sequence is illustrated in Fig. 4.2. In this general scheme, the major desaturation pathways and preferred substrates are Δ9-desaturation of 16:0 and 18:0 to monounsaturated fatty acids, Δ6-desaturation of 18:2n-6 and 18:3n-3 to 18:3n-6 and 18:4n-3, and Δ5-desaturation of 20:3n-6 and 20:4n-3 to 20:4n-6 and 20:5n-3. With few exceptions, most fatty acid isomers are not good substrates for these desaturases, and their desaturated products in tissues are present either at or below detection limits when diets contain adequate levels of 18:2n-6.

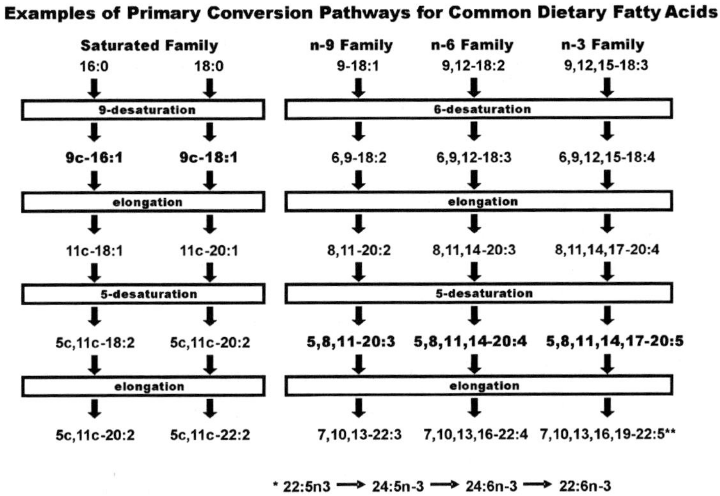

Fig. 4.2. Primary conversion pathways for common dietary fatty acids.

Monounsaturated Isomers

In vitro results show that most *trans* and *cis* positional C_{18} fatty acid isomers can be desaturated and elongated within the context of the conversion pathways shown in Fig. 4.2. EFA deficiency enhances desaturate activity, and desaturated products can be detected for most *trans* 18:1 positional isomers when incubated with rat liver microsomes from EFA deficient rats (Pollard et al., 1980). In general, *trans* 18:1 isomers with double bonds located nearer to the methyl and carboxyl ends of the fatty acid acyl chain are better substrates for Δ9 desaturase than fatty acid isomers with double bonds in the center portion of the fatty acid chain (Mahfouz et al., 1980b; Pollard et al., 1980). This is the opposite of the specificity observed for elongation of *t*-18:1 positional isomers to *t*-20:1 products (Kameda et al., 1980). The main desaturase products were the expected *t,c*-18:2 isomers although some studies report that *c,c*-18:2 products were detected also (Riisom and Holman, 1981; Mahfouz et al., 1980b; Pollard et al., 1980; Miller et al., 2003). The percent desaturation of the 8*t*-, 9*t*-, and 10*t*-18:1 isomers was less than 1%, which contrasts with the over 10% desaturation of the 6*t*- and 13*t*-18:1 isomers. Interestingly, 4–5% of the 11*t*-, 12*t*-, 13*t*- and 14*t*-18:1 isomers were converted also to *c,c*-18:2 isomers with a conversion of 23% reported for the 5*t*-18:1 isomer (Mahfouz et al., 1980b). This suggests the possibility of a *trans* to *cis* isomerase. In a similar study, the desaturation of a series of both *cis* and *trans* 18:1 positional isomers by rat liver microsomes also found little Δ9-desaturation of the 8*t*-, 9*t*-, and 10*t*-18:1 isomers but over 40% Δ9-desaturation for the 11*t*-, 13*t*-, 14*t*-, and 15*t*-18:1 isomers (Pollard et al., 1980).

In contrast to *trans* 18:1 isomers, the *cis* 18:1 positional isomers were not good substrates for Δ9 desaturase (stearoyl-CoA desaturase), although small amounts of products were formed from the 13*c*-, 14*c*-, and 15*c*-18:2 isomers (Pollard et al., 1980). A more recent study compared the Δ9-desaturation of the *cis* and *trans* 7- and 11-18:1 isomers (Broadwater et al., 2002). In this study, desaturation of 7*t*-18:1 isomers produced three main products (7*t*,9*c*-, 7*t*,9*t*-, 7*t*,10*c*-18:2) but 11*t*-18:1 produced only 9*c*,11*t*-18:2 and no desaturation products were formed from the 7*c*- and 11*c*-18:1 isomers. Interestingly, Δ6-desaturation was observed also for most *cis* and *trans* 18:1 isomers with a double bond between the 8 and 13 positions and also for 14*c*- and 15*c*-18:1. In contrast, Δ5-desaturation was observed only for the 9*t*-, 10*t*-, 14*t*-, 15*t*-18:1, 8*c*-18:1 and 11*c*-18:1 isomers (Pollard et al., 1980). The experimental model can impact in vitro results. For example, significant conversion in vitro of 12*c*-18:1 to 18:2n-6 was reported for hen, pig, and goat but not for rat and hamster (Gurr et al., 1972) which suggests that in some species a second "Δ9 desaturase" is presence that may be able to use other isomers as substrates also.

Result from in vivo studies differ considerably from in vitro studies. The preferred substrates for Δ9 desaturase are 16:0 and 18:0 (Enoch et al., 1976; Elovson, 1965) and, in contrast to in vitro results for 18:1 isomers, substantial in vivo conversion of *trans* and *cis* 18:1 positional isomers to *c*,*t*- and *c*,*c*-18:2 isomers by Δ9-desaturation is generally not observed. For example, isotope studies in human subjects (Emken et al., 1979a; Emken et al., 1979b, 1980, 1983, 1985, 1986, 1989) and laying hens (Mounts et al., 1971; Lanser and Emken, 1981; Lanser and Emken, 1987) detected little or no formation of isotope-labeled desaturated products for most of the major *cis* and *trans* 18:1 positional isomers in PHVO but there are some exceptions. One is the desaturation in vivo of 11*t*-18:1 (Bauman et al., 2003). Results from recent diet studies suggest substantial conversion in vivo of 11*t*-18:1 to 9*c*,11*t*-18:2 in mice (Santora et al., 2000; Lin et al., 2004), rats (Banni et al., 2001; Corl et al., 2003; Adam et al., 2004), and humans (Salminen et al., 1998; Turpeinen et al., 2002). However, results from an isotope study in humans (Adlof et al., 2000) are not consistent with the estimated 19% conversion based on diet studies (Turpeinen et al., 2002). In this study a high (33%) isotopic enrichment was observed in the 9*c*,11*t*-18:2 Δ9-desaturated product formed from 11*t*-18:1-2H but the estimated % conversion based on actual weight data was low (ca. 1%) and conversion of labeled 11*c*-18:1 to 9*c*,11*c*-18:2 was ca. 0.1%. Another exception was the substantial conversion of labeled 10*t*-18:1 to a conjugated octadecadienoic acid metabolite in chickens (Lanser and Emken et al., 1987) which suggests the presence of a Δ12 desaturase.

In addition to desaturated fatty acid metabolites of *cis* and *trans* 18:1 isomers, significant levels of chain-shortened *t*-16:1 positional isomers and elongated *t*-24:1 have been identified in PHVO fed rats (Wood, 1979) and small amounts of *cis* and *trans* 14:1 and 16:1 positional isomers are present in human tissues (Adlof and Emken, 1986; Hudgins et al., 1991). Also identified have been other metabolites that cannot be explained by the fatty acid conversion pathways in Fig. 4.2. For example, 8*t*-18:1 was found in tissue lipids of rats that had not been fed

8*t*-18:1 (Wood, 1979) and labeled 9*t*,12*c*-18:2 was found in plasma lipids from subjects fed labeled 12*c*-18:1 (Emken et al., 1980). A plausible explanation for the formation of these metabolites is that they arise from isomerized β-oxidation intermediates that escape from the β-oxidation cycle (see β-oxidation section). A β-oxidation intermediate (4*t*-14:1 or 2*t*-12:1) could be elongated to 8*t*-18:1 and a 5*t*,8*c*-14:2 or 3*t*,6*c*-12:2 intermediate of 12*c*-18:1 could be elongated to 9*t*,12*c*-18:2.

Polyunsaturated Isomers

The predominant PUFA isomers in PHVO, heat-treated vegetable oils and dairy fats are 9*c*,12*t*-18:2, 9*t*,12*c*-18:2, 9*t*,12*c*,15*c*-18:3, 9*c*,12*c*,15*t*-18:3, 9*c*,11*t*-18:2 and 10*t*,12*c*-18:2. Most *trans* and positional C_{18} polyunsaturated fatty acid isomers can be desaturated and/or elongated to form isomeric analogs of the 18:2n-6 and 18:3n-3 metabolites. A number of studies are listed in Table 4.1 (in vitro studies) and Table 4.2 (in vivo studies) that illustrate the conversion of various methylene and non-methylene-interrupted polyunsaturated isomeric structures. The relative conversion rates of these isomers have been compared to 18:2n-6 and 18:3n-3 in several different in vitro and in vivo experimental models. The 9*t*,12*t*-18:2 isomer has also been well investigated (see review Kinsella et al., 1981) but studies related to the metabolism of this isomer are generally not

TABLE 4.1
In vitro Formation of Desaturated and Elongated Products from PUFA Isomers

Model	Isomer Substrates	Products	Reference
Rat liver microsomes	eight *c,c*-18:2 and six *c,c,c*-18:3 positional isomers	variable amounts of $C_{20:2}$ and $C_{20:3}$ elongated products	Ludwig and Sprecher, 1979
Perfused rat liver	*t,t*-18:2 and *c,t*-/*t,c*-18:2	no *c,t*/*t,c*—20:4 detected	Fukuda et al., 1993
Microsomes	5*t*,9*c*,12*c*-18:3	no desaturated product	Houtsmuller, 1981
Mouse liver microsomes	9*c*,11*t*-18:2	conjugated 18:3	Belury and Kempa-Steczko, 1997
Rat liver microsomes	9*c*,12*c*,15*t*-18:3	6*c*,9*c*,12*c*,15*t*-18:4	Chardigny et al., 1997
Perfused rat liver	9*c*,12*t*- and 9*t*,12*c*-18:2 9*c*,12*c*,15*t*- and 9*t*,12*c*, 15*c*-18:3	*t*-18:3n-6, 17*t*-20:5n-3, *t*-20:2n-6 and 17*t*-20:3n-3,	Bretillon et al., 1998a
Rat liver microsomes	9*t*,12*c*- and 9*c*,12*t*-18:2	6*c*,9*c*,12*t*-18:3; 11*c*,14*t*-20:2; 11*t*,14*c*-20:2	Berdeaux et al., 1998
Bovine endothelial cells	5*c*,8*c*,11*t*,14*c*,17*c*-20:5 5*c*,8*c*,11*c*,14*c*,17*t*-20:5 5*c*,8*c*,11*t*,14*c*,17*t*-20:5	All isomers elongated to *t*-22:5 and *t*-24:5	Loi et al., 2000b
Rat liver microsomes	9*c*,11*t*-18:2 and 8*c*,11*c*,13*t*-20:3	6*c*,9*c*,11*t*-18:3; 8*c*,11*c*,13*t*-20:4; 5*c*,8*c*,11*c*,13*t*-20:4	Berdeaux et al., 2002
Rat liver slices	9*c*,11*t*- and 10*t*,12*c*-18:2	6*c*,9*c*,11*t*- and 6*c*,10*t*,12*c*-18:3	Gruffat et al., 2003.

TABLE 4.2
In vivo Formation of Desaturated and Elongated Products from PUFA Isomer

Model	Isomer Substrates	Products[b]	Reference
Rat, EFAD[a]	9*t*,12*t*-18:2 and *t*-18:3 mix	14*t*-20:4 and *t*-20:5n-3	Blank and Privett, 1963
Rat, EFAD	16:2n-4; 16:3n-7; 17:2n-5; 17:4n-5; 19:2n-6 (all *cis*)	18:2n-6 and 18:3n-3 isomer analogs	Schlenk and Sand, 1967
Rat, EFAD	9*c*,15*c*- and 12*c*,15*c*-18:2	no Δ9 or Δ6-desaturated products	Sprecher et al., 1967
Rat, EFAD	11*c*,14*c*- and 11*t*,14*t*-20:2	*c*,*c*-20:4; no *t*,*t*-20:4	Stearns et al., 1967
Rat, EFAD	9*t*,12*t*-; 9*c*,12*t*-; 9*t*,12*t*-18:2	14*t*-20:4 and 9*t*-20:4; no *t*,*t*-20:4	Privett et al., 1967
Rat	*t*,*t*-18:2 and *c*,*t*-/*t*,*c*-18:2	*c*,*t*/*t*,*c*-20:4	Anderson et al., 1975
Rat	*t*,*t*-18:2 and *c*,*t*-/*t*,*c*-18:2	*c*,*t*/*t*,*c*-20:4	Cook, 1980
Rat	*trans*-18:3n-3 mixture	17*t*-20:5n-3 and 19*t*-22:6n-3	Grandgirard et al., 1989
Rat	*trans*-18:3n-3 mixture	*t*-20:5n-3	Blond et al., 1990
Mouse	9*c*,12*t*- and 9*t*,12*c*-18:2	14*t*-20:3, 11*t*-20:4 and 14*t*-20:4	Beyers and Emken, 1990
Rat	*trans*-18:3n-3 mixture	17*t*-20:5n-3 and 19*t*-22:6n-3	Grandgirard et al., 1994
Rat	12*c*-18:1 and 9*c*,13*t*-18:2	5*c*,8*c*,14*c*-20:3 and 5*c*,8*c*,11*c*,15*t*-20:4	Ratnayake et al., 1994
Lamb	CLA isomers	conjugated 18:3	Banni et al., 1996
Rat	9*c*,12*t*- and 9*t*,12*c*-18:2	14*t*-20:4 and 11*t* -20:4	Berdeaux et al., 1996
Rat	9*t*,12*c*,15*c*- and 9*t*,12*c*,15*t*-18:3 mixture	11*t* -20:5 and 11*t*,17*t*-20:5	Chardigny et al., 1996
Rat	9*c*,11*t*- and 10*t*,12*c*-18:2	13*t*-20:4; 12*t*-20:3 and 12*t*-20:4	Sébédio et al., 1997
Rat	*trans*-18:3 mixture	17*t*-20:5; 19*t*-22:5; 19*t*-22:6	Grandgirard et al., 1998
Rat, EFAD	9*c*,11*t*- and 10*t*,12*c*-18:2	13*t*-20:4 and 12*t*- 20:4	Sébédio et al., 1999
Rat	*trans*-18:3 mixture	17*t*-20:5 and 19*t*-22:5	Loi et al., 2000a
Rat	9*t*,12*t*-; 9*t*,12*c*-; 9*c*,12*t*-; 9*c*,13*t*-18:2	15*t*-20:4 and 14*t*-20:4	Chen et al., 2001
Rat	9*c*,11*t*- and 10*t*,12*c*-18:2	conjugated 16:2, 18:3, 20:3 isomers	Sébédio et al., 2001
Rat	CLA Isomers	conjugated 18:3, 20:3, 20:4 isomers	Banni et al., 2001b
Human	10*t*,12*c*-18:1	6*c*,10*t*,12*c*-18:3	Emken et al., 2002
Rat	9*c*,11*t*,13*t*-18:3	9*c*,11*t*-18:2 (biohydrogenated product)	Tsuyoshi et al., 2004

[a] *Abbreviation*: EFAD, essential fatty acid deficient

[b] Only the position of the trans double bonds are listed for the C_{20} and C_{22} PUFA metabolites. *Cis* double bonds not listed are at positions shown in Fig. 4.2 unless otherwise indicated.

specifically cited in this review because only tracers of this isomer are present in edible fats and vegetable oils.

The conversion rates for specific isomeric PUFA acids differ considerably and reflect the sensitivity of the Δ6 and Δ5 desaturases to double bond configuration and position. For example, Δ6-desaturation-elongation rates for conversion of 9*c*,12*t*-18:2 to 5*c*,8*c*,11*c*,14*t*-20:4 in vivo by mice and rats are nearly equal to conversion rates for 18:2n-6 to 20:4n-6 and are 4 – 5 times greater than conversion of 9*t*,12*c*-18:2 (Beyers and Emken, 1990; Berdeaux et al., 1996). These in vivo data are similar to in vitro results with rat liver microsomes (Berdeaux et al., 1998) and are consistent with in vivo qualitative results from EFAD rats (Blank and Privett, 1963). Formation of 6*c*,9*c*,12*t*-18:3 by Δ6-desaturation was 16 times greater for 9*c*,12*t*-18:2 than 18:2n-6 in perfused rat liver (Bretillon et al., 1998a). However, conversion in vivo and in vitro of 9*c*,12*t*-18:2 is much lower than for 18:2n-6 in rat brain (Cook, 1980; Cook, 1981). In rats fed PHVO, conversion of 9*c*,13*t*-18:2 to 5*c*,8*c*,11*c*,15*t*-20:4 was as high or higher than conversion of 9*c*,12*t*-20:4 to 5*c*,8*c*,11*c*,14*t*-20:4 (Ratnayake et al., 1994; Chen et al., 2001). Formation of desaturated products from 9*t*,12*t*-18:2 in vivo (Privett et al, 1967; Anderson et al., 1975; Ratnayake et al., 1994) and in vitro (Cook, 1981) was very low or not detected. No desaturated products were detected in vivo studies with 12*c*,15*t*-18:2 (Emken et al., 1987), 12*c*,15*c*-18:2 (Sprecher et al., 1967) and 9*c*,15*c*-18:2 (Sprecher et al., 1967).

A number of studies have reported the conversion of *trans* 18:3 isomers to *trans* 20:5 and *trans* 22:6 isomers (see Tables 4.1 and 4.2) Metabolites of *trans* 18:3 isomers identified in liver lipids from rats fed mixtures of *trans* 18:3 isomers include 5*c*,8*c*,11*c*,14*c*,17*t*-20:5 (17*t*-20:5), 5*c*,8*c*,11*t*,14*c*,17*c*-20:5 (11*t*-20:5) and 5*c*,8*c*,11*t*,14*c*,17*t*-20:5 (11*t*,17*t*-20:5) 7*c*,10*c*,13*c*,16*c*,19*t*-22:5 (19*t*-22:5), 4*c*,7*c*,10*c*,13*c*,16*c*,19*t*-22:6 (19*t*-22:6) (Blond et al., 1990; Grandgirard et al., 1989; Chardigny et al., 1996; Grandgirard et al., 1998; Loi et al., 2000a). Conversion of 18:3n-3 to 20:5n-3 by perfused rat liver was 5 – 13 times higher than conversion of 9*c*,12*c*,15*t*-18:3 and 9*t*,12*c*,15*c*-18:3 to 17*t*-20:5 and 11*t*-20:5 respectively (Bretillon et al., 1998a). Low levels (0.01 – 0.5%) of 17*t*-20:5 and 19*t*-22:6 were also detected in human platelets (Chardigny et al., 1993). Elongation of 9*t*,12*c*,15*c*-18:3 to 11*t*,14*c*,17*c*-20:3 also occurred and was 2.5 times greater than elongation of 9*c*,12*c*,15*t*-18:3 and 18:3n-3 to their respective 20:3 metabolites (Bretillon et al., 1998a). Bovine aortic endothelial cells were used to show that further elongation of the desaturation products, 17*t*-20:5, 11*t*,17*t*-20:5, and 11*t*-20:5, to their respective 22:4 products can occur (Loi et al., 2000b).

Conjugated linoleic acid (CLA) isomers are converted to 20:2, 20:3, and 20:4 conjugated isomers, presumably by the same desaturases and elongases responsible for conversion of 18:2n-6 to n-6 LC-PUFA. The characterization of CLA metabolites and a discussion of the metabolic pathways have been reviewed in detail (Banni and Martin, 1998; Sébédio and Christie, 2003). Identified metabolites (some tentatively) of 9*c*,11*t*-18:2 include 6*c*,9*c*,11*t*-18:3, 11*c*,13*t*-20:2, 8*c*,11*c*,13*t*-20:3, 5*c*,8*c*,11*c*,13*t*-20:4 and metabolites of 10*t*,12*c*-18:2 include 8*t*,10*c*-16:2, 6*c*,10*t*,12*c*-18:3, 12*t*,14*c*-20:2, 8*c*,12*t*,14*c*-20:3 5*c*,8*c*,12*t*,14*c*-20:4 (Banni et al., 1996; Berdeaux et al., 2002; Sébédio et al., 1997, 1999, 2001).

Three studies report an unexpected distinct difference between the relative % conversion of 9*c*,11*t*-18:2 and 10*t*,12*c*-18:2 in vivo to their respective desaturated, elongated, and chain-shortened products (Sébédio et al., 1997, 2002; Emken et al, 2002). In rat liver, the two main metabolites of 10*t*,12*c*-18:2 were 8*t*,10*c*-16:2 and 6*c*,10*t*,12*c*-18:3 and the main metabolite of 9*c*,11*t*-18:2 was 8*c*,11*c*,13*t*-20:3 (Sébédio et al., 2002;). In a study with adult women fed labeled 9*c*,11*t*- and 10*t*,12*c*-18:2, the only labeled CLA metabolite detected in measurable amounts was 6*c*,10*t*,12*c*-18:3-^{2}H and its elongated 8*c*,12*t*,14*c*-20:3 metabolite was not detected. (Emken et al, 2002). The 6*c*,10*t*,12*c*-18:3-^{2}H CLA metabolite was detected in plasma TAG but not in phospholipid samples which is consistent with results for rat liver lipids (Sébédio et al, 2002). These results suggest that CLA metabolites are probably poor substrates for the phospholipid acyltransferases. Other studies have demonstrated that both CLA isomers and their metabolites can be desaturated and elongated to conjugated LC-PUFA isomers by rodents (Nugteren, 1970; Belury and Kempa-Steczko, 1997; Sébédio et al., 1997; Loi et al., 2000b; Banni et al., 2001b; Sébédio et al., 2001; Berdeaux et al., 2002; Gruffat et al., 2003). Clearly, the presence of a conjugated diene structure perturbs the π electron density and spatial arrangement or conformation of CLA isomers compared to 9*c*,12*c*-18:2 and introduces interesting metabolic properties.

As a general rule, isomers that have low conversion rates to 20:4 and 20:5 metabolites tend to be converted to "dead-end" products, such as 5*c*,11*c*,14*t*-20:3, or elongated to 20:2 isomers. Isomers that are poor substrates for Δ6 desaturase also tend to inhibit 18:2n-6 and 18:3n-3 conversion and the presence of a 9*c*- or 12*c*- double bond generally facilitates Δ6-desaturation of methylene-interrupted 18:2 and 18:3 isomers.

Inhibition of Fatty Acid Desaturation/Conversion

Unsaturated fatty acids (isomeric and non-isomeric) inhibit Δ9-, Δ6-, and Δ5-desaturation via various mechanisms, such as substrate inhibition, product inhibition, and competitive inhibition. The degree of inhibition is related to fatty acid structure (double bond position, double bond configuration, and acyl chain-length). In general, n-6 PUFA inhibit Δ6-desaturation to a greater degree than 9*c*-18:1 and 18:3n-3 but the relative amount of inhibition is different for various organs or tissue types.

Inhibition in vivo

Diets containing PHVO generally depress Δ6- and Δ5 desaturase activities relative to non-PHVO diets and dietary intake of isomeric fatty acids tends to be associated with lower 20:4n-6 levels in tissues. However, dietary intake of isomeric fatty acids tend not to be associated with lower 20:5n-3 and 22:6n-3 tissue levels. For example, the % 20:4n-6 in liver phospholipid from female rats fed PHVO (Lawson et al., 1983) decreased by 35% but % 20:5n-3 and 22:6n-3 increased by 85 and 20%, respectively, and the decrease in % 20:4 inversely correlated with % 12*c*-18:1. In a related study, male rats fed diets supplemented with *t*-18:1 isomer concentrates, the level of 20:4n-6 and 22:6n-3 in rat liver phosphatidylcholine decreased by 50 and 15% but diets supplemented with *c*-18:1 isomers concentrates decreased 20:4n-6 and 22:6n-3 by 15 and 40%, respectively (Lawson et al, 1985). These results demonstrate that the influence of *cis* and *trans* 18:1 isomers on 20:4n-6 and 22:6n-3 is clearly different. However diets containing 45% 9*t*-18:1 with a n-6/n-3 ratio of 1.4 produced modest decreases of 20:4n-6 and 22:6n-3 (15–20%) in liver PC of female rats compared to zero trans control diets containing either palmitic or oleic acid (Astorg and Chevalier, 1988) which suggests *t*-18:1 positional isomers other than 9*t*-18:1 are major contributors to decreased LC-PUFA synthesis in vivo.

Fatty acid isomer acyl chain length has an influence on desaturation. Rats fed diets supplemented with 18:1 isomers depressed Δ6-desaturation but not Δ5-desaturation, whereas diets containing 20:1 and 22:1 isomers had a greater effect and depressed both Δ6- and Δ5-desaturation (Holmer et al., 1982; Kirstein et al., 1983). Results for rat brain and rat liver are markedly different. In young rat brain, 9*c*-18:1 is 2–4 times more effective inhibitor of Δ6-desaturation than 9*t*-18:1 or a mixture of *trans* 18:1 isomers and 18:2n-6 is also a stronger inhibitor of brain Δ6 desaturase (0.5–1.5 times) than *trans* 18:1 isomers (Cook, 1981).

Other factors appear as or more important than percent conversion of 18:2n-6 and 18:3n-3 for determining the levels of 20:4n-6, 20:5n-3 and 22:6n-3 in tissue lipids. For example, conversion of ^{14}C-18:2n-6 to ^{14}C-20:4n-6 was about three times higher in rats fed a high 9*c*-18:1 diet than in rats fed a PHVO diet but the PHVO (40% *trans*) diet did not significantly reduce 20:4n-6 levels in liver and heart phospholipids relative to the 9*c*-18:1 diet (Zevenbergen and Houtsmuller, 1989). These results were consistent with an earlier study that showed adequate levels of dietary 18:2n-6 prevented reduction of 20:4n-6 in heart and liver phospholipid of rats fed 20 en% *trans* diet (Zevenbergen et al., 1988). A more recent study also found no reduction in liver PL 20:4n-6 of rats fed diets containing 4.5–15% *trans* fatty acids and 10% 18:2n-6. In addition, these *trans* diets increased the % 22:6n-3 in liver PC and PE compared to the zero trans-10% 18:2n6 control diet. (Bysted et al, 1998). Levels of n-6 and n-3 LC-PUFA in liver microsomes were also not different in pregnant rats fed a 14% *trans* diet compared to a zero *trans* diet despite a 70% reduction in liver Δ6 desaturase activity (Larque et al., 2000). In a follow-up study using the same diets, no reduction in Δ6 desaturase activity and LC-PUFA were observed in non-pregnant rats (Larque et al., 2003). In the later study, increasing dietary *trans* to 28% significantly increased the *trans* fatty acid levels in liver microsomes but did not further decrease Δ6 desaturase activity and did not reduce LC-PUFA levels compared to controls.

However, *trans* 18:1 and *trans* 18:2 isomers do reduce phospholipid 20:4n-6 levels if intake of 18:2n-6 is low. For example, a 30% reduction in Δ6 desaturase activity correlated with a 25% reduction in 20:4n-6 levels in liver phospholipids of rats fed essential fatty acid deficient diets containing PHVO and hydrogenated coconut oil, (Hill et al., 1982). In a related study, Δ9 desaturase activity in liver microsomes from rats fed diets supplemented with a mixture of *trans* isomers (14% 9*t*-18:1 + 51% *t,t*-18:1 + 18% *ct*/*tc*-18:2) was greatly stimulated. In addition, Δ6 desaturase activity was depressed and correlated with a reduction in the 20:4n-6 and 22:5n-3 content of liver microsomes (De Schrijver and Privett, 1982). These results were likely due to the high *t,t*-18:2 content of the diet supplement and are consistent with results from other studies that show *t,t*-18:2 is a strong Δ6 desaturase inhibitor. In general, results from in vivo rat studies are consistent with the concept that inhibition of Δ6 desaturase by *trans* 18:1 isomers does not significantly compromise n-6 and n-3 LC-PUFA in liver phospholipid if intake of 18:2n-6 and 18:3n-3 is adequate.

An in vivo study in rats suggest that mono-*trans* 18:2 and 18:3 isomers do not inhibit Δ6 desaturase. Diets containing 20% levels of a mixture of *c,t*/*t,c*-18:2 fed to rats for 6 weeks did not reduce 20:4n-6 levels in liver (Anderson et al., 1975) and suggests no inhibition of Δ6 desaturase. The effect of mono-*trans* 18:3 isomers (*t,c,c*- and *c,c,t*-18:3) on inhibition of Δ6 and Δ5-desaturation was explored by measuring desaturase activity in liver microsomes from rats fed LSO and heated-LSO diets for 6 weeks (Blond et al., 1990). Results showed no affect of diet on Δ6 desaturase activity but Δ5 desaturase activity was about 50% higher in livers of heated-LSO fed rats, suggesting *trans*-18:3 isomers stimulated Δ5-desaturation. An alternative possibility is that the four-fold higher 18:3n-3 content of the LSO diet or the approximately 50% higher levels of 20:5n-3 and 22:5n-3 in the LSO fed rat liver lipids depressed Δ5 desaturase activity.

In contrast to methylene-interrupted C_{18} polyunsaturated mono-*trans* isomers, non-methylene-interrupted C_{18} polyunsaturated mono-*trans* isomers (i.e., CLA isomers) appear to influence desaturation of 18:2n-6 in vivo, although results from human diet studies are inconsistent. Decreased Δ9- and Δ6-desaturation and increased Δ5-desaturation was suggested based on changes in the proportions of fatty acids in total serum phospholipid reported for women fed a CLA supplemented diet for 84 days (Smedman and Vessby, 2001). A decrease in plasma phospholipid 20:4n-6 from subjects given CLA for 56 days is consistent with inhibition of Δ6-desaturation (Mougios et al., 2001). In contrast, no significant change in plasma, platelet, and mononuclear cells fatty acid composition occurred in women fed a CLA supplemented diet for 63 days (Benito et al., 2001a, 2001b; Kelley et al., 2001). Decreased synthesis of 18:2n-6 metabolites in mammary gland has been reported also for rats fed CLA diets (Banni et al., 1999) and supports an early suggestion, based on enhanced EFA deficiency symptoms in rats, that CLA isomers inhibit 18:2n-6 desaturation (Holman and Lyman, 1951).

Inhibition in vitro

In vitro studies have been used to assess the influence of a large number of individual isomers on Δ9, Δ6, and Δ5 desaturase inhibition. For example, the effect of the Δ5–16 positional *trans* 18:1 isomers on inhibition of desaturase activities was investigated using rat liver microsomes (Mahfouz et al., 1980a). Relative to 18:0, Δ9-desaturation of 16:0 was reduced about 10% by the 8*t*- and 9*t*-18:1 isomers, 15–25% by the 11*t*- and 14*t*-18:1 isomers, and by about 50% for all other isomers; Δ6-desaturation of 18:2n-6 was reduced 25–30% by most isomers except 14*t*-18:1 (<5% reduction); Δ5-desaturation of 20:3n-6 was reduced about 10–20% by the 6*t*-, 7*t*- and 8*t*-18:1 isomers, about 40% by the 9*t*-, 10*t*-, 13*t*-, and 14*t*-18:1 isomers, and the 11*t*- and 12*t*-18:1 had little effect. In a similar manner, the effect of a series of *cis* 18:1 positional isomers Δ3–Δ17 on rat liver microsomal Δ9, Δ6, and Δ5 desaturases was investigated (Mahfouz et al., 1981). Relative to 18:0, Δ9-desaturation of 16:0 was reduced 35% by the 9*c*-18:1 and 10*c*-18:1 and by 10–15% for all other *cis* 18:1 isomers; Δ6-desaturation of 18:2n-6 was reduced 35% by the 8*c*-18:1 isomers and by 10–20% for all other *cis* 18:1 isomers; Δ5-desaturation of 20:3n-6 was reduced ca. 45% by the 8*c*-18:1 isomer and approximately 15–25% by all the other *cis* 18:1 isomers. In vitro rat liver microsomes showed that 9*c*,12*t*-18:2 is a much more potent Δ6 desaturase inhibitor than 9*t*,12*c*-18:2, which correlated inversely with the much higher rate of 9*c*,12*t*-18:2 conversion to *t*-20:4 (Berdeaux et al., 1998)

Results from rat and human brain studies are considerably different from rat liver microsomal results. A study using rat brain homogenates showed that 9*t*- and 9*c*-18:1 were equally poor inhibitors of Δ9 and Δ6 desaturase and that 18:2n-6 and 18:3n-3 were five times better inhibitors of Δ6 desaturase than 9*t*-18:1 (Cook, 1981). The effect of 11 *cis* and *trans* 18:1 positional isomers (Δ8 to Δ13) and three 18:2 isomers (9*t*,12*c*-, 9*c*,12*t*-, 12*c*,15*t*) on Δ6- and Δ5-desaturation by cultured human brain glioma cells were markedly different for 18:2n-6 and 18:3n-3. All the *t*-18:1 isomers inhibited 18:2n-6 conversion to 20:4n-6 by about 50% relative to 9*c*-18:1 (Cook and Emken, 1990). The 12*c*-18:1 isomer enhanced 18:2 conversion to 20:4n-6 by about 30% in contrast to no affect for all other *c*-18:1 positional isomers. The 11*t*- and 12*t*-18:1 isomers were the only *trans* and *cis* 18:1 positional isomers that inhibited 18:3n-3 conversion to 20:5n-3 (about 30–40% relative to 9*c*-18:1). Relative to 18:2n-6, the 9*c*,12*t*-18:2, 12*c*,15*t*-18:2, 9*t*,12*c*-18:2 and 9*t*,12*t*-18:2 isomers inhibited conversion of 18:2n-6 to 20:4n-6 by about 25, 50, 70, and 85% respectively (Cook and Emken, 1990). Interestingly, the 9*t*,12*c*-18:2 and 12*c*,15*t*-18:2 isomers enhanced 18:3n-3 conversion to 20:5n-3 by about two-fold and 9*c*,12*t*-18:2 enhanced conversion by about 40% relative to 18:2n-6. In contrast, the 9*t*,12*t*-18:2 isomer totally blocked 18:3n-3 conversion to 20:5n-3. When compared to controls with no added fatty acids, 18:2n-6 inhibited 18:3n-3 conversion to a much greater extent than the *c,t*- and *c,t*-18:2 isomers which is consistent with human in vivo results that show 18:2n-6 inhibits 18:3n-3 conversion to n-3 LC-PUFA (Emken et al., 1994).

The effect of *cis* and *trans* 9-18:1 and 11-18:1 isomers on Δ9- and Δ6-desaturation by human skin fibroblasts also differ from liver and brain tissue results (Rosenthal and Whitehurst, 1983). Results show that the 9*c*- and 11*c*-18:1 isomers, but not 9*t*- and 11*t*-18:1 isomers, inhibited Δ9 desaturase by 22 and 12%, respectively, relative to 18:0. In fact, 9*t*-18:1 stimulate Δ9-desaturation of 18:0 by 37%, but 11*t*-18:1 had no effect. 9*c*-18:1 and 11*c*-18:1 inhibited Δ6-desaturation by about 40% whereas 9*t*-18:1 and 11*t*-18:1 inhibited Δ6-desaturation by a factor of 4 and 2, respectively. Relative to 18:2n-6, 9*c*-18:1 inhibition of Δ6 desaturase was about 50% less and 11*t*-18:1 inhibition was similar.

Inhibition of Δ9 desaturase by the CLA isomers (9*c*,11*t*-18:2 and 10*t*,12*c*-18:2) has been reported in rat liver microsomes (Bretillon et al., 1999), mouse liver microsomes (Park et al., 2000) cultured HepG2 cells (Choi et al., 2001), two human breast cancer cell lines (Choi et al., 2002), and in liver and mammary gland of mice (Lin et al., 2004). Elongation of 18:2n-6 to 20:2n-6 was strongly (ca. 50%) inhibited by 9*c*,11*t*- and 10*t*,12*c*-18:2, but not by 9*t*,11*t*- and 9*c*,12*c*-18:2 in yeast cells containing human elongase (Chuang et al., 2001). In rat liver microsomes, 10*t*,12*c*-18:2, but not 9*c*,11*t*-18:2, inhibited Δ9-desaturation of 18:0, in mouse liver microsomes, 10*t*,12*c*-18:2 reduced Δ9 desaturase by 70% compared to 9*c*,11*t*-18:2, 18:2n-6, 9*t*,12*c*-18:2, 9*t*,11*t*-18:2, 9*c*-18:1, and the 9*t*-, 10*t*-, 11*t*-18:1 positional isomers (Park et al., 2000). Both CLA isomers reduced Δ9-desaturation of 16:0 in human breast cancer cell lines, and 10*t*,12*c*-18:2 was a stronger inhibitor of Δ9-desaturation than 9*c*,11*t*-18:2 in mouse mammary gland. An in vitro study with rat liver microsomes indicates about 30% more inhibition of Δ6 desaturase activity by 9*c*,11*t*-18:2 than 10*t*,12*c*-18:2 (Bretillon et al., 1999). Inhibition of Δ6- and Δ5-desaturation of 18:2n-6 and 18:3n-3 by high levels of 10*t*,12*c*-18:2 but not by 9*c*,11*t*-18:2 was observed in human liver cell cultures (Eder et al., 2002).

The inhibition results summarized are based on two general types of studies: in vivo diet studies that use accumulation of n-6 and n-3 metabolites in tissues as a measure of desaturase activity, and in vitro studies that measure desaturase activity. There are inherent problems with both types of studies, and conclusions based on results from both types of studies can be misleading because conversion of 18:2n-6 and 18:3n-3 to n-6 and n-3 LC-PUFA is a complex issue. Many factors other than desaturase activity can influence accretion of n-6 and n-3 metabolites in tissue lipids, and desaturase activity is regulated by factors other than dietary fatty acids. Examples are peroxisomal oxidation, dietary protein, animal model, age, sex, pregnancy, and genetics. Conversely, control and regulation mechanisms are disrupted in cell homogenates, isolated enzyme, and cell culture studies.

Impaired LC-PUFA synthesis related to reduced Δ6-desaturation has been evoked by many authors as an explanation for the inverse correlation observed between levels of 20:4n-6 and 22:6n-3 and *trans* C_{18} fatty acids in various human cord blood and umbilical tissue lipid classes (Koletzko, 1992; Desci, 1995; Elias and Innis, 2001; Decsi et al, 2001, 2002). However, results from in vivo and invitro metabolism studies indicate that inhibition of Δ6-desaturase is an unlikely explanation for reduced n-6 and n-3 LC-PUFA levels in tissues. Inhibition by the modest *trans* fatty acid levels (3–6%) in typical U.S. diets and the low % *trans* (0.3 – 3%) in tissue phospholipids would be trivial compared to inhibition exerted by the much higher levels of n-6 and n-3 fatty acids in diets and tissue phospholipids. Based on the current levels of *trans* fatty acids in diets, it seems unlikely that inhibition of Δ6 desaturase activity by *trans* fatty acid isomers would have a significant metabolic and physiological impact in normal human subjects.

Acyltransferase Specificity

Acyltransferases are responsible for phospholipid and cholesterol esterification. They are quite sensitive to fatty acid structure and they have a critical role in regulation of lipid class fatty acid composition. The focus of this section is to compare the relative specificity of the phospholipid acyltransferases, lecithin: cholesterol acyltransferase (LCAT), and acyl-CoA:cholesterol acyltransferase (ACAT) for isomeric and non-isomeric fatty acid structures.

Phospholipid Acyltransferase

Phospholipid acyltransferases have an inherent specificity for different fatty acid structures and they are responsible for the selective incorporation of specific fatty acids into different phospholipid lipid classes. The specificity of the acyltransferase is a major mechanism responsible for controlling tissue fatty acid composition and maintaining cell membrane fluidity and cell function. Acyltransferase selectivity data are available for a large number of monounsaturated and polyunsaturated fatty acid isomers and illustrate that distribution and incorporation of fatty acid isomers are well controlled by the acyltransferases. In this review, only representative examples of the extensive literature available are cited since there are several general reviews that include comprehensive coverage of fatty acid isomer acylation and incorporation (Emken and Dutton, 1979; Emken, 1984, 1995, 1997; Senti, 1985; Sébédio and Christy, 1998; Sébédio et al., 2003).

Phosphatidylcholine and phosphatidylethanolamine acyltransferase selectivity values for a series (Δ8 to Δ13) of stable-isotope labeled *cis* and *trans* 18:1 positional isomers fed to adult human subjects are summarized in Fig. 4.3 and 4.4. Also included are values for two CLA isomers (9*c*,11*t*-18:2, 10*t*,12*c*-18:2), and 12*c*,15*t*-18:2. Values for four non-isomers (16:0, 18:0, 18:2n-6, 18:3n-3) are included for comparison. The values represent incorporation over a 2–3 day time period and all values are relative to an internal control (9c-18:1) that was included in all studies (Emken et al., 1979a, 1980, 1983, 1985, 1986, 1987, 1989, 1993, 1994, 1998, 1999, 2002). Note: the selectivity values in Fig. 4.3 and 4.4 for PC,

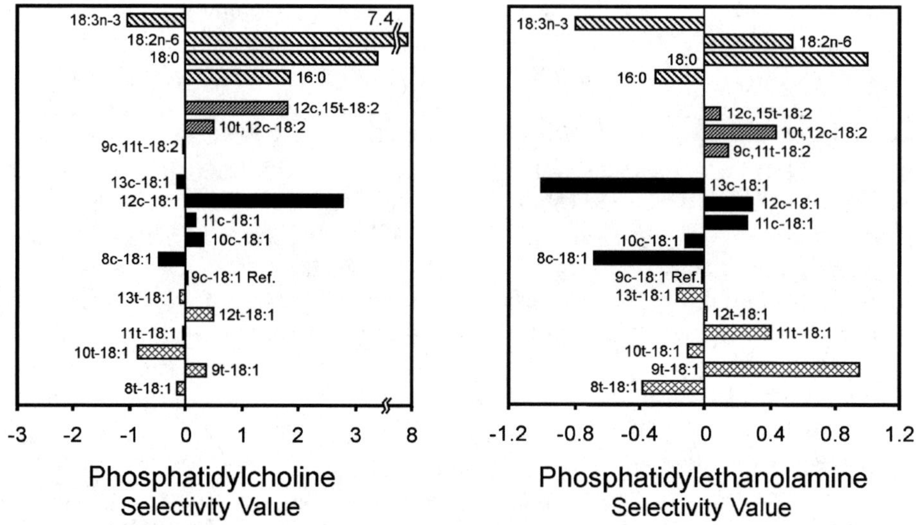

Fig. 4.3. Selectivity values for human plasma phosphatidylcholine and phosphatidylethanolamine. Positive selectivity values indicate preferential incorporation and negative values indicate discrimination relative to 9*c*-18:1. Selectivity values are relative values and are equivalent to percent difference divided by 100.

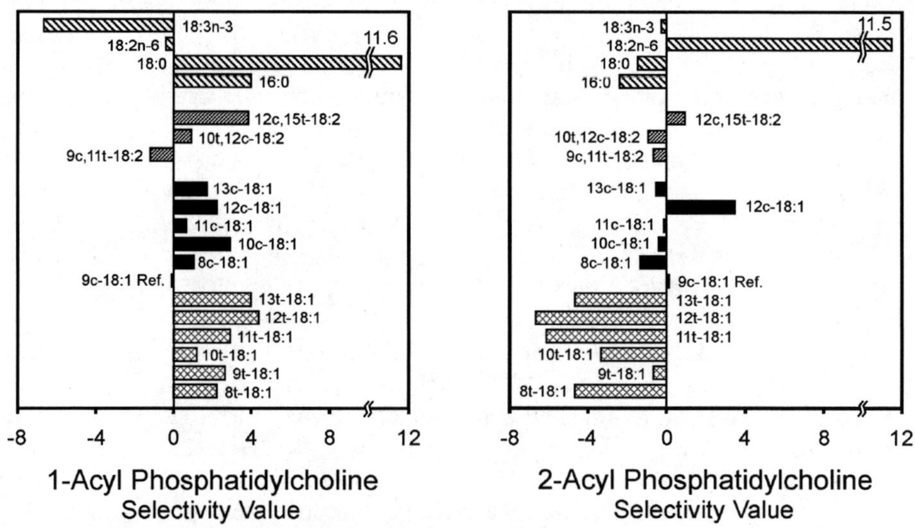

Fig. 4.4. Selectivity values for human plasma 1-Acyl and 2-Acyl phosphatidylcholine. Positive selectivity values indicate preferential incorporation and negative values indicate discrimination relative to 9*c*-18:1. Selectivity values are relative values and are equivalent to percent difference divided by 100.

PE, PC-1, and PC-2 reflect acyltransferase activity relative to 9*c*-18:1 and are not a measure of absolute activity. The selectivity values for total PC and PE (Fig. 4.3) are equivalent to percent differences from 9*c*-18:1 and illustrate the large differences in acyltransferase specificity for various dietary fatty acids and the large difference between phospholipid classes. For example, a positive selectivity value of 1.0 indicates a 100% or 2-fold higher incorporation of the fatty acid isomer relative to 9*c*-18:1. Conversely, a negative 1.0 selectivity value indicates 9*c*-18:1 incorporation is 100% or 2-fold higher than the fatty acid isomer. For fatty acid isomers, the positive values for 12*c*-18:1 and 12*c*,15*t*-18:2 incorporation in PC and the negative values for incorporation of 13*c*-18:1 and 8*c*-18:1 in PE are clear examples of the influence of double bond position and phospholipid class. The selectivity values for 9*t*-18:1, 18:2n-6 and 18:3n-3 plotted in Fig. 4.1, 4.3, and 4.5 are consistent with values calculated from results of a recent study in adult women fed these carbon-13 labeled fatty acids (McCloy et al., 2004).

Fatty acid composition data for nearly all tissues and lipid classes from rats fed a PHVO (51% *trans*) diet for 4 weeks provide additional information for the overall metabolism of many 18:1 isomers (Wood, 1979). Most of the rat composition data are consistent with the human plasma selectivity values and suggest a similarity between the specificity of human and rat acyltransferase. For example, 12*c*-18:1 distribution between the 1-acyl and 2-acyl positions of PC and PE was similar, *t*-18:1 isomers were almost exclusively incorporated in the 1-acyl position of PC and PE, 10*t*-18:1 incorporation into all tissue lipids was low, and the 12*c*-18:1 content was much higher than other isomers in all tissues except brain (Wood, 1979). In another well designed study with rats fed diets containing 4.5–15% PHVO, the observed discrimination against 8*t*- and 10*t*-18:1 and the preferential incorporation of 9*t*- and 12*t*-18:1 into liver and heart PC and PE (Bysted and Holmer, 1998) is consistent with human plasma PC and PE selectivity values. The results also show that 18:0 is the only fatty acid that is replaced by the *t*-18:1 isomers in PC and PE. In addition to the influence of lipid class on incorporation, rat studies results show that tissue type influences total 18:1 isomer accumulation and individual isomer accumulation patterns (Wood, 1979). For example, *cis* and *trans* 18:1 isomers are strongly excluded from brain tissue lipids except for sphingomyelin, which accumulated the *t*-18:1 elongated product, *t*-24:1. In general, there is no obvious correlation between physical properties (i.e., melting point) or structural features (i.e., double bond configuration or position) and acyltransferase specificity.

Comparison of the 1-acyl and 2-acyl phosphatidylcholine selectivity values in Fig. 4.4 for human plasma show a clear preference for incorporation of the *cis* and *trans* 18:1 positional isomers into the 1-acyl PC position and a discrimination against incorporation into the 2-acyl position relative to 9*c*-18:1. The notable exception is the high selective incorporation of 12*c*-18:1 into PC-2. Note that the average of the PC-1 plus PC-2 values does not equal the total PC data because these are ratio values and 9*c*-18:1 is not evenly distributed between both positions. The selectivity values for incorporation of the *trans* isomers into the 1-acyl PC position are also considerably less than values for 18:0, and the *cis* 18:1 isomers values are another order of magnitude smaller. In fact, the PC-1 selectivity values for the 18:1 isomers more closely resemble those for 16:0 than 18:0. These selectivity values reflect acyltransferase specificity and demonstrate that the structural conformations of the *trans* 18:1 isomers and 18:0 are clearly different. The *cis* 18:1 positional isomer values are an indication that 9*c*-18:1 is a preferred substrate.

The high 2-acyl PC selectivity values for 9*c*,12*c*-18:2 and the relatively high values for 12*c*-18:1 and 12*c*,15*t*-18:2 suggest that the 12*c* bond enhances acyltransferase selectivity. However, this suggestion is not consistent with the relatively low selectivity value for 9*c*,12*c*,15*c*-18:3 but is consistent for 18:3 isomers with a *trans* bond in the Δ9 or Δ15 position. For example, 9*c*,12*c*,15*t*-18:3 distribution between the 1- and 2-acyl positions in rat cardiolipin is similar to 18:2n-6 (Wolf et al., 1993) and in perfused rat liver, the selectivity for incorporation of 9*c*,12*c*,15*t*-18:3 into the PC-1 and PC-2 positions was the same as for 18:2n-6 and higher than for 18:3n-3 (Houtsmuller, 1981). For total PC and PE from perfused liver, there was no difference between incorporation of 9*t*,12*c*,15*c*-18:3 or 9*c*,12*c*,15*t*-18:3 and less 9*t*,12*c*-18:2 was incorporated than 18:2n-6. Reaction rates for esterification of 2-acyl PC with *c*,*t*- and *t*,*c*-18:2 have been compared to other fatty acids in vitro (Waku and Lands, 1968; Jezyk and Lands, 1968). The combined results from these two studies indicate the following order for the relative reaction rates: 18:2n-6 > 9*t*,12*c*-18:2 > 9*c*,12*t*-18:2 > 9*c*-18:1 > 9*t*-18:1 = 9*t*,12*t*-18:2. The rate for the 9*t*,12*c*-18:2 was about three times higher than the 9*c*,12*t*-18:2 rate and about 15% lower than 18:2n-6 rate. The order for acylation of 1-acyl PC was approximately the reverse of the 2-acyl PC rates. The double bond positions of *c*,*c*-18:2 isomers also affect the specificity of Acyl CoA synthetase and acyltransferase (Suzue and Marcel, 1972; Reitz et al., 1968). Results for 10*c*,13*c*-, 9*c*,12*c*-, 8*c*,11*c*-, 7*c*,10*c*-, and 6*c*,9*c*-18:2 indicate that only 9*c*,12*c*-18:2 and 6*c*,9*c*-18:2 are good substrates for acyl CoA synthetase and have high 2-acyl PC acyltransferase rates.

The human plasma phospholipid selectivity values in Fig. 4.3 and 4.4 for 9*c*,11*t*-18:2 and 10*t*,12*c*-18:2 clearly indicate that acyltransferase differentiates between the structures of these isomers. The PC acyltransferase selectivity values for 9*c*,11*t*-18:2 and 10*t*,12*c*-18:2 are more like *cis* 18:1 isomers than 18:0 or 18:2n-6 and the PE selectivity values are intermediate between 9*c*-18:1 and 18:2n-6. Interestingly, the values for the CLA isomers tend to be similar to values for 18:1 isomers that have double bonds in the 9, 10, 11, and 12 positions. For example, the PC selectivity value of −0.03 for 9*c*,11*t*-18:2 is close to the internal control, 9*c*-18:1, and the value of 0.5 for 10*t*,12*c*-18:2 is similar to 10*c* and 9*t*-18:1 and essentially the same as 12*t*-18:1. Also, the PE selectivity values for 9*c*,11*t*-18:2 and 10*t*,12*c*-18:2 are not greatly different than the 11*t*-, 11*c*-, and 12*c*-18:1 PE selectivity values. These relationships indicate that conjugation of dienoic fatty

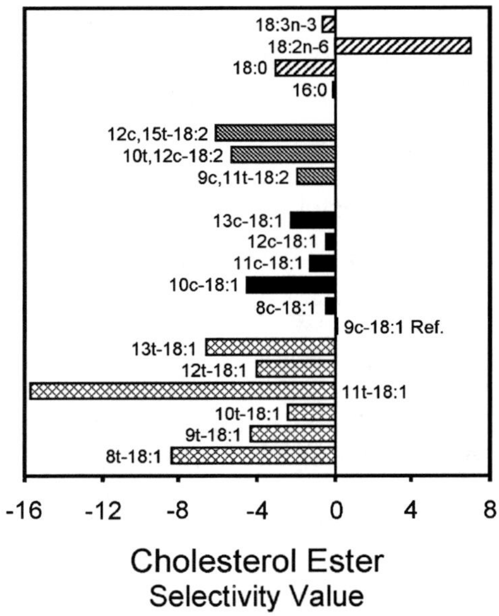

Fig. 4.5. Selectivity values for human plasma cholesterol ester.
Positive selectivity values indicate preferential incorporation and negative values indicate discrimination relative to 9*c*-18:1. Selectivity values are relative values and are equivalent to percent difference divided by 100.

acid structures results in a π electron density that resembles monoenoic fatty acids. Also of interest is the unusual negative PC-1 selectivity value of −1.17 for 9*c*,11*t*-18:2 which is the only *trans* fatty acid where discrimination has been observed to occur at the 1-acyl PC position.

In general, the 18:1 isomer selectivity values plotted in Fig. 4.3 and 4.4 for human plasma phospholipid are consistent with a plethora of results from in vitro (Reitz et al., 1969; Okuyama et al., 1972, 1977; Murphy and Spence, 1977; Marchand and Beare-Rogers, 1978) and in vivo animal studies (McMillan et al., 1963; Cook, 1978; Reichwald-Hacker et al., 1980; Moore and Dhopeshwarkar, 1981, 1982; Masuzawa et al., 1987; Pettersen and Opstvedt, 1989; Rocquelin et al., 1989; Holman et al., 1991). In addition, other studies with *trans* 18:2, *trans* 18:3, and CLA isomers show that most polyunsaturated fatty acid isomers also tend to be preferentially incorporated into the 1-acyl PC position, excluded from the 2-acyl PC position, and do not accumulate excessively in tissues and lipid classes (Sprecher et al., 1967; Jezyk and Lands, 1968; Reitz et al., 1968; Waku and Lands, 1968; Suzue and Marcel, 1972; Cook, 1980; Iversen et al., 1985; Pettersen and Ostvedt, 1992; Wolff et al., 1993; Banni et al., 1998, 2000; Bretillon et al., 1998a; Yurawecz et al., 1999; Loi et al., 2000a; Sébédio et al, 2001; Gruffat et al, 2003). Overall, a high acyltransferase specificity for 16:0 and 18:0 limits incorporation of fatty acid isomers into the 1-acyl position and a high specificity for 9*c*,12*c*-18:2 limits isomer incorporation into the 2-acyl

PC position, although 12*c*-18:1, 9*c*,12*c*,15*t*-18:3 and 9*t*,12*c*-18:2 isomers are exceptions. The result is that the biological impact of dietary fatty acid isomers is muted by strong competition from the non-isomeric fatty acids (16:0, 18:0, 9*c*-18:1, 9*c*,12*c*-18:2) in dietary fat.

Cholesterol Acyltransferase

The transfer of fatty acids from the 2-acyl phosphatidylcholine position to cholesterol by lecithin:cholesterol acyltransferase (LCAT), and acyl-CoA: cholesterol acyltransferase (ACAT) is responsible for CE synthesis in tissues and plasma (Glomset, 1979; Spector et al., 1979). However, the fatty acid specificity of these two enzymes are very different. The preferred substrate for LCAT is 18:2n-6, whereas 9*c*-18:1 is the preferred substrate for ACAT. There is also a large species differences. ACAT activity is high in rat liver and very low in human liver. Both ACAT and LCAT contribute to CE synthesis in rat plasma but most (>90%) of the CE in human plasma is synthesized by LCAT (Glomset, 1979; Spector et al., 1979). Because of these differences, the fatty acid composition of cholesterol ester from rat and human is different and rat liver CE data may not be particularly relevant to human plasma CE data.

The preference for cholesterol esterification by several isomeric and non-isomer fatty acids incubated with isolated rat liver microsomes was illustrated using a combination of experimental designs (Sgoutas, 1970). Reaction rates for 9*c*-18:1 were 3.5 times higher than for 9*t*-18:1 and 18:0. Based on the combined results presented, the relative order of preference for ACAT appears to be 9*c*-18:1 = 9*c*,12*t*-18:2 = 9*t*,12*c*-18:2 > 16:0 > 18:2n-6 > 9*t*-18:1 = 9*t*,12*t*-18:2 = 11*t*-18:1 = 11*c*-18:1 = 18:0.

The selectivity values for human plasma CE (Fig. 4.5) reflect LCAT specificity and, surprisingly, the relative preferences of LCAT for isomeric fatty acids are not greatly different from ACAT. Both the human plasma and rat liver results show a strong discrimination against the incorporation of *t*-18:1 isomers relative to 9*c*-18:1. Discrimination by LCAT is more extreme than ACAT and in general, LCAT discrimination is greater for the *trans* 18:1 isomers than for the *cis* 18:1 isomers. Within their respective groups, the 11*t*-18:1 and 10*c*-18:1 isomers exhibit the largest negative selectivity values. The relative order of LCAT preference for some selected fatty acids is: 18:2n-6 > 9*c*-18:1 = 16:0 > 18:3n-3 > 11*c*-18:1 = 9*c*,11*t*-18:2 > 18:0 > 9*t*-18:1 = 10*t*,12*c*-18:2 > 11*t*-18:1.

The negative CE selectivity values shown in Fig. 4.5 for the *c*-18:1 positional isomers are not the result of selective hydrolysis of CE. In fact, cholesterol hydrolase activity reported for a series of Δ2-Δ17 *c*-18:1 containing cholesterol esters was highest for cholesterol esterified with 9*c*-18:1 and activity decreased as the double bond position was moved toward either end of the acyl chain (Goller et al., 1970). Cholesterol hydrolysis rates were also lower for 9*t*-18:1-CE than for 9*c*-18:1-CE (Sgoutas, 1968).

The cholesterol acyltransferase selectivity values for human plasma CE show strong discrimination against incorporation of 12*c*,15*t*-18:2 and both CLA isomers but exclusion from CE was about three fold less for 9*c*,11*t*-18:2 than 10*t*,12*c*-18:1. This difference is consistent with results observed for rats fed diets supplemented with pure forms of these isomers (Sébédio et al., 2001). The large positive selectivity value for 9*c*,12*c*-18:2 is consistent with the recognized preference of LCAT for 18:2n-6 and is much higher than values for the other common dietary fatty acids including 9*c*,12*c*,15*c*-18:3.

The selectivity values for plasma CE are consistent with the low levels of *trans* fatty acids in human plasma CE. For example, less than 1% *trans* was present in plasma CE from subjects fed diets containing 40% *trans* (Zock and Katan, 1992), CE from subjects consuming habitual US diets contained 0.2% *trans* (Ohlrogge et al., 1981) and 10*c*-18:1 was strongly excluded from human tissues (Ohlrogge et al., 1981, 1982). Strong discrimination against incorporation of *trans* 18:3 isomers into human plasma CE and PL has been observed also. Plasma CE and PL from subjects fed diets supplemented with 1.4 g/d of *trans* 18:3 isomers, contained less than 0.2% and 0.1% *trans* 18:3, respectively (Sébédio et al., 2000).

Summary

All the same pathways and enzymes required for the metabolism of non-isomeric fatty acids are used also for the metabolism of the isomeric fatty acids in PHSO. Most fatty acid isomers, as well as non-isomers, are well absorbed, readily oxidized, and distributed selectively into lipid classes and between tissue types. Essentially all fatty acids are desaturated and elongated but the extent to which these reactions occur is dependent on fatty acid structure and varies enormously. Fatty acids that are biologically important have higher desaturation rates than non-biologically important fatty acids. There are many discernable differences between the metabolism of individual fatty acid isomers but the magnitude of the differences are no larger than the differences between individual non-isomeric fatty acids. In fact, the same mechanisms are used to regulate and control isomeric and non-isomeric fatty acid metabolism. Both double bond configuration and position influence the metabolic fate of specific isomers and are responsible for the metabolic diversity between individual isomers. In general, the oxidation rates for fatty acid isomers are higher than non-isomers, the large variation in desaturation rates is related to specific structural features, and isomers that are good substrates for $\Delta 6$ desaturase are generally better inhibitors. The incorporation and distribution of fatty acid isomers in phospholipids is intermediate between stearic and oleic acid, and they are poor substrates for cholesterol acyltransferase. Overall, the metabolic differences between non-isomeric fatty acids and isomeric fatty acids do not appear to present a metabolic challenge.

References

Adam L.L.; B.A. Corl; D.M. Barbano; et al. The Anticarcinogenic Effect of trans-11 18:1 is Dependent on its Conversion to cis-9,trans-11 CLA by Δ9-desaturase in Rats. *J Nutr* **2004,** *134*, 2698–2704.

Adlof R.O.; E.A. Emken. Distribution of Hexadecenoic, Octadecenoic and Octadecadienoic Acid Isomers in Human Tissue Lipids. *Lipids* **1986,** *21*, 543–547.

Adlof R.O.; S. Duval; E.A. Emken. Biosynthesis of Conjugated Linoleic Acid in Humans. *Lipids* **2000,** *35*, 131–135.

Anderson R.; R. Coots. The Catabolism of the Geometric Isomers of Uniformly 14C-labeled delta 9-octadecenoic Acid and Uniformly 14C-labeled delta 9,12-Octadecadienoic Acid by the Fasting Rat. *Biochim Biophys Acta* **1967,** *144*, 525–531.

Anderson R. Oxidation of the Geometric Isomers of delta 9-octadecenoic Acid by Rat-Liver Mitochondria. *Biochim Biophys Acta* **1967,** *144*, 18–24.

Anderson R.L.; C.S. Fullmer, Jr; E.J. Hollenbach. Effects of the trans Isomers of Linoleic Acid on the Metabolism of Linoleic Acid in Rats. *J Nutr* **1975,** *105*, 393–400, .

Aoyama T.; K. Fukui; K. Tanguchi; et al. Effect of Dietary Calcium on the Absorption of Triglycerides Esterified at 1,2 and 1,3 Positions of the Glycerol with Long Chain Saturated Fatty Acids in Rats. *Nutr Res* **1995,** *15*, 1005–1018.

Astorg P.O.; J. Chevalier. Effects of Elaidic Acid on Polyunsaturated Fatty Acids of Heart and Liver Phosphatidylcholine and Phosphatidylethanolamine in Rats Fed High or Low Levels of Linolenic Acid. *Nutrition Reports International* **1988,** *38*, 885–895.

Baer D.J.; J.T. Judd; P.M. Kris-Etherton; et al. Stearic Acid Absorption and its Metabolizable Energy Value are Minimally Lower Than Those of Other Fatty Acids in Healthy Men Fed Mixed Diets. *J Nutr* **2003,** *133*, 4129–4134.

Banni S.; G. Carta; M.S. Contini; et al. Characterization of Conjugated Diene Fatty Acids in Milk, Dairy Products, and Lamb Tissues. *J Nutr Biochem* **1996,** *7*, 150–155.

Banni S.; J.C. Martin. Conjugated Linoleic Acid and Metabolites. In *Trans Fatty Acids in Human Nutrition*. Sébédio J.-L.; Christie W.W., Eds. The Oily Press: Dundee, Scotland, 1998; pp 261–302.

Banni S.; E. Angioni; V. Casu; et al. Decrease in Linoleic Acid Metabolites as a Potential Mechanism in Cancer Risk Reduction by Conjugated Linoleic Acid. *Carcinogenese* **1999,** *20*, 1019–1024.

Banni S.; E. Angioni; E. Murru; et al. Vaccenic Acid Feeding Increases Tissue Levels of Conjugated Linoleic Acid and Suppresses Development of Premalignant Lesions in Rat Mammary Gland. *Nutr Cancer* **2001a,** *41*, 91–97.

Banni S.; G. Carta; M.S. Contini; et al. Distribution of Conjugated Linoleic Acid and Metabolites in Different Lipid Fractions in the Rat Liver. *J Lipid Res* **2001b,** *42*, 1056–1061.

Bauman D.E.; B.A. Corl; D.G. Peterson. The Biology of Conjugated Linoleic Acids in Ruminants. In *Advances in Conjugated Linoleic Acids Research, Vol 2*. Sébédio J.-L.; W.W. Christie; R.O. Adlof, Eds. AOCS Press: Champaign, IL, 2003; pp. 146–173.

Becker W. Distribution of 14C After Oral Administration of [1-14C] Linoleic Acid in Rats Red Different Levels of Essential Fatty Acids. *J Nutr* **1984,** 114, 1690–1696.

Belury M.A.; A. Kempa-Steczko. Conjugated Linoleic Acid Modulates Hepatic Lipid Composition in Mice. *Lipids* **1997,** 32, 199–204.

Benito P.; G.J. Nelson; D.S. Kelly; et al. The Effect of Conjugated Linoleic Acid on Plasma Lipoproteins and Tissue Fatty Acid Composition in Humans. *Lipids* **2001a,**

36, 229–236.

Benito P.; G.J. Nelson; D.S. Kelly; et al. The Effect of Conjugated Linoleic Acid on Platelet Function, Platelet Fatty Acid Composition, and Blood Coagulation in Humans. *Lipids* **2001b,** *36*, 221–227.

Berdeaux O.; J.-L. Sébédio; J.M. Chardigny; et al. Effect of trans n-6 Fatty Acids on the Fatty Acid Profile of Tissues and Microsomal Metabolism in the Rat. *Grasa. Aceit* **1996,** 47, 86–99.

Berdeaux O.; J.P. Blond; L. Bretillon; et al. In vitro Desaturation or Elongation of Mono-trans Isomers of Linoleic Acid by Rat Liver Microsomes. *Mol Cell Biochem* **1998,** *185*, 17–25.

Berdeaux O.; S. Gnadig; J.M. Chardigny; et al. In vitro Desaturation and Elongation of Rumenic Acid by Rat Liver Microsomes. *Lipids* **2002,** *37*, 1039–1045.

Bernard A.; B. Echinard; H. Carlier. Differential Intestinal Absorption of Two Fatty Acid Isomers: Elaidic and Oleic Acids. *Amer J Physiol* **1987,** *253*, G751–G759.

Beyers E.C.; E.A. Emken. Metabolites of cis,trans, and trans,cis Isomers of Linoleic Acid in Mice and Incorporation Into Tissue Lipids. *Biochem Biophys Acta* **1990,** *1082*, 275–284.

Blank M.L.; O.S. Privett. Studies on the Metabolism of cis, trans Isomers of Methyl Linoleate and Linolenate. *J Lipid Res* **1963,** *4*, 470–476.

Blond J.P.; C. Henchiri; P. Precigou; et al. Effect of 18:3 n-3 Geometrical Isomers of Heated Linseed Oil on the Biosynthesis of Arachidonic Acid in Rat. *Nutr Res* **1990,** *10*, 69–79.

Borgstrom B. Digestion and Absorption of Lipids. *International Review of Physiology Gastrointestinal Physiology II*, **1977,** *12*, 305–323.

Bretillon L.; J.M. Chardigny; J.P. Noel; et al. Desaturation and Chain Elongation of [1-14C] mono-trans Isomers of Linoleic and alpha-linolenic Acids in Perfused Rat Liver. *J Lipid Res* **1998a,** *39*, 2228–36.

Bretillon L.; J.M. Chardigny; J.-L. Sébédio; et al. Oxidative Metabolism of [1-14C] mono-trans Isomers of Linoleic and alpha-linolenic Acids in the Rat. *Biochim Biophys Acta* **1998b,** *1390*, 207–214.

Bretillon L.; J.M. Chardigny; S. Gregoire; et al. Effects of Conjugated Linoleic Acid Isomers on the Hepatic Microsomal Desaturation Activities in vitro. *Lipids* **1999,** *34*, 965–969.

Bretillon L.; J.M. Chardigny; J.-L. Sébédio; et al. Isomerization Increases the Postprandial Oxidation of Linoleic Acid but not alpha-linolenic Acid in Men. *J Lipid Res* **2001,** *42*, 995–997.

Brindley D.N. The Intracellular Phase of Fat Absorption. In *Biomembranes, Vol. 4B*. Smyth D.H., Ed. Plenum Press: 1974; pp 621–671.

Brink E.J.; E. Haddeman; N.J. de Fouw; et al. Positional Distribution of Stearic Acid and Oleic Acid in a Triacylglycerol and Dietary Calcium Concentration Determines the Apparent Absorption of These Fatty Acids in Rats. *J Nutr* **1995,** *125*, 2379–2387.

Broadwater J.; B.J. Laundre; B.G. Fox. Desaturation of trans-octadecenoyl-acyl Carrier Protein by stearoyl-acyl Carrier Protein Δ9 desaturase. *Inorg Biochem* **2002,** *78*, 7–14.

Bysted A.; G. Holmer; P. Lund. Influence of Moderate Amounts of trans Fatty Acids on the Formation of Polyunsaturated Fatty Acids. *J Am Oil Chem Soc* **1998,** *75*, 225–234.

Chardigny J.M.; J.P. Blond; L. Bretillon ; et al. Conversion of 18:3 delta 9cis,12cis,15trans in Rat Liver Microsomes. *Lipids* **1997,** *32*, 731–5.

Chardigny J.M.; J.-L. Sébédio; P. Juaneda; et al. Occurrence of n-3 trans Polyunsaturated Fatty Acids in Human Platelets. *Nutr Res* **1993,** *13*, 1105–1111.

Chardigny J.M.; J.-L. Sébédio; A. Grandgirard; et al. Identification of Novel trans Isomers of 20:5n-3 in Liver Lipids of Rats Fed a Heated Oil. *Lipids* **1996,** *31*, 165–168.

Chardigny J.M.; E. Masson; J.P. Sergiel; et al. The Position of Rumenic Acid on Triacylglycerols Alters its Bioavailability in Rats. *J Nutr.* **2003,** *133*, 4212–4214.

Chen Z.Y.; K.Y. Kwan; Y. Huang. Accumulation and Apparent Oxidation of cis,trans-18:2 Isomers Relative to Linoleic Acid in Rats. *Br J Nutr* **2001,** *86*, 249–255.

Choi Y.; Y. Park; J.M. Storkson; et al. Regulation of Stearoyl-CoA Desaturase Activity by trans-10,cis-12 Isomer of Conjugated Linoleic Acid in HepG2 Cells. *Biochem Biophysical Res Comm* **2001,** *284*, 689–693.

Choi Y.; Y. Park; J.M. Storkson; et al. Inhibition of Stearoyl-CoA Desaturase Activity by the cis-9,trans-11 Isomer and the trans-10,cis-12 Isomer of Conjugated Linoleic Acid in MDA-MB-231 and MCF-7 Human Breast Cancer Cells. *Biochem Biophys Res Comm* **2002,** *294*, 785–790.

Chuang L.T.; A.E. Leonard; J.W. Liu; et al. Inhibitory Effect of Conjugated Linoleic Acid on Linoleic Acid Elongation in Transformed Yeast with Human Elongase. *Lipids* **2001,** *36*, 1099–1103.

Clifford A.J.; L.M. Smith; R.K. Creveling; et al. Effects of Dietary Triglycerides on Serum and Liver Lipids and Sterol Excretion of Rats. *J Nutr* **1986,** *116*, 944–956.

Cook H.W. Incorporation, Metabolism and Positional Distribution of trans-unsaturated Fatty Acids in Developing and Mature Brain. Comparison of Elaidate and Oleate Administered Intracerebrally. *Biochim Biophys Acta* **1978,** 531, 245–256.

Cook H.W. Metabolism and Positional Distribution of Isomers of 18-carbon Dienoic Fatty Acids Containing trans and cis Double Bonds by Developing Rat Brain. *Can J Biochem* **1980,** *58*, 121–127.

Cook H.W. The Influence of trans Acids on Desaturation and Elongation of Fatty Acids in Developing Brain. *Lipids* **1981,** *16*, 920–926.

Cook H.C.; E.A. Emken. Geometrical and Positional Fatty Acid Isomers Interact Differently with Desaturation and Elongation of Linoleic and Linolenic Acids in Cultured Glioma Cells. *Biochem Cell Biol* **1990,** *68*, 653–660.

Coots R.H. A Comparison of the Metabolism of Elaidic, Oleic, Palmitic, and Stearic Acids in the Rat. *J Lipid Res* **1964a,** *5*, 468–472.

Coots R.H. A Comparison of the Metabolism of cis,cis-linoleic, trans,tras-linoleic, and a Mixture of cis,trans- and trans,cis-linoleic Acids in the Rat. *J Lipid Res* **1964b,** *5*, 473–476.

Corl B.A.; D.M. Barbano; D.E. Bauman; et al. cis-9,trans-11 CLA Derived Endogenously From trans-11 18:1 Reduces Cancer Risk in Rats. *J Nutr* **2003,** *133*, 2893–2900.

Craig-Schmidt M.C. Fatty Acid Isomers in Foods. In *Fatty Acids in Foods and Their Health Implications*. Chow C.K. Ed. Marcel Decker Inc.: New York, 1992; pp 365–398.

Cuebas D.; H. Schultz. Evidence for a Modified Pathway of Linoleate Degradation. *J Biol Chem* **1982,** *257*, 14140–14144.

De Schrijver R.; O.S. Privett. Interrelationship Between Dietary trans Fatty Acids and the 6- and 9- desaturases in the Rat. *Lipids* **1982,** *17*, 27–34.

Decsi T. Do trans Fatty Acids Impair Linoleic Acid Metabolism in Children? *Ann Nutr Metab* **1995,** *39*, 36–41.

Decsi T.; I. Burus; S. Molnár; et al. Inverse Association Between trans Isomeric and Long-chain Polyunsaturated Fatty Acids in Cord Blood Lipids of Full-term Infants. *Am J Clin Nutr* **2001,** *74*, 364–368.

Decsi T.; G. Boehm; H.M.R. Tjoonk; et al. Trans Isomeric Octadecenoic Acids are Related Inversely to Arachidonic Acid and DHA and Positively Related to Mead Acid in Umbilical Vessel Wall Lipids. *Lipids* **2002,** *37*, 959–965.

Delany J.P.; M.M. Windhauser; C.M. Champagne; et al. Differential Oxidation of Individual Dietary Fatty Acids in Humans. *Am J Clin. Nutr* **2000,** *72*, 905–911.

Demizieux L.; P. Degrace; J. Gresti; et al. Conjugated Linoleic Acid Isomers in Mitochondria: Evidence for an Alteration of Fatty Acid Oxidation. *J Lipid Res* **2002,** *43*, 2112–1122.

Du Plessis L.M. Isomerization of the Double Bonds of a Conjugated Fatty Acid During beta-oxidation. *Lipids* **1979,** *14*, 943–948.

Eder K.; N. Slomma; K. Becker. trans-10,cis-12 Conjugated Linoleic Acid Suppresses the Desaturation of Linoleic and Linolenic Acids in HepG2 Cells. *J Nutr* **2002,** *132*, 1115–1121.

Elias S.L.; S.M. Innis. Infant Plasma trans, n-6, and n-3 Fatty Acids and Conjugated Linoleic Acids are Related to Maternal Plasma Fatty Acids, Length of Gestation, and Birth Weight and Length. *Am J Clin Nutr* **2001,** *73*, 807–814.

Elovson J. Bioconversion of Palmitic and Stearic Acid in the Intact Rat. *Biochim Biophys Acta* **1965,** *106*, 291–303.

Emken E.A.; H.J. Dutton, Eds. *Geometrical and Positional Fatty Acid Isomers*. AOCS Press, Champaign, IL; 1979.

Emken E.A.; W.K. Rohwedder; H.J. Dutton; et al. Incorporation of Deuterium-labeled cis- and trans-9-octadecenoic Acid in Humans: Plasma, Erythrocyte, and Platelet Neutral Lipids. *Metabolism* **1979a,** *28*, 575–583.

Emken E.A.; W.K. Rohwedder; H.J. Dutton; et al. Incorporation of Deuterium-labeled cis- and trans-9-octadecenoic Acids in Humans: Plasma, Erythrocyte, and Platelet Phospholipids. *Lipids* **1979b,** *14*, 547–554.

Emken E.A.; H.J. Dutton; W.K. Rohwedder; et al. Distribution of Deuterium-labeled cis- and trans-12-octadecenoic Acids in Human Plasma and Lipoprotein Lipids. *Lipids* **1980,** *15*, 864–871.

Emken E.A. Human Studies with Deuterium Labeled Octadecenoic Acid Isomers. In *Dietary Fats and Health*. Perkins E.G.; W.J. Visek Eds. AOCS Press: Champaign, IL, 1983; pp 302–319.

Emken E.A.; R.O. Adlof; W.K. Rohwedder; et al. Incorporation of Deuterium-labeled trans- and cis-13-octadecenoic Acids in Human Plasma Lipids. *J Lipid Res* **1983,** *24*, 34–46.

Emken E.A. Nutrition and Biochemistry of trans and Positional Fatty Acid Isomers in Hydrogenated Oils. *Ann Rev Nutr* **1984,** *4*, 339–376.

Emken E.A.; W.K. Rohwedder; R.O. Adlof; et al. In vivo Distribution and Turnover and trans- and cis-10-octadecenoic Acid Isomers in Human Plasma Lipids. *Biochim Biophys Acta* **1985,** *836*, 233–245.

Emken E.A.; W.K. Rohwedder; R.O. Adlof; et al. Adsorption and Distribution of Deuterium-labeled trans- and cis-11-octadecenoic Acid in Human Plasma and

Lipoprotein Lipids. *Lipids* **1986,** *21*, 589–595.

Emken E.A. Absorption of deuterium-labeled cis- and trans-octadecenoic Acid Positional Isomers in Man. In *Fat Absorption, Vol. 1.* Kuksis A. Ed. CRC Press: Boca Raton, FL, 1986; pp 159–175.

Emken E.A.; W.K. Rohwedder; R.O. Adlof; et al. Metabolism in Humans of cis-12,trans-15-octadecadienoic Acid Relative to Palmitic, Stearic, Oleic, and Linoleic Acids. *Lipids* **1987,** *22*, 495–504.

Emken E.A.; W.K. Rohwedder; R.O. Adlof; et al. Incorporation of trans-8- and cis-8-octadecenoic Acid Isomers in Human Plasma and Lipoprotein Lipids. *Lipids* **1989,** *24*, 61–69.

Emken E.A. Do trans acids have adverse health consequences? In *Health Effects of Dietary Fatty Acids*. Nelson G. Ed. AOCS Press: Champaign, IL, 1991; pp 245–263.

Emken E.A.; R.O. Adlof; W.K. Rohwedder; et al. Influence of Linoleic Acid on Desaturation and Uptake of Deuterium-labeled Palmitic and Stearic Acids in Humans. *Biochim Biophys Acta* **1993,** *1170*, 173–181.

Emken E.A.; R.O. Adlof; R.M. Gulley. Dietary Linoleic Acid Influences Desaturation and Acylation of Deuterium-labeled Linoleic and Linolenic Acids in Young Adult Males. *Biochim Biophys Acta* **1994,** *1213*, 277–288.

Emken E.A. Metabolism of Dietary Stearic Acid Relative to Other Fatty Acids in Human Subjects. *Am J Clin Nutr* **1994,** *60*, 1023S–1028S.

Emken E.A. Physiological Properties, Intake and trans Fatty Acids and Coronary Heart Disease Risk. Kris-Etherton P.M. (Guest ed). *Am J Clin Nutr* **1995,** *62*, 659S–678S.

Emken E.A. (ed). Trans Fatty Acids: Infant and Fetal Development. *Am J Clin Nutr* **1997,** *66*, 715S–736S.

Emken E.A.; R.O. Adlof; S.M. Duval; et al. Influence of Dietary Arachidonic Acid on Metabolism of Deuterated Linoleic Acid by Adult Male Subjects. *Lipids* **1998,** *33*, 471–480.

Emken E.A.; R.O. Adlof; S.M. Duval; et al. Effect of Dietary Docosahexaenoic Acid on Desaturation and Uptake in vivo of Isotope Labelled Oleic, Linoleic and Linolenic Acids by Male Subjects. *Lipids* **1999,** *34*, 785–791.

Emken E.A.; R.O. Adlof; S. Duval; et al. Effect of Dietary CLA on Metabolism of Isotope-labeled Oleic, Linoleic and Conjugated Linoleic Acid Isomers in Women. *Lipids* **2002,** *37*, 741–750.

Emken E.A.; R.O. Adlof; S. Duval; et al. Effect of Triacylglycerol Structure on Absorption and Metabolism of Isotope Labeled Palmitic and Linoleic Acid by Humans. *Lipids* **2004,** *39*, 1–9.

Enoch H.G.; A. Catala; P. Stritmatter. Mechanism of Rat Liver Microsomal Stearyl-CoA Desaturase. Studies of the Substrate Specificity, Enzyme-substrate Interactions and Function of the Lipid. *J Biol Chem* **1976,** *251*, 5095–5103.

Flatmark T.; A. Nilsson; J. Kvannes; et al. On the Mechanism of Induction of the Enzyme System for Peroxisomal β-oxidation of Fatty Acids in Rat Liver by Diets Rich in Partially Hydrogenated Fish Oil. *Biochim Biophys Acta* **1988,** *962*, 122–130.

Friedman H.I.; B. Nylund. Intestinal Fat Digestion, Absorption, and Transport. *Am J Clin Nutr* **1980,** *33*, 108–1139.

Fukuda N.; N. Igari; T. Etoh; et al. A Comparison of the Metabolism of cis,cis-, cis,trans/trans,cis- and trans,trans-9,12-octadecadienoic Acids in Rat Liver. *Nutr Res* **1993,** *13*, 779–786.

Giudetti A.M.; A.C. Beynen; A.G. Lemmens; et al. Hepatic Fatty Acid Metabolism in Rats Fed Diets With Different Contents of C18:0, C18:1 cis and C18:1 trans Isomers. *Br J Nutr* **2003,** *90*, 887–893.

Glomset J.A. Lecithin: Cholesterol Acyltransferase. *Prog Biochem Pharmacol* **1979,** *15*, 41–66.

Goller H.J.; D.S. Sgoutas; I.A. Ismail; et al. Dependence of Sterol Ester Hydrolase Activity on the Position of Ethylenic Bond in Cholesteryl cis-octadecenoates. *Biochem* **1970,** *9*, 3072–3076.

Grandgirard A.; A. Piconneaux; J.-L. Sébédio; et al. Occurrence of Geometrical Isomers of Eicosapentaenoic and Docosahexaenoic Acids in Liver Lipids of Rats Fed Heated Linseed Oil. *Lipids* **1989,** *24*, 799–804.

Grandgirard A.; J.M. Bourre; F. Julliard; et al. Incorporation of trans Long-chain n-3 Polyunsaturated Fatty Acids in Rat Brain Structures and Retina. *Lipids* **1994,** *29*, 251–258.

Grandgirard A.; A. Piconneaux; J.-L. Sébédio; et al. Trans Isomers of Long-chain n-3 Polyunsaturated Fatty Acids in Tissue Lipid Classes of Rats Fed with Heated Linseed Oil. *Reprod Nutr Dev* **1998,** *38*, 17–29.

Gruffat D.; A. De La Torre; J.-M. Chardigny; et al. In vitro Comparison of Hepatic Metabolism of 9cis-11trans and 10trans-12cis Isomers of CLA in the Rat. *Lipids* **2003,** *38*, 157-163.

Gurr M.I.; M.P. Robinson; A.T. James; et al. The Substrate Specificity of Desaturases: the Conversion of cis-12-octadecenoic Acid into Linoleic Acid in Different Animal and Plant Species. *Biochim Biophys Acta* **1972,** *280*, 415–421.

Gurr M.I.; A.T. James. *Lipid Biochemistry.* Chapman and Hall: London, 1975.

Guzman, M.; W. Klein; T. Gomez del Pulgar; et al. Metabolism of trans Fatty Acids by Hepatocytes. *Lipids* **1999,** *34*, 381–386.

Heimermann W.H.; R.T. Holman; D.T. Gordon; et al. Effect of Double Bond Position in Octadecenoates Upon Hydrolysis by Pancreatic Lipase. *Lipids* **1973,** *8*, 45–47.

Hill E.G.; S.B. Johnson; L.D. Lawson; et al. Perturbation of the Metabolism of Essential Fatty Acids by Dietary Partially Hydrogenated Vegetable Oil. *Proc Natl Acad Sci USA* **1982,** *79*, 953–957.

Hiltunen J.K.; H. Osmundsen; J. Bremer. β-oxidation of Polyunsaturated Fatty Acids Having Double Bonds at Even-numbered Positions in Isolated Rat Liver Mitochondria. *Biochim Biophys Acta* **1983,** *752*, 223–232.

Holman R.T.; C.M. Lyman. Metabolism of Isomers of Linoleic and Linolenic Acids. *Proc Soc Exp Biol Med* **1951,** *76*, 100.

Holman R.T.; F. Pusch; B. Svingen; et al. Unusual Isomeric Polyunsaturated Fatty Acids in Liver Phospholipids of Rats Fed Hydrogenated Oil. *Proc Natl Acad Sci* **1991,** *88*, 4830–4834.

Holmer G.; C.E. Hoy; D. Kirstein. Influence of Partially Hydrogenated Vegetables and Marine Oils on Lipid Metabolism in Rat Liver and Heart. *Lipids* **1982,** *17*, 585–593.

Houtsmuller U.M.T. Columbinic Acid, a New Type of Essential Fatty Acid. *Progr. Lipid Res* **1981,** *20*, 889–896.

Hudgins L.C.; J. Hirsch; E.A. Emken. Correlation of Isomeric Fatty Acids in Human Adipose Tissue with Clinical Risk Factors for Cardiovascular Disease. *Am. J. Clin. Nutr.* **1991,** *53*, 474–484.

Hsu C.M.L.; F.A. Kummerow. Influence of Elaidate and Erucate on Heart Mitochondria. *Lipids* **1977,** *12*, 486–494, .

Ide T.; M. Sugano. Oxidation and Esterification of cis- and trans- isomers of Octadecenoic and Octadecadienoic Acids in Isolated Rat Liver. *Biochim Biophys Acta* **1984,** *794*, 281–291.

Ide T.; M. Sugano. Strain Dependence of the Metabolism of cis- and trans-isomers of 9-octadecenoic Acid in Perfused Liver and Cell-free Preparation in Rats. *Biochim Biophys Acta* **1986,** *877*, 104–111.

Ide T.; M. Watanabe; M. Sugano; et al. Activities of Liver Mitochondrial and Peroxisomal Fatty Acid Oxidation Enzymes in Rats Fed trans Fat. *Lipids* **1987,** *22*, 6–10.

Iversen S.A.; P. Cawood; M.J. Madigan; et al. A diene-conjugated Isomer of Linoleic Acid, 18:2(9,11), in Human Plasma Phospholipid. *Life Chem Rep* **1985,** *3*, 45–48.

Jensen R.G.; J. Sampugna; R.L. Pereina. Pancreatic Lipase. Lipolysis of Syntheric Triglycerides Containing a trans Fatty Acid. *Biochim Biophys Acta* **1964,** *84*, 481–482.

Jezyk P.; W.E.M. Lands. Specificity of Acyl-CoA: Phospholipid Acyltransferases: Solvent and Temperature Effects. *J Lipid Res* **1968,** *9*, 525–531.

Kameda K.; A.J. Valicenti; R.T. Holman. Chain Elongation of trans-octadecenoic Acid Isomers in Rat Liver Microsomes. *Biochim Biophys Acta* **1980,** *618*, 13–17.

Karmen A.; M. Whyte; D.S. Goodman. Fatty Acid Esterification and Chylomicron Formation During Fat Absorption: I. Triglycerides and Cholesterol Esters. *J Lipid Res* **1983,** *4*, 312–321.

Kelly D.S.; V. Simon; P.C. Taylor; et al. Dietary Supplementation with Conjugated Linoleic Acid Increased its Concentration in Human Peripheral Blood Mononuclear Cells, but Did Not Alter Their Function. *Lipids* **2001,** *36*, 669–674.

Kinsella J.E.; G. Bruckner; J. Mai; et al. Metabolism of trans Fatty Acids with Emphasis on the Effects of trans,trans-octadecadienoate on Lipid Composition, Essential Fatty Acid, and Prostaglandins: An Overview. *Amer J Clin Nutr* **1981,** *34*, 2307–2318.

Kirstein D.; C.E. Hoy; G. Holmer. Effect of Dietary Fats on the delta6- and delta5-Desaturation of Fatty Acids in Rat Liver Microsomes. *Brit J Nutr* **1983,** *50*, 749–756.

Koba K.; A. Akahoshi; M. Yamasaki; et al. Dietary Conjugated Linolenic Acid in Relation to CLA Differently Modifies Body Fat Mass and Serum and Liver Lipid Levels in Rats. *Lipids* **2002,** *37*, 343–350.

Koletzko B. trans Fatty Acids May Impair Biosynthesis of Long-chain Polyunsaturates and Growth In Man. *Acta Paediatr* **1992,** *81*, 302–306.

Kubow S. The Influence of Positional Distribution of Fatty Acids in Native, Interesterified and Structure-specific Lipids on Lipoprotein Metabolism and Atherogenesis. *J Nutr Biochem* **1966,** *7*, 531–541.

Lanser A.C.; E.A. Emken. Competitive Deposition of trans-12- and cis-9-octadecenoates Into Egg Yolk Lipids. Lipids **1981,** *16*, 15–19.

Lanser A.C.; E.A. Emken; J.B. Ohlrogge. Oxidation of Oleic and Elaidic Acids in Rat and Human Heart Homogenates. *Biochim Biophys Acta* **1986,** *875*, 510–515.

Lanser A.C.; E.A. Emken. Comparative Deposition of trans-10 and cis-9-octadecenoates in Egg and Tissue Lipids of the Laying Hen. *J Agric Food Chem.* **1987,** *35*, 248–252.

Larque E.; P.A. Garcia-Ruiz; F. Perez-Liamas; et al. Dietary trans Fatty Acids Alter the

Compositions of Microsomes and Mitochondria and the Activities of Microsome Δ6-fatty Acid Desaturase and Glucose-6-phosphatase in Livers of Pregnant Rats. *J Nutr* **2003,** *133*, 2526–2531.

Larque E.; F. Perez-Llamas; V. Puerta; et al. Dietary trans Fatty Acids Affect Docosahexaenoic Acid Concentrations in Plasma and Liver But Not Brain of Pregnant and Fetal Rats. *Pediat. Res* **2000,** *47*, 278–283.

Lawson L.D.; R.T. Holman. Beta-oxidation of the Geometric and Positional Isomers of Octadecenoic Acid by Rat Heart and Liver Mitochondria. *Biochim Biophys Acta* **1981,** *665*, 60–65.

Lawson L.D.; E.G. Hill; R.T. Holman. Suppression of Arachidonic Acid in Lipids of Rat Tissues by Dietary Mixed Isomeric cis and trans Octadecenoates. *J Nutr* **1983,** *113*, 1827–1835.

Lawson L.D.; E.G. Hill; R.T. Holman. Dietary Fats Containing Concentrates of cis or trans Octadecenoates and the Patterns of Polyunsaturated Fatty Acids of Liver Phosphatidylcholine and Phosphatidylethanolamine. *Lipids* **1985,** *20*, 262–267.

Liang Y.W.; R.E. Ensenauer; L. Sweetman; et al. Leaky β-oxidation of a trans-fatty Acid: Incomplete β-oxidation of Elaidic Acid is Due to the Accumulation of 5-trans-Tetradecenoyl-CoA and Its Hydrolysis and Conversion to 5-trans-tetradecoylcarnitine in the Matrix of Rat Mitochondria. *J Biol Chem* **2004,** *279*, 52160–52167.

Lin X.; J.J. Loor; J.H. Herbein. Trans10,cis12-18:2 is a More Potent Inhibitor of de novo Fatty Acid Synthesis and Desaturation Than cis9,trans11-18:2 in the Mammary Gland of Lactating Mice. *J Nutr* **2004,** *134*, 1362–1368.

Loi C.; J.M. Chardigny; S. Alamanza; et al. Incorporation and Metabolism of Dietary trans Isomers of Linolenic Acid Alter the Fatty Acid Profile of Rat Tissues. *J Nutr* **2000a,** *130*, 2550–2555.

Loi C.; J.M. Chardigny; C. Cordelet; et al. Incorporation and Metabolism of trans 20:5 in Endothelial Cells. Effect on Prostacyclin Synthesis. *Lipids* **2000b,** *35*, 911–918.

Lovejoy J.C.; S.R. Smith; C.M. Champagne; et al. Effects of Diets Enriched in Saturated (palmitic), Monounsaturated (oleic), or trans (elaidic) Fatty Acids on Insulin Sensitivity and Substrate Oxidation in Healthy Adults. *Diabetes Care* **2002,** *25*, 1283–1288.

Ludwig S.A.; H. Sprecher. Substrate Specificity Studies on the Malonyl-CoA-dependent Chain Elongation of All cis Polyunsaturated Fatty Acids by Rat Liver Microsomes. *Arch Biochem Biophys* **1979,** *197*, 333–341.

Luo, M.J.; T.E. Smeland; K. Shoukry; et al. $\Delta^{3,5}$,$\Delta^{2,4}$-Dienoyl-CoA Isomerase From Rat Liver Mitochondria. *J Biol Chem* **1994,** *269*, 2384–2388.

Mahfouz M.; S. Johnson; R.T. Holman. Inhibition of Desaturation of Palmitic, Linoleic and eicosa-8,11,14-trienoic Acids in vitro by Isomeric cis-octadecenoic Acids. *Biochim Biophys Acta* **1981,** *663*, 58–68.

Mahfouz M.M.; S. Johnson; R.T. Holman. The Effect of Isomeric trans-18:1 Acids on the Desaturation of Palmitic, Linoleic and eicosa-8,11,14-trienoic Acids by Rat Liver Microsomes. *Lipids* **1980a,** *15*, 100–107.

Mahfouz M.M.; A.J. Valicenti; R.T. Holman. Desaturation of Isomeric trans-octadecenoic Acids by Rat Liver Microsomes. *Biochim Biophys Acta* **1980b,** *618*, 1–12.

Malewiak M.I.; R. Rozen; X.L. Liepvre; et al. Oleate Metabolism and Endogenous Triacylglycerol Hydrolysis in Isolated Hepatocytes From Rats Fed a High-fat Diet. *Diabete and Metabolism (Paris)* **1988,** *14*, 270–276.

Mani P.; M. Liebman; D.C. Rule. Effects of a Tristearate-containing Lipid and Canola Oil on Plasma and Tissue Lipids in Rats. *Plant Foods for Human Nutrition* **1994,** *45*, 203–212.

Marchand C.M.; J.L. Beare-Rogers. The Acylation of 1-palmityglycerol 3-phosphate with cis and trans C-16 to C-22 Monoenoic Fatty Acids in Rat Liver Microsomes. Lipids **1978,** *13*, 329–333.

Martin J.C.; J.-L. Sébédio; C. Caselli; et al. Lymphatic Delivery and in vitro Pancreatic Lipase Hydrolysis of Glycerol Esters of Conjugated Linoleic Acids in Rats. *J Nutr* **2000,** *130*, 1108–1114.

Masters C.; D. Crane. The Role of Peroxisomes in Lipid Metabolism. *Trends Biochem Sci.* **1984,** *9*, 314–317.

Masuzawa Y.; M.R. Prasad; W.E.M. Lands. Distribution of Dietary trans-octadecenoate Among acyl-CoA and Other Lipid Fractions of Rat Liver and Heart. *Biochim Biophys Acta* **1987,** *919*, 297–306.

McCloy U.; M.A. Ryan; P.B. Pencharz; R.J. Ross; S.C. Cunnane. A Comparison of the Metabolism of Eighteen-carbon 13C-unsaturated Fatty Acids in Healthy Women. *J. Lipid Res.* **2004,** *45*, 474–485.

McMillan G.C.; M.D. Silver; B.I. Weigensberg. Elaidinized Olive Oil and Cholesterol Atherosclerosis. *Arch Path* **1963,** *76*, 106–112.

Mead J.F.; R.B. Alfin-Slater; D.R. Howton; et al. Eds. *LIPIDS, Chemistry, Biochemistry and Nutrition*. Plenum Press: New York, NY, 1986.

Menon N.; G. Dhopeshwarkar. Differences in the Fatty Acid Profile and beta-oxidation by Heart Homogenates of Rats Fed cis and trans Octadecenoic Acids. *Biochim Biophys Acta* **1983,** *751,* 14–20.

Miller A.; E. McGrath; C. Stanton; et al. Vaccenic Acid (t11-18:1) is Converted to c9,t11-CLA in MCF-7 and SW480 Cancer Cells. *Lipids* **2003,** *38*, 623–632.

Moore C.E.; G. Dhopeshwarkar. Positional Specificity of trans Fatty Acids in Fetal Lecithin. *Lipids* **1981,** *16*, 479–484.

Moore C.E.; G. Dhopeshwarkar. Trans Fatty Acids: Transport and Positional Specificity in Rat Placental Lecithin. *Biol. Neonate* **1982,** *42*, 138–144.

Mougios V.; A. Matsakas; A. Petridou; et al. Effect of Supplementation with Conjugated Linoleic Acid on Human Serum Lipids and Body Fat. *J. Nutritional Biochem.* **2001,** *12*, 585–594.

Mounts T.L.; E.A. Emken; W.K. Rohwedder; et al. Metabolism of Labeled Isomeric Octadecenoates by the Laying Hen. *Lipids* **1971,** *6*, 912–918.

Murphy M.G.; M.W. Spence. Effect of Fatty-acid Double-bond Position on the Selectivity of Rat-brain Enzymes: The Incorporation of Oleic and cis-vaccenic Acids Into Lysolecithin in vitro. *J Neurochemistry* **1977,** *29*, 251–259.

Neat C.E.; M.S. Thomassen; H. Osmundsen. Effects of High Fat Diets on Hepatic Fatty Acid Oxidation in Rats. *Biochem J.* **1981,** *196*, 149–159.

Nugteren D.H. Inhibition of Prostaglandin Biosynthesis by 8-cis, 12-trans, 14-cis-Eicosatrienoic Acid and 5-cis, 8-cis, 12-trans, 14-cis-eicosatetraenoic Acid. *Biochim Biophys Acta* **1970,** *210*, 171–176.

Ohlrogge J.B.; E.A. Emken; R.M. Gully. Human Tissue Lipids: Occurrence of Fatty Acid Isomers From Dietary Hydrogenated Oils. *J Lipid Res* **1981,** *22*, 955–960.

Ohlrogge J.B.; R.M. Gulley; E.A. Emken. Occurrence of Octadecenoic Fatty Acid Isomers From Hydrogenated Fats in Human Tissue Lipid Classes. *Lipids* **1982,** *17*,

551–557.

Okuyama H.; W.E.M. Lands; F.D. Gunstone; et al. Selective Transfers of trans-ethylenic Acids by Acyl Coenzyme A. Phospholipid Acyltransferases. *Biochem* **1972,** *11*, 4392–4398.

Okuyama H.; K. Yamada; Y. Kameyama; et al. Regulation of Membrane Lipid Synthesis in Escherichia Coli After Shifts in Temperature. *Biochem* **1977,** *16*, 2668–2673.

Ono K.; D. Fredrickson. The Metabolism of 14C-labeled cis and trans Isomers of Octadecenoic and Octadecadienoic Acids. *J Biol Chem* **1964,** *239*, 2482–2488.

Osmundsen H.; J. Bremer; J.I. Pedersen. Metabolic Aspects of Peroxisomal β-oxidation. *Biochim Biophys Acta* **1991,** *1085*, 141–158.

Park Y.; J.M. Storkson; J.M. Ntambi; et al. Inhibition of Stearoyl-CoA Desaturase Activity by trans-10,cis-12 Conjugated Linoleic Acid and Its derivatives. *Biochim Biophys Acta* **2000,** *1486*, 285–292.

Pettersen J.; J. Opstvedt. Trans Fatty Acids. 3. Fatty Acids Composition of the Brain and Other Organs in trans the Newborn Piglet. *Lipids* **1989,** *24*, 616–624.

Pettersen J.; J. Ostvedt. Trans Fatty Acids.5. Fatty Acid Composition of Lipids of the Brain and Other Organs in Suckling Piglets. *Lipids* **1992,** *27*, 761–769.

Pollard M.R.; F.D. Gunstone; A.T. James; et al. Desaturation of Positional and Geometric Isomers of Monoenoic Fatty Acids by Microsomal Preparations From Rat Liver. *Lipids* **1980,** *15*, 306–312.

Privett O.S.; E.M. Stearns, Jr; E.C. Nickell. Metabolism of the Geometric Isomers of Linoleic Acid in the Rat. *J Nutr* **1967,** *92*, 303–310.

Ratnayake W.M.N.; Z.Y. Chen; G. Pelletier; et al. Occurrence of 5c,8c,11c,15t-eicosatetraenoic Acid and Other Unusual Polyunsaturated Fatty Acids in Rats Fed Partially Hydrogenated Canola Oil. *Lipids* **1994,** *29*, 707–714.

Reddy J.K.; T. Hashimoto. Peroxisomal β-oxidation and Peroxisomal Proliferator-Activator Receptor: An Adaptive Metabolic System. *Ann Rev Nutr* **2001,** *21*, 193–230.

Reichwald-Hacker I.; S. Grosseoetringhau; I. Kiewitt; et al. Molecular Structure of Glycerolipids in Rats Fed Partially Hydrogenated Triacylglycerols. *J Nutr* **1980,** *110*, 1122–1129.

Reitz R.C.; W.E.M. Lands; W.W. Christie; et al. Effects of Ethylenic Bond Position Upon Acyltransferase Activity With Isomeric cis,cis-octadecadienoyl Coenzyme A Thiol Esters. *J Biol Chem* **1968,** *243*, 2241–2246.

Reitz R.C.; M. El-Sheikh; W.E.M. Lands; et al. Effects of Ethylenic Bond Position Upon Acyltransferase Activity With Isomeric cis-octadecenoyl Coenzyme A Thiol Esters. *Biochim Biophys Acta* **1969,** *176*, 480–490.

Ren Y.; H. Schultz. Metabolic Functions of the Two Pathways of Oleate β-oxidation Double Bond Metabolism During the β-oxidation of Oleic Acid in Rat Heart Mitochondria. *J Biol Chem* **2003,** *278*, 111–116.

Renaud S.C.; J.C. Ruf; D. Petithorny. The Positional Distribution of Fatty Acids in Palm Oil and Lard Influences Their Biological Effects in Rats. *J Nutr* **1995,** *125*, 229–237.

Riisom T.; R.T. Holman. Incorporation Into Lipid Classes of Products From Microsomal Desaturation of Isomeric trans-octadecenoic Acids. *Lipids* **1981,** *16*, 647.

Rocquelin G.; L. Guenot; P.O. Astorg; et al. Phospholipid Content and Fatty Acid Composition of Human Heart. *Lipids* **1989,** *24*, 775–780.

Rosenthal M.D.; M.C. Whitehurst. Selective Effects of Isomeric cis and trans Fatty Acids on Fatty Acyl Δ9 and Δ6 Desaturation by Human Skin Fibroblasts. *Biochim Biophys Acta* **1983,** *753*, 450–459.

Royce S.M.; R.P. Holmes. The Saturation and Isomerization of Dietary Fatty Acids and the Respiratory Properties of Rat Heart Mitochondria. *Biochem Biophys Acta* **1984,** *792*, 371–375.

Sakono, M.; F. Miyanaga; S. Kawahara; et al. Dietary Conjugated Linoleic Acid Reciprocally Modifies Ketogenesis and Lipid Secretion by Rat Liver. *Lipids* **1999,** *34*, 997–1000.

Salminen I.; M. Mutanen; M. Jauhiainen; et al. Dietary trans Fatty Acids Increase Conjugated Linoleic Acid Levels in Human Serum. *J Nutr Biochem* **1998,** *9*, 93–98.

Santora J.; D.L. Palmquist; K.L. Roehrig. Tans-vaccenic Acid is Desaturated to Conjugated Linoleic Acid in Mice. *J Nutr* **2000,** *130*, 208–215.

Schlenk H.; D.M. Sand. A New Group of Essential Fatty Acids and Their Comparison With Other Polyenoic Fatty Acids. *Biochim Biophy. Acta* **1967,** *144*, 305–320.

Sébédio J.-L.; P. Juaneda; G. Dobson; et al. Metabolites of Conjugated Isomers of Linoleic Acid (CLA) in the Rat. *Biochim Biophys Acta* **1997,** *1345*, 5–10.

Sébédio J.-L.; W.W. Christie Eds. Trans Fatty Acids in Human Nutrition. The Oily Press: Dundee, Scotland, 1998

Sébédio J.-L.; P. Juaneda; S. Gregoire; et al. Geometry of Conjugated Double Bonds of CLA Isomers in a Commercial Mixture and in Their Hepatic 20:4 Metabolites. *Lipids* **1999,** *34*, 1319–25.

Sébédio J.-L.; P. Juaneda; S. Gregoire; et al. Geometry of Conjugated Double Bonds of CLA Isomers in a Commercial Mixture and in Their Hepatic 20:4 Metabolites. Lipids **1999,** *34*, 1319–1325.

Sébédio J.-L.; S.H.F. Vermunt; J.M. Chardigny; et al. The Effect of Dietary trans α-Linolenic Acid on Plasma Lipid Platelet Fatty Acid Composition: the TransLinE Study. *Euro J Clin Nutr* **2000,** *54*, 104–113.

Sébédio J.-L.; E. Angioni; J.M. Chardigny; et al. The Effect of Conjugated Linoleic Acid Isomers on Fatty Acid Profiles of Liver and Adipose Tissues and Their Conversion to Isomers of 16:2 and 18:3 Conjugated Fatty Acids in Rats. *Lipids* **2001,** *36*, 575–582.

Sébédio J.-L., W.W. Christie; R.O. Adlof, Eds. *Advances in Conjugated Linoleic Acids Research, Vol 2.* AOCS Press: Champaign, IL, 2003.

Senti F.R. Ed. Health Aspects of Dietary trans Fatty Acids. Life Sciences Research Office, Federation of American Societies for Experimental Biology, 1985.

Sergiel J.P.; J.M. Chardigny; J.-L. Sébédio; et al. Beta-oxidation of Conjugated Linoleic Acid Isomers and Linoleic Acid in Rats. *Lipids* **2001,** 36, 1327–1329.

Sgoutas D. Hydrolysis of Synthetic Cholesterol Esters Containing trans Fatty Acids. *Biochim Biophys Acta* **1968,** *164*, 317–326.

Sgoutas D.S. Effect of Geometry and Position of Ethylenic Bond Upon Acyl Coenzyme A-cholesterol-O-acyltransferase. *Biochem* **1970,** *9*, 1826–1833.

Smedman A.; B. Vessby. Conjugated Linoleic Acid Supplementation in Humans – Metabolic Effects. *Lipids* **2001,** *36*, 773–781.

Spector A.A.; S.N. Mathur; T.L. Kaduce. Role of Acylcoenzyme A: Cholesterol O-Acyltransferase in Cholesterol Metabolism. *Prog Lipid Res* **1979,** *18*, 31–53.

Sprecher H.W.; H.J. Dutton; F.D. Gunstone; et al. Metabolic Conversions and the Positional Distributions in Liver Lecithin of Some Unnatural Dienoic Acids in the Rat. Lipids **1967,** *2*, 122–126.

Stearns, Jr E.M.; J.A. Rysavy; O.S. Privett. Metabolism of cis-11, cis-14- and trans-11, trans-14-eicosadienoic Acids in the Rat. *J Nutr* **1967,** *93*, 485–490.

Sugano M.; A. Tsujita; M. Yamasaki; et al. Lymphatic Recovery, Tissue Distribution, and Metabolic Effects of Conjugated Linoleic Acid in Rats. *J Nutr Biochem* **1997,** *8*, 38–43.

Surawicz C.M.; D.R. Saunders; J. Sillery; et al. Linolenate Transport by Human Jejunum: Presumptive Evidence for Portal Transport at Low Absorption Rates. *Am. J. Physiol* **1981,** *240*, G157–G162.

Suzue G.; Y.L. Marcel. Specificity of Long-chain Acyl Coenzyme a Synthetase From Rat Liver Microsomes. Influence of the Position of Double Bonds in Octadecadienoic Acids. *Biochem* **1972,** *11*, 1704–1708.

Tserng K.-Y.; S.-J. Jin. NADPH-Dependent Reductive Metabolism of cis-5 Unsaturated Fatty Acids. A Revised Pathway for the β-oxidation of Oleic Acid. *J Biol Chem* **1990,** *266*, 11614–11620.

Tsuyoshi T.; Y. Tokuyama; M. Igarashi; et al. α-Eleostearic acid (9Z11E13E-18:3) is Quickly Converted to Conjugated Linoleic Acid (9z11e-18:2) in Rats. *J Nutr* **2004,** *134*, 2634–2639.

Turpeinen A.M.; M. Mutanen; A. Aro; et al. Bioconversion of Vaccenic Acid to Conjugated Linoleic Acid in Humans. *Am. J Clin Nutr* **2002,** *76*, 504–510.

Waku K.; W.E.M. Lands. Control of Lecithin Biosynthesis in Erythrocyte Membranes. *J Lipid Res* **1968,** *9*, 12–18.

Willebrands A.F.; K.J. Van der Veen. The Metabolism of Elaidic Acid in the Perfused Rat Heart. *Biochim Biophys Acta* **1966,** *116*, 583–585.

Wisneski J.A.; E.W. Gertz; R.A. Neese; et al. Myocardial Metabolism of Free Fatty Acids Studies with 14C-labeled Substrates in Humans. *J Clin Invest* **1987,** *79*, 359–366.

Wolff R.L.; N.A. Combe; B. Entressangles; et al. Preferential Incorporation of Dietary cis-9,cis-12,trans-15 18:3 Acid Into Rat Cardiolipins. *Biochim Biophys Acta* **1993,** *1168*, 285–291.

Wood R. Distribution of dietary geometrical and positional isomers in brain, heart, kidney, liver, lung, muscle, spleen, adipose and hepatoma. In Emken EA, Dutton HJ (eds). Geometrical and Positional Fatty Acid Isomers. AOCS Press: Champaign, IL, 1979; pp.213–281.

Yurawecz M.P.; J.K.G. Kramer; M.E.R. Dugan; et al. Incorporation of conjugated fatty acids into biological matrices. In *Advances in Conjugated Linoleic Research Vol 1.* Yurawecz M.P.; M.M. Mossoba; J.K.G. Kramer; et al, Eds. AOCS Press: Champaign, IL, 1999; pp 238–254.

Zevenbergen J.L.; U.M. Houtsmuller; J.J. Gottenbos. Linoleic Acid Requirements of Rats Fed trans Fatty Acids. *Lipids* **1988,** *23*, 178–186.

Zevenbergen J.L.; U.M.T. Houtsmuller. Effect of Dietary Fats on Linoleic Acid Metabolism. A Radiolabel Study in Rats. *Biochem Biophys Acta* **1989,** *1002*, 312–323.

Zock P.L.; M.B. Katan. Hydrogenation Alternatives: Effects of trans Fatty Acids and Stearic Acid versus Linoleic Acid on Serum Lipids and Lipoprotein in Humans. *J Lipid Res* **1992,** *33*, 399–410.

Chapter 5

Ruminant *Trans* Fatty Acids: Composition and Nutritional Characteristics

Peter J. Huth, PhD

National Dairy Council, 10255 West Higgins Road, Suite 900, Rosemont, IL 60018-5616

Introduction

The fat component of the diet consists of a complex mixture of saturated and unsaturated fatty acids from a variety of food sources. Saturated fatty acids differ primarily by carbon chain length and, in general, range from C_4-C_{22} in the diet and are derived predominantly from animal food sources. Unsaturated fatty acids, in addition to variations in carbon chain length, vary in the number, position, and geometric configuration of their double bonds. Furthermore, it is now well known that fatty acids (FA) having the same carbon chain length but differing in the number and geometric position of double bonds can have dramatically different physical and biological properties. Familiar examples in the diet include oleic acid (*cis*-9,18:1), linoleic acid (*cis*-9,12-18:2), and α-linolenic acid (*cis*-9,12,15-18:3) that are derived principally from plant based sources. Other FA that have both the same carbon chain length and number of double bonds but have the hydrogens of the double bond positioned in the *cis* or *trans* configuration are termed geometric fatty acid isomers. Dietary examples include oleic acid ($C_{18:1}$ *cis*, n-9) and elaidic acid ($C_{18:1}$ *trans*, n-9). The predominant source of elaidic acid in the diet is from partially hydrogenated vegetable oils (PHVO) and makes up approximately 20–30% of the total $C_{18:1}$ *trans* isomers of PHVO. Still other FA have the same carbon chain length but contain multiple conjugated positional and geometric double bonds. Examples include conjugated linoleic acid (CLA), which is a collective term used to describe one or more of the C_{18} octadecadienoic positional and geometric fatty acid isomers of the essential fatty acid, linoleic acid (18:2, n-6). The primary source of naturally occurring CLA in the human diet is contributed by dairy products and ruminant meats.

Cis and *trans* 18-carbon octadecenoic isomeric fatty acids and the conjugated isomers of linoleic acid have been shown to have distinctly different biologic and metabolic effects. For example, oleic acid has been shown to modestly lower LDL-cholesterol (LDL-C) and raise HDL-cholesterol (HDL-C) compared to 18:1-*trans* monounsaturated fatty acids derived from partially hydrogenated vegetable oils. Similarly, unlike linoleic acid, which has been shown to stimulate carcinogenesis in animal models, CLA has been shown to protect against carcinogenesis in vitro in a wide range of human cancer cell lines and in vivo in a number of animal models of chemically induced skin papillomas, forestomach neoplasias, and preneoplastic lesions and tumors of the colon and mammary gland. Also, geometric isomers of CLA that are predominant in ruminant fats (i.e., $C_{18:2}$, *cis*-9, *trans*-11) appear to have different effects on body composition and lipid metabolism than other CLA isomers derived from non-ruminant sources (i.e., $C_{18:2}$, *trans*-10, *cis*-12).

The distinct forms and levels of *trans*-fatty acids (TFA) found in ruminant animal fats and their physiological and metabolic effects may be distinctly different from PHVO that have major dietary guidance and nutrition labeling implications.

Origin of Ruminant *Trans* Fatty Acids

The understanding of lipid metabolism in the rumen along with major improvements in analytical techniques for quantifying the intermediates of ruminant FA metabolism has led to major advances in understanding the origins, types and levels of TFA in ruminant milk and meat fats. Ruminant TFA result from the partial biohydrogenation of feed-derived polyunsaturated fatty acids (PUFA) by rumen bacteria. The diets of dairy cows are typically low in total fat and generally consist of about 4–5% lipid. The major PUFA in the diet of dairy cows are linoleic (LA) and α-linolenic (ALA) acid. LA is derived principally from oilseeds and concentrates fed to dairy cows; while ALA is provided principally from the consumption of forage crops. When dietary triglycerides, phospholipids, and glycolipids enter a dairy cow's rumen, which contains 40–50 liters of fluid with 10^{10}–10^{11} bacteria and 10^6 protozoa per milliliter, the ester linkages of these lipids are first hydrolyzed by rumen bacteria hydrolases followed by biohydrogenation. Hydrolysis is generally extensive (>85%) and the rate and degree of hydrolysis can be affected by a number factors (for reviews, see Doreau and Ferlay, 1994; Doreau and Chilliard, 1997; Harfoot and Hazlewood, 1997).

Biohydrogenation of unsaturated fatty acids is the second major change that these fatty acids undergo in the rumen and, in order to proceed, requires that they be in the free fatty acid (FFA) form. Hence, the rate of rumen bacteria biohydrogenation of fatty acids is dependent on the rate of lipid hydrolysis. Additionally, biohydrogenation generally increases with increasing levels of fatty acid unsaturation as the rumen bacteria carry out these reactions primarily as a protective mechanism against the toxic effects of PUFA (Bauman et al., 2003a). Most of the LA (70–95%) and ALA (85–100%) undergo biohydrogenation through established metabolic pathways by two classes of ruminant bacteria: Group A bacteria hydrogenate LA and ALA to $C_{18:1}$ *trans* FA, and Group B bacteria that completely hydrogenate $C_{18:1}$ *trans* FA to stearic acid (Fig. 5.1) (Harfoot and Hazlewood, 1997; Palmquist et al., 2004). Thus, as pointed out from studies by Kemp and Lander (1984), no single species of rumen bacteria carry out the complete series of metabolic reactions for the conversion of LA and ALA to stearic acid.

The FA products of rumen biohydrogenation that reach the small intestine

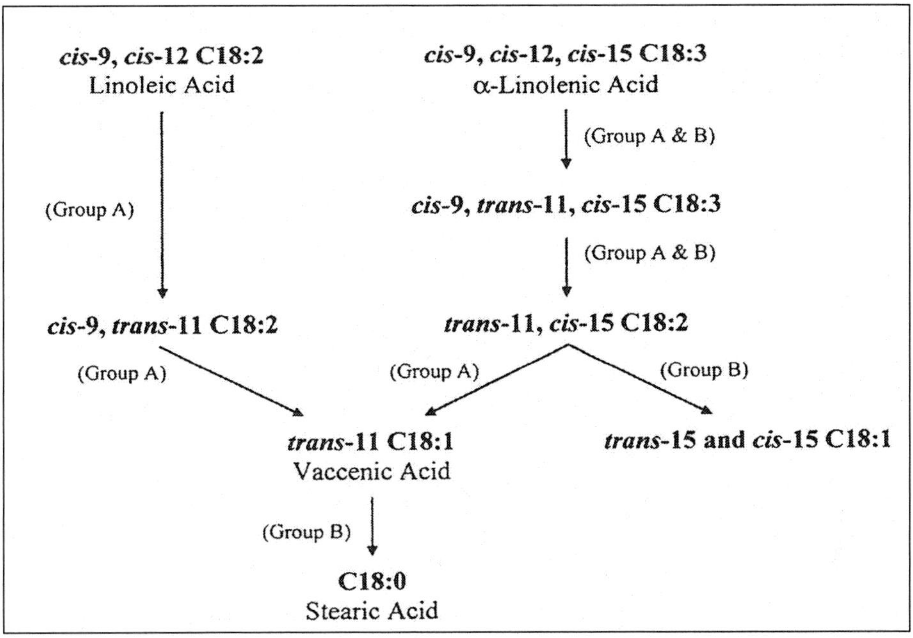

Fig. 5.1. Biochemical pathways for the biohydrogenation of linoleic and linolenic acid in the rumen. (from Lock and Bauman, 2004)

are primarily saturated FFA along with smaller amounts of biohydrogentation intermediates including $C_{18:1}$, *trans*-11 (*trans*-vaccenic acid) derived from LA and ALA and *cis*-9, *trans*-11 CLA from LA. However, it is clear that the biohydrogenation process in the rumen is substantially more complex and that many other pathways of biohydrogenation exist based on the wide range of $C_{18:1}$ *trans*, CLA and non-conjugated 18:2 *trans* isomers that have been identified in ruminant milk and meat fat.

Trans Fatty Acids in Milk Fat and Dairy Products

Bovine milk fat consists predominantly of triglycerides (96–98%) (TG) and small amounts of phospholipids, cholesterol, and cholesterol esters found in the surrounding milk fat globular membrane (McGuire and Bauman, 2002). The composition of milk fat TG consists of over 400 FA that are mostly intermediates of rumen lipid metabolism and generally occur only in trace amounts (Jensen, 2002). The predominant FA in milk fat consists of saturated FA ($C_{4:0}$–$C_{18:0}$), palmitoleic, oleic, linoleic, and smaller amounts of *trans* fatty acids which occur in a wide range of positional and geometric isomers. Unlike partially hydrogenated vegetable oils where the TFA levels can vary widely and may account for as little as 10% to as much as 60% of the total FA content, the TFA of ruminant fats in beef and dairy products account for about 2–5% of total fatty acids. The TFA

TABLE 5.1
The *Trans* Fatty Acid Content (wt% of Fatty Acids) of Bovine Milk Fats From Different Countries[a]

Mean	SD	N	Technique	Year	Country	*trans*-FA
3.4		3	GC	1983	USA[d]	C18:1
3.33	0.99	31	Ag-TLC/GC	1994	Austria[e]	C18:1
3.4		10	GC	1982	Germany[f]	C18:1
2.75		10	GC	1993	Germany[g]	C18:1
3.62	1.22	1756	Ag-TLC/GC/TR[b]	1995	Germany[h]	C18:1
3.83	1.34	100	Ag-TLC/GC	1995	Germany[h]	C18:1
3.19		1	Ag-TLC/GC/TR	1995	Belgium[h]	C18:1
4.21	0.60	4	Ag-TLC/GC/TR	1995	Denmark[h]	C18:1
4.04	0.30	10	Ag-TLC/GC/TR	1995	Spain[h]	C18:1
3.8		24	Ag-TLC/GC	1994	France[h]	C18:1
4.47	0.92	10	Ag-TLC/GC/TR	1995	France[h]	C18:1
4.01	0.15	4	Ag-TLC/GC/TR	1995	Greece[h]	C18:1
4.14	1.30	12	Ag-TLC/GC/TR	1995	Italy[h]	C18:1
5.91	0.92	22	Ag-TLC/GC/TR	1995	Ireland[h]	C18:1
3.51		1	Ag-TLC/GC/TR	1995	Luxembourg[h]	C18:1
4.09	0.91	24	Ag-TLC/GC/TR	1995	Netherlands[h]	C18:1
4.78	0.91	23	Ag-TLC/GC/TR	1995	U.K.[h]	C18:1
3.56		37	GC	2004	All Regions USA[c]	C18:1
4.47		5	GC	2004	Southeast USA[c]	C18:1
3.67		7	GC	2004	Mountain USA[c]	C18:1
3.48		7	GC	2004	Northeast USA[c]	C18:1
3.41		7	GC	2004	Midwest USA[c]	C18:1
3.20		6	GC	2004	Northern USA[c]	C18:1
3.63		5	GC	2004	Pacific USA[c]	C18:1

[a] Adapted from D. Precht and J. Molkentin, 1995.
[b] Derived from triglyceride formulae based on Ag-TLC/GC.
[c] Total *trans* – CLA, adapted from International Dairy Foods Association (2004), http://www.idfa.org/reg/transfat.cfm.
[d] Enig et al., 1983.
[e] Henninger and Ulberth, 1994.
[f] Renner and Yoon, 1982.
[g] Pfalzgraf et al., 1993.
[h] Precht and Molkentin, 1996.

of milk and meat fats consists of a variety of positional and geometric fatty acid isomers which result from the partial biohydrogenation of feed PUFA by rumen bacteria.

Except for CLA isomers, the TFA in milk fat are predominantly positional isomers of $C_{18:1}$ *trans*, with smaller quantities of *trans*-14:1, *trans*-15:1, *trans*-16:1, and *trans*-17:1 having been reported (Hay and Morrison, 1970; Precht and Molkentin, 1995, 1996, 2000a; Jakobsen et al., 2006). Additionally, the presence of an array of non-conjugated *trans*-18:2 positional and geometric isomers have been recorded at levels ranging from 0.1 – 1.4% in milk fat acquired during the winter and summer seasons, respectively (Precht and Molkentin, 2000a).

Positional Isomers of $C_{18:1}$-Trans *in Milk Fat*

Precht and Molkentin (1995, 1996) analyzed the level of total TFA including *trans* -18:1, and 13 positional isomers of *trans*-18:1 between *trans*-Δ4 and *trans*-Δ16 in milk fat samples from Germany and 10 other EU countries. Using a combination of argentation thin-layer chromatography and gas chromatography, total TFA ranged from 1.29–6.75 wt% with a mean level of 3.62 wt% in 1756 German milk samples (927 winter, 236 transition, 593 summer samples) while those from the 10 EU countries ranged from 3.19–5.91 wt% with an average value of 4.23 wt% (Table 5.1). The total TFA level in U.S. milk fat was recently determined in 37 whole raw milk samples obtained from milk processors across six regions of the country to obtain a representative sample of commercially available dairy products in the United States. Using gas chromatography (AOAC 996.06), TFA

TABLE 5.2
***Trans*-Octadecenoic Acid Isomers From 1756 German Milk Fats**

	mean	
trans position	wt% of fatty acids	% of C18:1-*trans*
Δ4	0.05	1.6
Δ5	0.05	1.5
Δ6–8	0.16	4.7
Δ9	0.23	6.9
Δ10[a]	0.17	4.7
Δ11[a]	1.72	43.2
Δ12	0.21	6.3
Δ13/14	0.49	14.2
Δ15	0.28	7.9
Δ16	0.33	9.8
trans-18:1 (total)	3.69	

[a]Based on 1707 milk fat samples

Adapted from: Precht and Molkentin, 1995.

levels (minus CLA) in milk fat ranged from 3.20–4.47 wt% with a weighted average of 3.56 wt% (Table 5.1) (International Dairy Foods Association, 2004). The lowest levels of TFA (3.20 wt% milk fat) occurred in milk samples from the Northern region of the United States (Minnesota, Wisconsin) while the highest (4.47 wt%) occurred in the Southeastern region (Florida, Virginia). Since milk samples were obtained during the winter months of December–February these differences are most likely due to feeding conditions whereby more pasture feeding, which is associated with higher TFA levels, would occur in the Southeast whereas lower levels are associated with barn feeding of concentrates and would occur in the Northern region during these winter months (Precht and Molkentin, 1997a).

The individual TFA positional isomers of 18:1 (i.e. *trans*-18:1 Δ4–Δ16) expressed as the percent of the total amount of $C_{18:1}$-*trans* are shown in Table 5.2. The most predominant isomer, *trans*-vaccenic acid (18:1-*trans*, Δ11) accounted for 43.2% of the total 18:1-*trans* isomers (Precht and Molkentin, 1995). This is somewhat lower than that reported by others: 58.2% and 50.4% Wolff (1994), and 60.5%, Parodi (1976a). In other reports, analysis of 10 samples of Danish butter, the mean level of total TFA was 4.8 g/100 g milk fat with18:1-*trans* constituting 64% of total *trans* fat while $C_{14:1}$ *trans*, $C_{16:1}$ *trans*, C18:2 *trans*, and CLA each accounted for 8–10% of total TFA (Jakobsen et al., 2006). In most studies, the predominant18:1-*trans* isomer was Δ11 (*trans*-vaccenic acid) and generally accounts for about 50% or more of the total 18:1-*trans* isomers.

The distribution pattern of 18:1-*trans* isomers and the predominance of the Δ11 isomer in bovine milk fat is clearly different than that from PHVO where the predominant TFA isomer is Δ9 (elaidic acid) along with considerable amounts of *trans* Δ10, *trans* Δ11 and *trans* Δ6–8 (Fig. 5.2) (Precht and Molkentin, 1995). Earlier studies have frequently regarded elaidic acid as the most important of the 18:1-*trans* isomers of partially hydrogenated vegetable oils (Bruckner, 1995) and many nutritional studies have been based on this premise (e.g., Nestel et al., 1992). However, by a combination of Ag-TLC with GC using a 100 m column, a separation of 18:1-*trans* positional isomers into 10 single peaks has been achieved (Molkentin and Precht, 1995) while others have achieved comparable results (Wolff and Bayard, 1995; Parodi, 1976b). The predominant 18:1-*trans* isomers in PHVO includes: Δ9 (elaidic) at 21.9% of total 18:1-*trans* isomers followed closely by Δ10 (20.7%), and Δ6-8 (17.5%) and lesser amounts of Δ11 (14.8%), Δ12 (12.0%) and Δ13/14 (9.9%) (Precht and Molkentin, 1995; Molkentin and Precht, 1995; Wolff and Bayard, 1995; Parodi, 1976b).

Effects of Feeding Conditions on $C_{18:1}$-Trans *in Milk Fat*

Parodi (1976a) reported little seasonal variation in the distribution of *trans*-18:1 isomers. However, the effect of feeding conditions has been shown to play a major role in altering the levels of 18:1-*trans* isomers in bovine milk fat. In 1,756 german milk fat samples acquired over the course of a year, 18:1-*trans* levels ranged

A

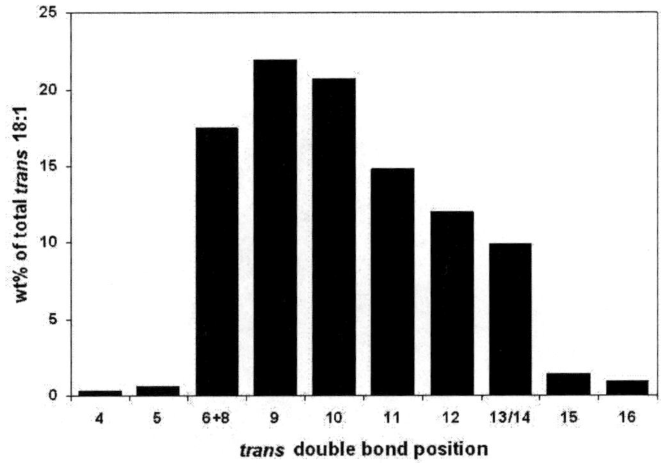

B

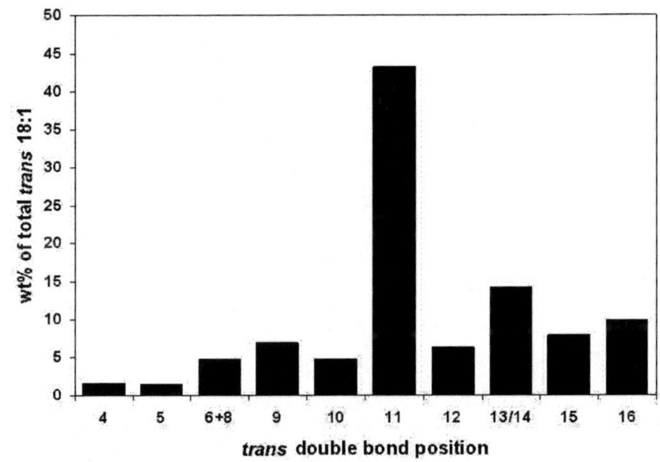

Fig. 5.2. Isomeric distribution of *trans*-octadecenoic acids (mean wt% of *trans*-18:1 fatty acids) in (A) partially hydrogenated vegetable oil (margarine, $n = 46$ samples) and (B) bovine milk fat ($n = 1765$ samples). (adapted from Precht and Molkentin, 1995).

from 2.65 wt% in winter during barn feeding (high concentrates, starch-rich fodder) to 5.08 wt% in the summer (pasture feeding) with intermediate levels, 3.80 wt%, during transition periods (barn to pasture feeding) (Precht and Molkentin, 2000a). Fig. 5.3 shows the weekly results for the contents of $C_{18:1}$-*trans*, $C_{18:2}$-*trans* (minus CLA) and $C_{18:2}$, *cis*-9, *trans*-11 (CLA) from bulk milk samples from the same herds analyzed over a one year period (Precht and Molkentin, 1997b). The qualitative pattern of the data are nearly identical for each of the fatty acids shown and are reflective of feeding conditions incurred over the year. The levels are lowest in winter, between the 48th week and the 19th week of the following

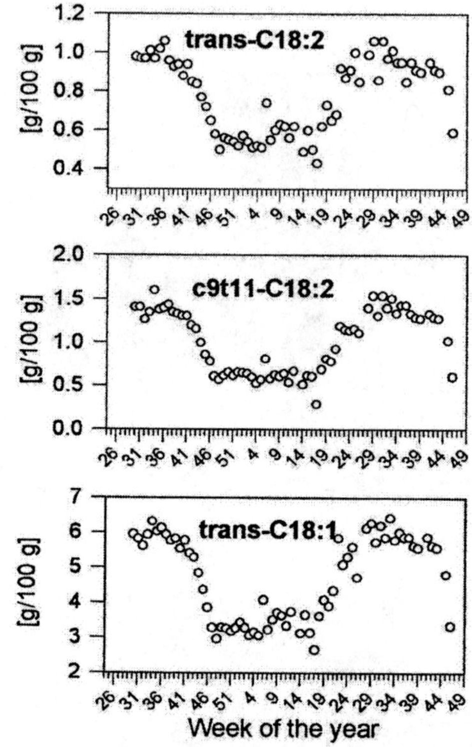

Fig. 5.3. Weekly variation in the contents of *trans*-$C_{18:2}$, the main CLA isomer *cis*-9 *trans*-11-C18:2, and *trans*-C18:1 in milk fat (from Precht and Molkentin, 2000a).

year during which feed rations are high in concentrates and low in pasture grass. Further seasonal work on the relative distributions of $C_{18:1}$-*trans* isomers showed that *trans*-vaccenic acid was the predominant isomer accounting for 35.1%, 46.4%, and 55.3 % over the winter, transition, and summer feeding periods. This is the result of increasing PUFA levels in the bovine feed with lower amounts in winter and higher amounts in summer pasture feeding (Precht and Molkentin, 1997a). These workers also demonstrated that the seasonal influences of feeding that were observed for *trans* Δ11 (*trans*-vaccenic acid) also occurs for all other *trans*-18:1 positional isomers in milk fat (i.e., Δ4, Δ5, Δ6–Δ16) — except *trans* Δ10 which remained unchanged (Precht and Molkentin, 1997a). Moreover, in comparison to the average total $C_{18:1}$-*trans* levels of 3.62% observed in the German milk fats, the relatively high mean level of 4.24% seen in 111 milk fats from 10 other EU countries are indicative of different climatic conditions and feeding practices (Precht and Molkentin, 2000a; Precht and Molkentin, 1996). In countries such as Ireland with *trans*-18:1 levels of 5.19% and where pasture

TABLE 5.3
***Trans*-Octadecadienoic Acid Isomers in Milk Fat**

Non-conjugated 18:2[a,b]			Conjugated 18:2 (CLA)	
Isomer	wt% of fatty acids	# of total C18:2-*trans*	Isomer	% of total CLA isomers
trans-9,*cis*-12	0.07	7.1	*trans*-7,*cis*-9	1.20–8.89
trans-9,*trans*-12	0.09	9.1	*trans*-7,*trans*-9	0.02–2.39
cis-9,*trans*-12	0.10	10.1	*trans*-8,*cis*-10	0.06–1.47
trans-11,*cis*-15	0.33	33.3	*trans*-8,*trans*-10	0.19–0.37
cis-9,*trans*-13	0.11	11.1	*cis*-9,*trans*-11	72.56–91.16
trans-8,*cis*-13	0.11	11.1	*trans*-9,*trans*-11	0.77–2.87
other isomers	0.18	18.1	*trans*-10,*cis*-12	0.03–1.51
trans-18:2 (total)	0.99	100	*trans*-10,*trans*-12	0.28–1.31
			cis-11,*trans*-13	0.18–4.70
			trans-11,*cis*-13	0.07–8.00
			trans-11,*trans*-13	0.28–4.24
			cis-12,*trans*-14	0.04–0.80
			trans-12,*trans*-14	0.33–2.76
			cis,*cis* isomers	0.06–4.80

[a]Based on 100 milk fat samples
[b]Adapted from: Precht and Molkentin, 1997c.

feeding predominates corresponds closely with German milk fat from summer feeding periods (Precht and Molkentin, 2000a).

Milk fat has also been shown to contain several other *trans* fatty acid isomers, including *trans*-14:1, *trans*-15:1, *trans*-16:1 and positional isomers of *trans*-18:2, which altogether account for about 2–3% of total fatty acids.

Trans-Octadecadienoic Acids ($C_{18:2}$-*Trans*) in Milk Fat

Although $C_{18:1}$-*trans* isomers account for most of the *trans* isomers in milk fat (Precht and Molkentin, 1995), small amounts of $C_{18:2}$-*trans* are also present. The total *trans*-18:2 fatty acids in milk fat consists of both conjugated-18:2 (CLA) and non-conjugated-18:2 isomers each have having an array of positional and geometric isomers.

Non-Conjugated Trans-*octadecadienoic Acids*

Precht and Molkentin (2000a) demonstrated the presence of an array of positional and geometric isomers of non-conjugated (NC) 18:2-*trans* dienes in milk fat. These and other workers have also confirmed that the predominant NC-

18:2-*trans* isomer in milk fat is *trans*-11, *cis*-15 (Precht and Molkentin, 1997c; Ulberth and Henninger, 1994). The type and levels of NC-18:2-*trans* isomers identified in milk fat by Precht and Molkentin (1997c) include the *trans*-11, *cis*-15 (0.33%), *cis*-9, *trans*-13 (0.11%), and probably *trans*-8, *cis*-13 (0.11%) as well as smaller amounts of other isomers (Table 5.3). Additionally, several *trans,trans*-non-methylene interrupted dienes (t,t-NMID) were also reported. The level of NC-*trans*-18:2 isomers (minus CLA and t,t-NMID) recorded in 1,756 milk fat samples averaged 0.63%. These levels were shown to increase from a winter feeding level of 0.46% to 0.66% during transition feeding periods in spring and late autumn to 0.87% during summer pasture feeding (Fig. 5.3) (Precht and Molkentin, 1997b).

Conjugated Trans-*Octadecadienoic Acids (CLA)*

Conjugated linoleic acid (CLA) is a collective term used to describe one or more of the C_{18} dienoic positional and geometric fatty acid isomers of the essential fatty acid, linoleic acid (18:2, n-6), which have at least one *trans* double bond. For a complete history of the discovery of the CLA fatty acid isomers in milk fat and their structural determination the reader is referred to an excellent review by Parodi (1999).

Ruminant foods are the primary sources of CLA, with dairy products accounting for approximately 75% of the estimated 100–300 mg CLA consumed per day (Ritzenthaler et al., 2001; Parodi, 2003; Fritsche and Steinhart, 1998). The content of CLA in milk fat from commercial dairy products, such as milk, cheese and cultured products varies considerably ranging from 1.0–29.0 mg/g fat and is independent of the type of dairy product but depends almost exclusively on the level contained in the starting milk fat (Parodi, 2003). Regional and season differences in the CLA content of milk fat have been noted. For example, concentrations of CLA in milk fat have been reported to vary markedly between European countries from 0.13–1.89 wt% and is likely due to differences in ruminant feed composition (Precht and Molkentin, 2000b).

There is substantial interest in CLA because of the potential health benefits that these fatty acids may confer. Unlike linoleic acid which has been shown to stimulate carcinogenesis in animal models, CLA has consistently been shown to protect against carcinogenesis in vitro in a wide range of human cancer cell lines and in vivo in a number of animal models of chemically induced skin papillomas, forestomach neoplasias, and preneoplastic lesions and tumors of the colon and mammary gland (Banni et al.,2003; Belury, 2002; and Ip et al., 2003). Other potential benefits of CLA include antidiabetogenic, anti-atherogenic, immunomodulation, anti-obestity, and modulation of bone growth (Belury, 2002; Watkins et al., 2003; Cook et al., 2003).

Isomers of CLA

Research has established that the predominant CLA fatty acid in ruminant fat is the *cis*-9, *trans*-11 isomer which represents about 75–90% of the total CLA isomers (Parodi, 2003; Bauman et al., 2003b). The name "rumenic acid" has been proposed as the common name for this isomer because of its unique ruminant origin (Kramer et al., 1998). Like the ruminant *trans*-18:1 isomers, the content of *cis*-9, *trans*-11 CLA in milk fat has been shown to increase from winter concentrate feeding levels of 0.45–0.76% during the spring transition feeding period with the highest levels (1.20%) observed during summer pasture feeding (Fig. 5.3) (Precht and Molkentin, 1995). With the development of improved analytical techniques including Ag-TLC/gas chromatography and tandem-column Ag-high performance liquid chromatograpy, (reviewed in Parodi, 2002), numerous *cis-trans*, *trans-cis* and *trans-trans* isomers of CLA have been identified at positions Δ7,9; Δ8,10; Δ9,11; Δ10,12; Δ11,13; and Δ12,14 of the 18 carbon fatty acid chain (Table 5.3) (Precht and Molkentin, 1997c; Bauman et al., 2000; Rickert et al., 1999; Piperova et al., 2002). Bauman et al. (2000) showed that control bovine butter contained CLA levels of 5.3 mg/g of fatty acids (all isomers) that consisted of 85.8% *cis-trans*, 9.4% *trans-trans* and 4.8% *cis-cis* isomers. The *cis*-9, *trans*-11 isomer constituted 76.5% of the total CLA isomers followed by 6.7% of *cis*-7, *trans*-9 and 1.1% of *trans*-10, *cis*-12 and minor amounts of other geometric and positional isomers. After feeding cows a low forage diet supplemented with sunflower oil for 6–10 days, CLA levels were increased 7–8 times compared to control butter (41 versus 5.3 mg/g fatty acids) with the *cis*-9, *trans*-11 isomer accounting for 90.8% of the total CLA isomers. In general, compared to control butter, the CLA-enhanced butter had less medium chain fatty acids (3.56 versus 6.86 g/100g FA), less $C_{12:0}$–$C_{16:0}$ saturated fatty acids (26.9 versus 45.98 g/100 g FA), higher levels of $C_{18:1}$-*cis* unsaturated fatty acids (oleic; 24.17 versus 18.11 g/100g FA) and a predominant

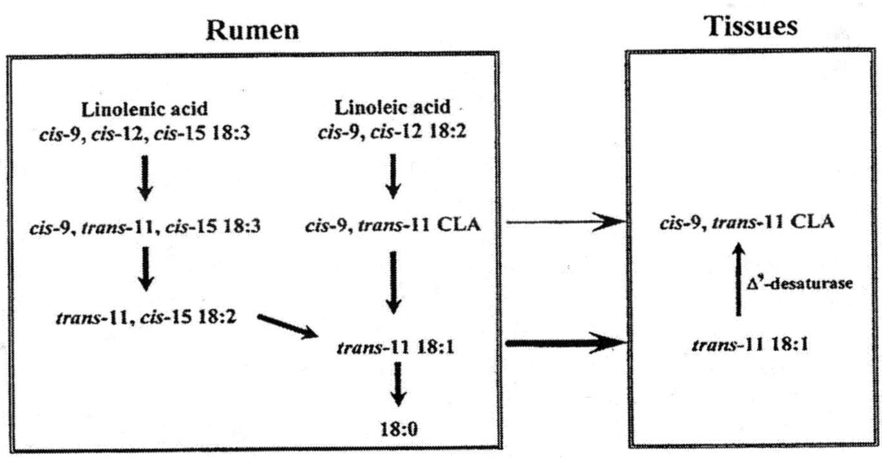

Fig. 5.4. Pathways for ruminal and endogenous synthesis of *cis*-9,*trans*-11 CLA in the dairy cow. (From Lock and Bauman, 2004b)

TABLE 5.4
***Trans* Fatty Acid Content of Ruminant Meat Products**

		Mean g/100g meat fatty acids (g/100g meat)			
Meat Cut	n[a]	C16:1 *trans*	C18:1 *trans*	C18:2 *trans*	Total *trans*
Beaf					
thick flank	8	0.4 (0.02)	1.6 (0.08)	—	2.0 (0.10)
outside round	8	0.1 (0.01)	1.4 (0.08)	—	1.5 (0.09)
chuck	8	0.5 (0.05)	2.9 (0.32)	—	3.4 (0.37)
rump	8	0.2 (0.02)	1.4 (0.17)	—	1.6 (0.19)
brisket	7	0.2 (0.03)	2.8 (0.43)	0.1 (0.02)	3.1 (0.48)
Veal					
thick flank	4	—	2.3 (0.07)	—	2.3 (0.07)
outside round	4	—	2.0 (0.09)	—	2.0 (0.09)
chuck	4	0.1 (0.01)	5.5 (0.43)	—	5.6 (0.44)
rump	4	—	2.3 (0.20)	—	2.3 (0.20)
brisket	4	0.1 (0.02)	4.4 (0.64)	0.00 (0.00)	4.5 (0.66)
Lamb					
shoulder	11	0.8 (0.08)	4.7 (0.45)	0.00 (0.00)	5.5 (0.53)
leg	9	0.8 (0.11)	3.7 (0.52)	0.09 (0.02)	4.6 (0.65)
loin	11	0.9 (0.14	3.9 (0.63)	0.09 (0.02)	4.9 (0.79)

[a]Number of samples
Adapted from: Leth et al., 1998.

increase in $C_{18:1}$-*trans* fatty acids 15.04 versus 5.04 g/100g FA that consisted predominantly of the $C_{18:1}$, *trans*-11 isomer (*trans*-vaccenic acid; 48.3 versus 24.5% of total $C_{18:1}$ *trans* isomers). In an assessment of the CLA content of other dairy products, analysis of 16 cheese samples (curd cheese, processed cheese, Edamer, Tilsiter, Brie, Emmenthal, Camembert, Mozzarella, Feta) showed a total CLA content (all isomers) of 9.3 mg/g fat that consisted of 93.2% *cis-trans*, 6.2% *trans-trans* and 0.7% *cis-cis* isomers (Rickert et al., 1999). The *cis*-9, *trans*-11 and *trans*-10, *cis*-12 isomers (not separated) constituted 83.5% of the total CLA isomers followed by 4.7% *cis*-11, *trans*-13 and 3.7% *cis*-7, *trans*-9. The primary *trans*/*trans* isomers were 11,13 (2.3%), 9,11 (1.5%), and 12,14 (0.9%).

To date, research on the biological effects of CLA have been principally focused on the *cis*-9, *trans*-11 and *trans*-10,*cis*-12 isomers because of the natural predominance of the *cis*-9, *trans*-11 isomer in milk fat and the occurrence of both isomers in significant amounts in synthetic preparations from vegetable oils. With respect to dairy products, studies have clearly demonstrated the anti-carcinogenic and antiatherogenic effects of the *cis*-9, *trans*-11 CLA isomer (Ip et

TABLE 5.5
Distribution of $C_{18:1}$ *Trans* Isomers in Beef Meat Fat and Tallow

<table>
<tr><th></th><th colspan="5">Origin of Fat</th></tr>
<tr><th>trans Position</th><th>Perirenal[a]</th><th colspan="2">Subcutaneous[a,b]</th><th>Meat[c]</th><th>Tallow[c]</th></tr>
<tr><td>Δ6</td><td>—[d]</td><td>0.3</td><td>0.2</td><td rowspan="4">8.7</td><td rowspan="4">11.6</td></tr>
<tr><td>Δ7</td><td>1.0</td><td>0.5</td><td>0.3</td></tr>
<tr><td>Δ8</td><td>2.1</td><td>1.6</td><td>1.5</td></tr>
<tr><td>Δ9</td><td>5.0</td><td>8.9</td><td>13.6</td></tr>
<tr><td>Δ10</td><td>11.9</td><td>5.4</td><td>6.4</td><td rowspan="2">66.9</td><td rowspan="2">63.8</td></tr>
<tr><td>Δ11</td><td>46.9</td><td>68.6</td><td>64.4</td></tr>
<tr><td>Δ12</td><td>6.0</td><td>2.7</td><td>2.4</td><td>5.9</td><td>4.7</td></tr>
<tr><td>Δ13</td><td>6.6</td><td>2.6</td><td>2.3</td><td rowspan="2">9.5</td><td rowspan="2">10.5</td></tr>
<tr><td>Δ14</td><td>7.4</td><td>3.6</td><td>3.6</td></tr>
<tr><td>Δ15</td><td>5.5</td><td>2.6</td><td>2.3</td><td>4.1</td><td>4.6</td></tr>
<tr><td>Δ16</td><td>7.6</td><td>3.2</td><td>3.0</td><td>4.9</td><td>4.7</td></tr>
</table>

Adapted from: [a]Parodi, 1976b; [b]Hay and Morrison, 1973; [c]Wolff, 1995. [d]Not reported.

al., 2003; Kritchevsky, 2003). Other studies have demonstrated that both isomers of CLA (i.e. *cis*-9, *trans*-11 CLA and *trans*-10, *cis*-12 CLA) appear to be equally effective in inhibiting tumors in rat mammary carcinogenesis (Ip et al., 2003) and in the inhibition of atherosclerosis in rabbit models (Kritchevsky, 2003). Since *trans*-10, *cis*-12 CLA only represents approximately 1–2 % of the total CLA isomers in milk fat whereas the *cis*-9, *trans*-11 CLA isomer predominates at levels ranging from 72–91%, technical efforts to increase the CLA content of dairy foods effectively means alterations in this isomer only.

Origin of CLA in Bovine Milk Fat

It is now well established that the presence of CLA in milk fat involves ruminant biohydrogenation of feed and pasture derived LA, ALA, and γ-linolenic acids through the major metabolic pathways shown in Fig. 5.4. The initial steps in the biohydrogenation process involves the hydrolysis of feed derived triglycerides by rumen microbial lipases to free fatty acids. As noted in Fig. 5.4, *cis*-9, *trans*-11 CLA (rumenic acid) is formed as the first intermediate in the pathway, exclusively from LA through an isomerase from the rumen bacteria Butyrivbrio fibrisolvens which isomerizes the *cis*-12 double bond of LA (*cis*-9, *cis*-12 18:2) to the *trans* configuration on carbon-11. However, rumenic acid is only a transitory intermediate that is further biohydrogenated by rumen bacteria to *trans*-18:1 Δ11, *trans*-vaccenic acid, the predominant *trans*-18:1 fatty acid in milk fat that is a stable intermediate that tends to accumulate. Unlike LA, other major feed

TABLE 5.6
Daily Intake and Ruminant Contribution of *Trans* Fatty Acids in Men From Different Countries[a]

	United States	Belgium	Denmark	Finland	France	Germany	Greece	Iceland	Italy	Netherlands	Norway	Portugal	Spain	Sweden	United Kingdom
Total *trans*															
% energy	2.2	1.4	1.0	0.8	1.1	0.8	0.5	2.1	0.5	1.5	1.5	0.6	0.7	1.1	1.3
g/day	6.4	4.4	2.9	2.3	2.7	2.4	1.2	6.7	1.6	4.8	4.8	1.6	2.1	3.0	2.8
Ruminant *trans*															
% of foods	20–25	43	N/A	44.1	72.3	77.1	52.1	39.2	75.4	29	26.3	65.5	62.5	50.8	35
% energy	N/A[b]	1.4	1.0	0.8	1.1	0.8	0.5	2.1	0.5	1.5	1.5	0.6	0.7	1.1	1.3
g/day	1.28–1.6	1.89	N/A	1.01	1.95	1.85	0.6	2.61	1.2	1.39	1.26	1.04	1.31	1.52	0.98

[a]Source: Hulshof et al., 1999.
[b]N/A – not available

polyunsaturated fatty acids (ALA and γ-linolenic acids) cannot be converted in the rumen to rumenic acid. Rather, these highly unsaturated fatty acids are bioconverted to a series of intermediates in the rumen leading finally to *trans*-vaccenic acid, which passes from the rumen and is absorbed across the intestine into the circulatory system and is transported to the mammary gland and peripheral adipose tissue for triglyceride synthesis (Griinari and Bauman, 1999). Through a series of experiments involving infusions of *trans*-vaccenic acid to bypass rumen biohydrogenation and the use of inhibitors of Δ9-desaturase, Griinari et al. (2000) showed that endogenous synthesis accounted for the majority of rumenic acid present in milk fat with 70–93% generated from the desaturation of *trans*-vaccenic acid by Δ9-desaturase in the lactating mammary gland (Griinari and Bauman, 1999; Griinari et al., 2000; Bauman and Griinari, 2003).

Trans Fatty Acids in Ruminant Meat Products

The positional and geometric TFA isomers contained in ruminant meat fats are similar to those found in ruminant milk fat and consists of $C_{18:1}$-*trans*, $C_{16:1}$-*trans*, $C_{18:2}$-*cis,trans* conjugated dienes (CLA) and small amounts of $C_{18:2}$-*trans* unconjugated dienes. For a more in-depth review of the CLA content of ruminant meat products, the reader is directed to the excellent review by Parodi (2003).

Like dairy products, meats from ruminant animals contain relatively low levels of TFA compared to PHVO but are distinctly higher in TFA than meat from nonruminants. The fatty acid composition of ruminant meat products from 14 different European countries showed that the mean $C_{18:1}$-*trans* content for beef was 2.84% of fatty acids (range: 1.30–7.00%) and for lamb somewhat higher at 4.57% (range: 2.72–6.70%) whereas substantially lower levels were observed in nonruminant meats such as pork at 0.45% (range: 0.17–1.60%) and chicken 0.47% (range: 0.16–1.03%) (Aro et al., 1998). In some countries including Belgium and The Netherlands, beef or lamb meat fat appeared to contain more

TABLE 5.7
Daily Intake of Ruminant *Trans* Fatty Acids Among 3098 Danish Males and Females[a,b]

		Percentiles					
		10		50		90	
		g/day	(en%[c])	g/day	(en%)	g/day	(en%)
All 1–6 years	551	0.9	(0.5)	1.4	(0.8)	2.1	(1.0)
All 7–14 years	710	1.0	(0.5)	1.6	(0.6)	2.4	(0.9)
All 15–29 years	789	0.9	(0.5)	1.8	(0.7)	2.9	(0.9)
All 30–80 years	1348	0.9	(0.5)	1.8	(0.7)	2.9	(1.0)

[a]Source: Jakobsen et al., 2006.
[b]Includes conjugated linoleic acid
[c]en%, % of energy intake

TABLE 5.8
Daily Intake of *Trans* Fatty Acids From Ruminant Food Sources Among 3098 Danish Males and Females[a,b]

Product	Age Group	Number	Ruminant TFA[c] Intake, g/day		
			Percentiles		
			10	50	90
Dairy Products	1–6 years	551	0.8	1.2	1.9
	7–14 years	710	0.8	1.4	2.1
	15–29 years	489	0.7	1.4	2.5
	30–80 years	1348	0.8	1.5	2.7
Meat Products	1–6 years	551	0.0	0.1	0.1
	7–14 years	710	0.0	0.1	0.3
	15–29 years	489	0.1	0.2	0.5
	30–80 years	1348	0.0	0.1	0.2
Ice Cream and Chocolate	1–6 years	551	0.0	0.1	0.2
	7–14 years	710	0.0	0.1	0.2
	15–29 years	489	0.0	0.1	0.2
	30–80 years	1348	0.0	0.1	0.1

[a]Source: Jakobsen et al., 2006.
[b]Includes conjugated linoleic acid
[c]TFA = *trans* fatty acids

trans fatty acids than in most other countries. This likely reflects differences in ruminant feeding practices between countries. For example, evaluation of feedlot rations versus grass feeding showed there was a 2- and 3-fold increase in $C_{18:1}$ TFA in grain-fed Australian beef cuts compared to beef from cattle fed grass (Ponnampalam et al., 2006).

Also, like dairy products, the $C_{18:1}$-*trans* isomers account for the majority of TFA in ruminant meats followed by much lower levels of $C_{16:1}$- and $C_{18:2}$-*trans* (Leth et al., 1998 Parodi, 1976b; Hay and Morrison, 1973; Wolff, 1995). The $C_{18:1}$-*trans* islomers accounted for 80–93% of the total TFA followed by $C_{16:1}$-*trans* (6–20%) while $C_{18:2}$-*trans* were not detected in most cuts evaluated (Table 5.4). The level of TFA also varies across ruminant species. The reported mean level of $C_{18:1}$-*trans* across different cuts of beef, veal, and lamb was 2.1, 4.0, and 4.5 g/100 g meat fatty acids, respectively—significantly lower in beef fat than in veal and lamb fat whereas the difference between veal and lamb were not significant (Leth et al., 1998). The levels of $C_{16:1}$-*trans* was 0.24, 0.14, and 0.79 g/100 g fatty acids in beef, veal, and lamb fat, respectively. In all meat samples in which $C_{16:1}$-*trans* was found, the content was highest in lamb while in veal, the presence of $C_{16:1}$-*trans* was inconsistent and was found in only 2 out 5 samples analyzed.

The TFA content of ruminant meats also varied widely depending on the type of meat cut. The level of total *trans* fat in beef was lowest in the outside round (1.5 g/100 g meat fat) and highest for brisket and chuck (3.1 and 3.4 g/100 g meat fat), respectively (Table 5.4). A very similar pattern was observed for veal whereas in lamb, the levels of total *trans* fat across meat cuts were less variable at 4.6, 4.9, and 5.5 g/100 g meat fat for the leg, loin, and shoulder, respectively (Leth el al, 1998).

The distribution patterns of $C_{18:1}$-*trans* fatty acid isomers in ruminant meat fat and tallow are similar to those seen in ruminant milk fats with *trans*-vaccenic acid ($C_{18:1}$-*trans*, $\Delta 11$) being the predominant isomer. The distribution of $C_{18:1}$-*trans* isomers in beef meat fats reported by different authors are generally in good agreement (Hay and Morrison, 1973; Parodi, 1976; Wolff, 1995). Table 5.5 shows that in two separate analyses of subcutaneous fat from beef, *trans*-vaccenic acid accounted for 68.6 and 64.4% of total 18:1-*trans* isomers (Hay and Morrison, 1973 and Parodi, 1976b, respectively) and are consistent with those reported for meat fat at 66.9% and tallow at 63.8% (Wolff, 1995) while in perirenal fat vaccenic acid accounted for 46.9% (Parodi, 1976b).

Estimated Intake of *Trans* Fatty Acids from Ruminant Food Sources

The average daily consumption of total TFA in the United States was estimated at 2.6+0.02 % of total energy intake (en%; range 2.6–2.8 en% across all ages and gender) with a median intake of 4.7 g/d (Allison et al., 1999). This is consistent with total TFA intakes reported for men and women in Minnesota during the same time period (2.5 and 2.4 en%, respectively) (Harnack et al., 2003). The absolute total TFA intake reported for participants aged 20–49 yrs, was 6.7 g/d for men and 4.5 g/d for women (Harnack et al., 2003). These intakes are consistent with the amounts reported by Allison et al. (1999), with mean intake estimates of 6.6 g/d for men and 4.6 g/d for women. In Europe, a study involving 14 European countries estimated the intake of total TFA to range from 0.5 en% (Greece and Italy) to 2.1 en% (Iceland) with absolute levels of intake ranging from 1.2 g/d (Greece) to 6.7 g/d (Iceland) (Table 5.6) (Hulshof et al., 1999). Results from this study and a later survey of European countries (van de Vijver et al., 2000), clearly indicate that the TFA intakes in Western Europe, 0.52–1.47 en%, 1.37–3.62 g/d for men and 0.57–1.65 en%, 1.04–1.77 g/d for women (van de Vijver et al., 2000), are lower than previously assumed. In an earlier review of TFA intake in different European countries covering a period from 1979–1992, the total TFA intake for Finland was estimated at 5.6 g/d in 1983 and 1.5–3.0 g/d in 1991, while for Germany estimated intakes were 4.5–6.4 g/d in 1979, and 3.4–4.1 g/d in 1992 (see Precht and Molkentin, 1995). Differences in the study populations and differences in dietary survey methods partly explain the differences. However, during the time elapsing from the earlier surveys, the fat composition and TFA content of food products have changed dramatically. For example, in Iceland and The Netherlands, an updated analysis of the TFA levels in food products in the TRANSFAIR part 3 survey were substantially lower than those in TRANSFAIR part 2, resulting in an approximate 45% and 38% reduction in the total daily TFA intake (6.7 versus 3.62 g/d and 4.8 g/d versus 2.96 g/d in men, respectively) (Hulshof et al., 1999; van de Vijver et al., 2000).

Trans *Fatty Acid Intake From Dairy Products and Ruminant Meats*

Dairy products (e.g. milk, milk products, cheese, butter) and ruminant meats (e.g. beef, veal and lamb) are the principal sources of naturally occurring TFA in the diet. Heimbach et al. (1995) estimated that the mean intake of naturally occurring TFA in the U.S. population is about 1.1 g/d while Hunter and Applewhite (1991) estimated this amount to be 1.34 g/d. Their estimates suggested that ruminant sources of TFA account for approximately 20–25% of the total TFA consumed in the U.S. with the majority derived from foods containing partially hydrogenated vegetable oils (baked goods, frying shortenings, margarines). In European studies, however, the estimated contribution of ruminant food sources to the total TFA intake varied widely ranging from 26.3% (Norway) to 77% (Germany) with absolute daily intakes of ruminant *trans* ranging from 0.62 g/d (Greece) to 2.61 g/d (Iceland) (Table 5.6) (Hulshof et al., 1999). Ruminant foods included: milk and milk products, cheese, butter, and meat and meat products. Fully 7 of the 14 countries surveyed derived over 50% of their daily intake of TFA from ruminant food sources (Hulshof et al., 1999). In general, the countries with the highest daily intake of TFA had the lowest overall contribution from ruminant food sources and it is evident that as industrially produced TFA are removed from the food supply, the likelihood of further reductions seems limited because of the relatively consistent contributions of naturally occurring TFA from ruminant food sources.

Jakobsen et al. (2006) estimated the average daily intake of ruminant TFA in the Danish population to range from 1.4–1.8 g/d in subjects 1–80 years of age with a mean intake of 1.7 g/d (0.7% en%) of which 70% (1.2 g/d) was contributed by $C_{18:1}$ TFA (Table 5.7). Similarly, Wolff (1995) estimated the daily intake of ruminant $C_{18:1}$ TFA in member countries of the European Economic Community (EEC) to be 1.4 g/d, with Denmark having the highest in the EEC at 1.8 g/d (Leth et al., 1998). However, using newer consumption data from 1993 instead of 1983 and more dairy food samples collected from all seasons, the mean ruminant TFA intake for Denmark was recalculated to 1.4 g/d. An explanation for this slightly higher intake, when compared to 1.2 g/d reported by Jakobsen et al. (2006), may be that the estimated intake of $C_{18:1}$ TFA by Wolff is based on food availability data whereas the lower estimate is derived from actual food intake data (Jakobsen et al., 2006). Table 5.8 shows the estimated daily intake of ruminant TFA based on ruminant food sources for various age groups. Dairy products were the predominant source of ruminant TFA contributing 89, 85, 83, and 87% of the total intake among persons 1–6 years, 7–14 years, 15–29 years, and 30–80 years of age, respectively, with an average contribution of 86% among persons 1–80 years of age (Jakobsen et al., 2006). Meat and meat products contributed substantially less at approximately 11.7% of the total ruminant TFA intake (0.2 g/d) followed by ice cream and chocolate confections at about 5.9% (0.1 g/d). This is consistent with the estimates by Leth et al., (1998) showing that among meat products, beef and veal were the most important sources contributing approximately 0.17 g of TFA/d followed by lamb at approximately 0.03 g/d.

Ruminant *Trans* Fatty Acids and Risk of Coronary Heart Disease

Background

Numerous studies over the past 50 years have evaluated the effect of *trans* fat from partially hydrogenated vegetable oils (PHVO) on plasma lipids and lipoproteins. Early clinical studies produced conflicting results because of poor quality study designs that often lacked adequate power due to low subject numbers, short experimental periods and inadequate control for intakes of TFA, saturated, monounsaturated and polyunsaturated fatty acids each having their own independent cholesterolemic effects (reviewed by Emken, 1979). However, renewed interest in the role of TFA on coronary heart disease risk was generated when a well designed controlled feeding trial clearly demonstrated that substituting *cis*-monounsaturated (oleic acid) with an equivalent amount of *trans*-monounsaturated fatty acids from PHVO significantly raised total cholesterol (TC) and LDL-cholesterol (LDL-C) levels in adult male and female subjects (Mensink and Katan, 1990). Since then, several clinical studies have confirmed these findings that TFA from PHVO at sufficiently high dietary levels raise LDL-C levels to a similar degree as saturated fatty acids, and whereas saturated fatty acids raise HDL-cholesterol (HDL-C), TFA do not (Aro et al., 1997; Judd et al., 1994, 1998, 2002; Lichtenstein et al., 1993, 1999; Zock and Katan, 1992). The levels of PHVO derived TFA used in these studies ranged from 3–11% of total energy intake (en%). Using the results of these studies in a best-fit regression estimate, Ascherio (1999) reported a linear increase in LDL/HDL ratio with increasing intakes of TFA or saturated fat but with an approximate 2-fold greater effect for *trans* fat compared to an equivalent intake of saturated fat. In a reassessment of these clinical studies, Hunter (2005) estimated that the dietary level of TFA necessary to raise LDL-C levels appear to be approximately 4 en% or higher. The mean intake of TFA in the U.S. population aged 3 yrs and older is estimated at 2.6 en% (5.3 g/d) with a 90th percentile intake of 3.1 en% (9.4 g/d) (Allison, 1999). The impact of TFA at these more typical intake levels on plasma LDL-C and HDL-C have not been evaluated and is currently unknown.

In addition to the findings from metabolic studies on the effects of TFA on plasma LDL-C and HDL-C, several epidemiologic studies have reported a positive association between the intake of TFA and the risk of coronary heart disease (CHD). This effect was primarily accounted for by industrially produced PHVO whereas ruminant derived TFA were not associated with an increase in CHD risk.

TABLE 5.9
Characteristics of Epidemiological Studies on Ruminant and Industrial *Trans* Fatty Acid Intake and Risk of Coronary Heart Disease[a]

Reference (Study Name)	Number of Subjects[b]	Number of Cases	Definition of end Points	TFA Intake (g/d)		Relative Risk of CHD	
				Industrial	Ruminant	Industrial	Ruminant
Prospective cohort studies							
Willett et al. (1993) (Nurses' Health Study)	85,095 W	431	Non-fatal MI or coronary death	1.44–3.42	0.96–2.28	1.78 (p=0.009)[c]	0.59 (p=0.23)[c]
Pietinen et al. (1997) (ATBC Study)	21,930 M	635	Coronary death	0.1–5.1	0.6–2.5	1.23 (p=0.004)[c]	0.83 (p=0.035)[c]
Oomen et al. (2001) (Zutphen Cohort Study)	667 M	98	Fatal and non-fatal MI	5.0[d]	1.7[d]	1.05	1.17
Case-control studies							
Ascherio et al. (1994)	407 M 114W	239	MI	0.84–5.04	0.45–1.79	1.94 (p=0.0001)[c]	1.02 (p=0.57)[c]

[a]Source: adapted from Weggemans RM, Rudrum M, Trautwein EA. *Eur J Lipid Sci Technol* 106:390-397, 2004.
[b]W: women; M: men; MI: myocardial infarction.
[c]p for trend.
[d]g/d based on estimates by Weggemans et al., 2004 (above).

The consistent findings from metabolic and epidemiologic studies have resulted in dietary recommendations by government agencies and medical organizations to decrease the intake of TFA. Efforts to reduce the levels of TFA in food product formulations have typically replaced PHVO with liquid vegetable oils or, in some cases, with blends of liquid oils and tropical oils including palm and modified versions of palm oil. However, as noted by the American Heart Association (Lichtenstein et al., 2006), even if all PHVO were removed from the food supply, it is estimated that TFA would still represent ~1 en% in the diet because of the amount of naturally occurring TFA contributed from meat and dairy products. The U.S. Dietary Guidelines for Americans (2005) recommends keeping TFA intake as low as possible, while the American Heart Association (Lichtenstein et al., 2006) recommend that the intake of TFA be <1% of energy.

TFA in the food supply are derived from two primary sources: food products that are manufactured with PHVO including baked goods, margarines, shortenings and frying fats, and naturally occurring TFA from ruminant food products containing meat and milk fat. The predominant *trans* isomers of both PHVO and ruminant fats are *trans* octadecenoic acids ($C_{18:1}$-*trans*) that consist of a range of positional isomers ranging from Δ4–Δ16 (Precht and Molkentin, 1995). Although both ruminant and PHVO contain the same $C_{18:1}$ TFA isomers, their isomeric profiles are quite different (Fig. 5.2). The isomeric profile of PHVO depends on hydrogenation conditions including type of catalyst used, temperature and pressure whereas ruminant TFA isomers are influenced by biohydrogenation conditions including rumen pH and the level of unsaturated fatty acids in the ruminant diet. The predominant TFA isomer of ruminant fats is *trans*-$C_{18:1}$ Δ11 (*trans*-vaccenic acid) and ranges from 43% to greater that 60% of the total $C_{18:1}$ TFA isomers whereas in PHVO the predominant isomers are *trans*-$C_{18:1}$ Δ9 (eleidic acid), *trans*-$C_{18:1}$ Δ10 followed by lower but still significant levels of Δ6–8, Δ11, Δ12, and Δ13/14.. (Fig. 5.2) (Precht and Molkentin, 1995). In addition to their differences in $C_{18:1}$ TFA isomers, ruminant and PHVO TFA also differ in amounts of $C_{18:2}$ and $C_{16:1}$ TFA isomers. Ruminant *trans* fat contain significant amounts of 16:1 *trans* that can account for up to 20% of the total TFA isomers (Fritsche and Steinhart, 1998; Leth et al., 1998).

Ruminant versus Industrial Trans *and Coronary Heart Disease Risk*

Clearly one major difference between TFA from PHVO and ruminant fat is that unlike elaidic acid and *trans*-18:1 Δ10, *trans*-vaccenic can be bio-converted to rumenic acid by stearoyl-CoA desaturase (also commonly known as Δ9-desaturase) most notably by ruminant animals but also in non-ruminant animals (Ip et al. 1999; Glaser et al. 2000; Santora et al. 2000; Banni et al. 2003; Loor et al. 2002; Kraft, 2004) as well as in humans (Salminen et al. 1998; Adlof et al. 2000). Serum levels of rumenic acid were shown to increase in subjects fed a diet high in PHVO, which contained *trans*-vaccenic acid (Salminen et al. 1998). In a more rigorous intervention study, Turpeinen et al. (2002) demonstrated that the conversion of *trans*-vaccenic to rumenic acid in humans contributes significantly to the amount of rumenic acid available to the body with an average conversion rate of 19%. The proportion of *trans*-vaccenic acid in serum total fatty acids increased 94%, 307%, and 620% above baseline in subjects provided diets containing 1.5, 3.0, and 4.5 g *trans*-vaccenic/day for 9 days. This was accompanied by a 50, 169, and 193% increase in the proportion of plasma rumenic acid. Significant interindividual differences were observed in the conversion rate with 2- to 3-fold differences in total increases in plasma rumenic acid seen in all groups. These results are consistent with those of Kuhnt et al. (2006) in which the conversion rate of *trans*-vaccenic to rumenic acid was estimated in humans after consuming 3 g *trans*-vaccenic/ day for 7 or 42 days. Serum *trans*-vaccenic increased 5-fold after 7 days and 8-fold after 42 days while serum rumenic acid levels increased by 1.7- and 2-fold, respectively. The conversion rate of *trans*-vaccenic to rumenic acid was estimated at 24% in serum and 19% in red blood cell membranes in subjects consuming 3 g *trans*-vaccenic/day for 42 days. These authors concluded that the endogenous conversion of *trans*-vaccenic to rumenic acid contributes approximately 25% of the total human pool of rumenic acid.

Epidemiologic Evidence

To date, a total of five epidemiologic studies have evaluated and compared the effect of *trans* fat intake from PHVO and ruminant fat sources on the risk of CHD. Three of these were prospective studies (Willett et al. 1993; Pietinen et al., 1997; Oomen et al. 2001), one was a case-control study (Ascherio et al. 1994) and one a cross-sectional study (Bolton-Smith et al. 1996). Evidence from the prospective and case control epidemiologic studies indicated a direct positive association between the intake of total *trans* fat and the relative risk of CHD (Willett et al., 1993; Ascherio et al., 1994; Pietinen et al., 1997; Oomen et al., 2001). In three out of the four studies, the increased relative risk of CHD was accounted for entirely by the intake of PHVO (Willett et al., 1993; Ascherio et al., 1994; Pietinen et al., 1997) (Table 5.9). In the prospective Nurses Health Study, Willett et al. (1993) observed a significant positive association between *trans* fat intake from vegetable fat and CHD risk for the highest verses the lowest quintile of intake (relative risk: 1.78; 95% confidence interval: 1.12–2.83) whereas a non-significant inverse association was observed for *trans* fat intake from animal fat and the relative risk of CHD (relative risk: 0.59; 95% confidence interval: 0.30–1.17). Similarly, in the prospective Beta-Carotene, Cancer Prevention Study by Pietinen et al. (1997) the multivariate adjusted relative risk of CHD for the highest compared to the lowest quintile of *trans* fat intake was positive and significant for vegetable oil *trans* (relative risk: 1.23; 95% confidence interval: 0.97–1.55) and non-significant for *trans* fat from animal sources (relative risk: 0.83; 95% confidence interval: 0.62–1.11). In a case-control study reported by Ascherio et al. (1994), the overall association between energy adjusted *trans* fat intake and the relative risk of a myocardial infarction (MI) was almost entirely accounted for by *trans* fat from hydrogenated vegetable fats (relative risk: 1.94; 95% confidence interval: 0.93–4.04) for the highest compared to the lowest quintile of intake). No significant association, however, was observed between intake of *trans* fat from animal fat and MI (relative risk: 1.02; 95% confidence interval: 0.43–2.41). In further analysis of the contribution of individual food items as potential predictors, high intakes of margarine (>2.5 pats per day) was shown to be directly associated with risk of MI compared to those consuming <1 pat per day (relative risk: 3.22; 95% confidence interval: 1.63–6.38). This is consistent with case control studies from Greece (Tzonou et al. 1993) and Italy (Tavani et al., 1997), which indicated medium or high intakes of margarine was associated with an increased risk of CHD. Additionally, data from the Framingham Study after 21 years of follow-up showed a positive association between a rise in the incidence of CHD with increasing margarine intake, but not butter intake (Gillman et al., 1997). However, in a review of the data from epidemiologic studies that compared the intake of ruminant and industrial *trans* fat intake on CHD risk, Weggemans et al. (2004) suggested that there are no differences in risk of CHD between ruminant and industrial TFA for intakes up to 2.5 g/day. At higher intakes (more than 3 g/day) total and industrial TFA are associated with an increased risk of CHD but there is insufficient data on ruminant TFA intake at these levels and, hence, there is currently little data to support a distinction between ruminant and industrial TFA for dietary or regulatory decisions.

Clinical Studies

Numerous clinical studies over the past 16 years have demonstrated that TFA from PHVO raise LDL-cholesterol to a similar degree as saturated fat (Mensink and Katan, 1990; Aro et al., 1997; Judd et al., 1994, 1998, 2002; Lichtenstein et al., 1993, 1999; Zock and Katan, 1992). On the other hand, in a review of five clinical studies, isomers of CLA (*cis*-9, *trans*-11 and/or *trans*-10, *cis*-12) did not show any significant effect on plasma TC or LDL-C levels (Terpstra, 2004). Additionally, *cis*-9, *trans*-11 (rumenic acid) had no significant effect on plasma HDL-C or triglyceride levels. Only one study showed a significant HDL-C lowering effect of CLA when the *trans*-10, *cis*-12 isomer was administered (Riserus et al., 2002) but not when a mixture of the *trans*-10, *cis*-12 and *cis*-9, *trans*-11 was tested (Riserus et al., 2001). In summary, there were no significant effects CLA on plasma TC, LDL-C and HDL-C concentrations or on plasma triglyceride levels (Benito et al., 2001; Berven et al., 2000; Mougios et al., 2001; Smedman and Vessby, 2001; Noone et al., 2002). On the other hand, in a dose-response study with CLA isomers, *cis*-9, *trans*-11 CLA lowered LDL-C and had no effect on HDL-C whereas the *trans*-10, *cis*-12 CLA isomer raised LDL-C with no effect on HDL-C (Tricon et al. 2004). Taken together, these results indicate that the ruminant-derived *cis*-9, *trans*-11 CLA isomer has little effect or may actually lower plasma LDL-C and, similarly, has little effect on HDL-C or triglycerides.

Although conversion of *trans*-vaccenic to CLA (rumenic acid) may contribute to the potential neutral effect of ruminant TFA on plasma lipids, no studies have yet evaluated the independent effects of *trans*-vaccenic acid on plasma lipids and lipoprotein cholesterol in humans. However, in healthy middle-aged men who consumed dairy products (milk, butter, and cheese), naturally enriched in CLA and *trans*-vaccenic acid, for 6 weeks, no significant effects were observed on plasma TC, LDL-C, HDL-C or triglycerides (Tricon et al., 2006). Although *trans*-vaccenic acid was not tested independently in this study, it was the predominant TFA isomer in the CLA-enriched milk and milk products consumed by the subjects. During the CLA intervention phase, the dietary intake of rumenic acid was 1.42 g/day along with 4.69 g/day of *trans*-vaccenic acid for a grand total of 6.33 g/day of rumenic + *trans*-vaccenic acid. The total $C_{18:1}$-*trans* isomers in the control products contributed ~0.3 %en during the control diet, whereas $C_{18:1}$-*trans* isomers from the enriched dairy products contributed ~2.1 %en during the CLA intervention phase. Mensink et al. (2003) have suggested that for every 1% of energy increase in TFA raises LDL-C by ~0.04 mmol/L relative to carbohydrates. Thus, from the study by Tricon (2006) it may be predicted that the estimated 1.8 en% net increase in TFA from *trans*-vaccenic acid should have increased LDL-C by ~0.07 mmol/L (2.8 mg/dL). However, as noted, plasma LDL-C was not altered by the consumption of the rumenic and *trans*-vaccenic acid enriched dairy products, albeit perhaps not surprising since the predicted LDL-C effect is relatively small. Additionally, no effect of the rumenic and *trans*-vaccenic acid enriched dairy products was observed on other suspected biomarkers of CHD risk assessed in this study including LDL particle size, susceptibility of LDL to oxidation, and the inflammatory markers: interleukin 6, soluble vascular cell adhesion molecule-1, soluble intercellular adhesion molecule-1, soluble E-selectin, or C-reactive protein. Clearly, additional clinical research is needed to determine whether the intake of ruminant TFA and especially *trans*-vaccenic acid is similar or different from industrial sources of TFA with respect to the impact on coronary heart disease risk.

References

Adlof R.O.; S. Duval; E.A. Emken. Biosynthesis of Conjugated Linoleic Acid in Humans. *Lipids* **2000,** *35*, 131–135.

Allison D.B.; S.K. Egan; L.M. Barraj; et al. Estimated Intakes of *trans* Fatty and Other Fatty Acids in the US Oopulation. *J Am Diet Assoc* **1999,** *99*, 166–74.

Aro A.; M. Jauhiainin; R. Partanen; et al. Stearic Acid, *trans* Fatty Acids, and Dairy Fat: Effects on Serum and Lipoprotein Lipids, Apolipoproteins, Lipoprotein(a), and Lipid Transfer Proteins in Healthy Subjects. *Am J Clin Nutr* **1997,** *65*, 1419–1426.

Aro A.; J.M. Antoine; L. Pizzoferrota; O. Reykdal; et al. *Trans* Fatty Acids in Dairy and Meat Products from 14 European Countries: The TRANSFAIR study. *J Food Compos Anal* **1998,** *11*, 150–160.

Ascherio A.; C.H. Hennekens; J.E. Buring; et al. *Trans* Fatty Acids Intake and Risk of Myocardial Infarction. *Circulation* **1994,** *89*, 94–101, .

Ascherio A.; M.B. Katan; P.L. Zock; et al. *Trans* Fatty Acids and Coronary Heart Disease. *N Engl J Med* **1999,** *340* (25), 1994–1998.

Banni S.; E. Angioni; E. Murru; et al. Vaccenic Acid Feeding Increases Tissue Levels of Conjugated Linoleic Acid and Suppresses Development of Premalignant Lesions in Rat Mammary Gland. *Nutr Cancer* **2001,** *41*, 91–97.

Banni S.; S.D. Heys; K.W.J. Wahle. Conjugated Linoleic Acids as Anticancer Nutrients: Studies in vitro and Cellular Mechanisms. In *Advances in Conjugated Linoleic Acid Research, Volume 2.* Sebedio J.-L.; W.W. Christie; R.O. Adlof Eds. AOCS Press: Champaign, IL, 2003; pp 267–282.

Bauman D.E.; D.M. Barbano; D.A. Dwyer; et al. Technical Note: Production of Butter with Enhanced Conjugated Linoleic Acid for Use in Biomedical Studies With Animal Models. *J. Dairy Sci* **2000,** *83*, 2422–2425.

Bauman D.E.; J.W. Perfield; M.J. de Veth; et al. New perspectives on lipid digestion and metabolism in ruminants. In *Proceedings of the Cornell Nutritional Conference.* Cornell University: Ithaca, NY, 2003a; pp 175–189.

Bauman D.E.; B.A. Corl; D.G. Peterson. The Biology of Conjugated Linoleic Acids in Ruminants. In *Advances in Conjugated Linoleic Acid Research, Volume 2* Sebedio J.-L.; W.W. Christie; R.O. Adlof Eds. AOCS Press: Champaign, IL, 2003b; pp 146–173.

Bauman D.E.; J.M. Griinari. Nutritional Regulation of Milk Fat Synthesis. *Annu. Rev. Nutr* **2003,** *23*, 203–227.

Belury M.A. Dietary Conjugated Linoleic Acid in Health: Physiological Effects and Mechanisms of Action. *Annu. Rev. Nutr* **2002,** *22*, 505–531.

Benito P.; G.J. Nelso; D.S. Kelley; et al. The Effect of Conjugated Linoleic Acid on Plasma Lipoproteins and Tissue Fatty Acid Composition in Humans. *Lipids* **2001,** *36*, 229–236.

Berven G.; A. Be; O. Hals; et al. Safety of Conjugated Linoleic Acid (CLA) in Overweight or Obese Volunteers. *Eur J Lipid Sci Technol* **2000,** *102*, 455–462.

Bolton-Smith C.; M. Woodward; S. Fenton; et al. Does Dietary *trans* Fatty Acid Intake Related to the Prevalence of Coronary Heart Disease in Scotland? *Eur Heart J* **1996,** *17*, 837–845.

Bruckner J. *Trans*-fettsauren: Gesundheitliche Aspekte des Verzehrs. *Ernahrungs-Umschau* **1995,** *42*, 122–126.

Cook M.E.; D. Butz; G. Li; et al. Conjugated Linoleic Acid Enhances Immune Response but Protects Against the Collateral Damage of Immune Events. In *Advances in Conjugated Linoleic Acid Research, Volume 2.* Sebedio J.-L.; W.W. Christie; and R.O. Adlof Eds. AOCS Press: Champaign, IL, 2003; pp 283–291.

Doreau M.; A. Ferlay. Digestion and Utilization of Fatty Acids by Ruminants, *Anim. Feed Sci. Technol* **1994,** *45*, 379–396.

Doreau M.; Y. Chilliard. Digestion and Metabolism of Dietary Fat in Farm Animals. *Br. J. Nutr* **1997,** *8*, S15–S35.

Emken E.A. Utilization and Effects of Isomeric Fatty Acids in Humans. In *Geometrical and Positional Fatty Acid Isomers.* Emken E.A.; H.J. Dutton Eds. American Oil Chemists' Society: Champaign, IL, 1979; pp 99–129.

Enig M.G.; L.A. Pallansch; J. Sampugna; et al. Fatty Acid Composition of the Fat in Selected Food Items With Emphasis on *trans* Components. *J Am Oil Chem Soc* **1983,** *60*, 1788–1795.

Fritsche J.; H. Steinhart. Amounts of Conjugated Linoleic Acid (CLA) in German Foods and Evaluation of Daily Intake. *Z Lebensm Unters Forsch* **1998,** *206*, 77–82.

Gillman M.W.; L.A. Cupples; D. Gagnon; et al. Margarine Intake and Subsequent Coronary Heart Disease in Men. *Epidemiology* **1997,** *8*, 144–149.

Glaser K.R.; M.R.L. Scheeder; C. Wenk. Dietary C18:1 *trans* Fatty Acids Increase Conjugated Linoleic Acid in Adipose Tissue of Pigs. *Eur J Lipid Sci Technol* **2000,** *102*, 684–686.

Griinari J.M.; D.E. Bauman. Biosynthesis of Conjugated Linoleic Acid and its Incorporation into Meat and Milk in Ruminants. In *Advances in Conjugated Linoleic Acid Research, Volume 1*. Yurawecz M.P.; M.M. Mossoba; J.K.G. Kramer; M.W. Pariza; G.J. Nelson Eds. AOCS Press: Champaign, IL, 1999; pp 180–200.

Griinari J.M.; B.A. Corl; S.H. Lacy; et al. Conjugated Linoleic Acid is Synthesized Endogenously in Lactating Dairy Cows by Δ^9-desaturase. *J. Nutr* **2000,** *130*, 2285–2291.

Harfoot C.G.; G.P. Hazlewood. Lipid metabolism in the rumen. In *The Rumen Microbial Ecosystem,* 2nd Edn. Hobson P.N.; D.S. Stewart Eds. Chapman and Hall: London, 1997; pp 382–426.

Harnack L.; L. Lee; S.F. Schakel; et al. Trends in the *trans*-fatty Acid Composition of the Diet in a Metropolitan Area: the Minnesota Heart Survey. *J Am Diet Assoc* **2003,** *103* (9), 1166,

Hay J.D.; W.R. Morrison. Isomeric Monoenoic Fatty Acids in Bovine Milk Fat. *Biochim Biophys Acta* **1970,** *202*, 237–243.

Hay J.D.; W.R. Morrison. Positional Isomers of *cis* and *trans* Monoenoic Fatty Acids From Ox (steer) Perinephric Fat. *Lipids* **1973,** *8*, 94–95.

Heimbach J.T.; B.E. Sever; S.K. Egan. Estimation of Intakes of Naturally Occurring *trans* Fatty Acids in US Diets. *Paper presented at: Annual Meeting of the American Oil Chemists Society*. San Antonio, TX. May 7, 1995.

Henninger M.; F. Ulberth. *Trans* Fatty Acid Content of Bovine Milk Fat. *Milchwissenschaf* **1994,** *49*, 555–558.

Hulshof K.F.; M.A. van Erp-Baart; M. Anttolainen; et al. Intake of Fatty Acids in Western Europe with Emphasis on *trans* Fatty Acids: the TRANSFAIR Study. *Eur J Clin Nutr* **1999,** *53* (2), 143–57.

Hunter J.E.; T.H. Applewhite. Reassessment of *trans* Fatty Acid Availability in the US Diet. *Am J Clin Nutr* **1991,** *54*, 363–69.

Hunter J.E. Dietary Levels of *trans*-fatty Acids: Basis for Health Concerns and Industry Efforts to Limit Use. *Nutr Res*. **2005,** *25*, 499–513.

International Dairy Foods Association. http://www.idfa.org/reg/transfat.cfm Unpublished data, 2004.

Ip C.; S. Banni; E. Angioni; et al. Conjugated Linoleic Acid-enriched Butter Fat Alters Mammary Gland Morphogenesis and Reduces Cancer Risk in Rats. *J Nutr* **1999,** *129*, 2135–2142.

Ip M.M.; P.A. Masso-Welch; C. Ip. Prevention of Mammary Cancer With Conjugated Linoleic Acid: Role of the Stroma and Epithelium. *J Mammary Gland Biol Neoplasia* **2003,** *8*, 103–118.

Jakobsen M.U.; A. Bysted; N.L. Andersen; et al. Intake of Ruminant *trans* Fatty Acids in the Danish Population Aged 1–80 years. *Eur J Clin Nutr* **2006,** *60* (3), 312–318.

Jensen R.G. The Composition of Bovine Milk Lipids: January 1995 to December 2000. *J Dairy Sci* **2002,** *85*, 295–350.

Judd J.T.; B.A. Clevidence; R.A. Muesing; et al. Dietary *trans* Fatty Acids: Effects on

Plasma Lipids and Lipoproteins of Healthy Men and Women. *Am J Clinc Nutr* **1994,** *59*, 861–868.

Judd J.T.; D.J. Baer; B.A. Clevidence; et al. Effects of Margarine Compared with Those of Butter on Blood Lipid Profiles Related to Cardiovascular Disease Risk Factors in Normolipemic Adults Fed Controlled Diets. *Am J Clin Nutr* **1998,** *68*, 768–777.

Judd J.T.; D.J. Baer; B.A. Clevidence; et al. Dietary *cis* and *trans* Monounsaturated and Saturated FA and Plasma Lipids and Lipoproteins in Men. *Lipids* **2002,** *37*, 123–131.

Kemp P.; D.J. Lander. Hydrogenation in vitro of α-linoleic Acid to Stearic-acid by Mixed Cultures of Pure Strains of Rumen Bacteria, *J. Gen. Microbiol* **1984,** *130*, 527–533.

Kraft J. Incorporation of Conjugated Linoleic Acids into Body Lipids with Special Regard to the Isomeric Distribution, PhD thesis, Friedrich Schiller University, Jena, 2004.

Kramer J.K.G.; P.W. Parodi; R.G. Jensen; et al. Rumenic Acid: a Proposed Common Name for the Major Conjugated Linoleic Acid Isomer Found in Natural Products, *Lipids* **1998,** *33*, 835.

Kritchevsky D. Conjugated Linoleic Acids in Experimental Atherosclerosis, In *Advances in Conjugated Linoleic Acid Research, Volume 2.* Sebedio J.-L.; W.W. Christie; and R.O. Adlof Eds. AOCS Press: Champaign, IL, 2003; pp 293–301.

Kuhnt K.; J. Kraft; P. Moeckel; et al. *Trans*-11-18:1 is Effectively Δ9-desaturated Compared with *trans*-12-18:1 in Humans. *Brit J Nutr* **2006,** *95*, 752–761.

Leth T.; L. Ovesen; K. Hansen. Fatty Acid Composition of Meat From Ruminants, with Special Emphasis of *trans* Fatty Acids. *JAOCS* **1998,** *75*, 1001–1005.

Lichtenstein A.H.; L.M. Ausman; W. Carrasco; et al. Hydrogenation Impairs the Hypolipidemic Effect of Corn Oil in Humans. Hydrogenation, *trans* Fatty Acids, and Plasma Lipids. *Arterioslcer Thromb* **1993,** *12*, 154–161.

Lichtenstein A.H.; L.M. Ausman; S.M. Jalbert; et al. Effects of Different Forms of Dietary Hydrogenated Fats on Serum Lipoprotein Cholesterol Levels. *N Engl J med* **1999,** *340*, 1933–1940.

Lichtenstein A.H.; L.J. Appel; M. Brands; et al., Diet and Lifestyle Recommendations Revision 2006: a Scientific Statement From the American Heart Association Nutrition Committee. *Circulation* **2006,** *114*(1), 82–96.

Lock A.L.; D.E. Bauman. Modifying Milk Fat Composition of Dairy Cows to Enhance Fatty Acids Beneficial to Human Health. *Lipids* **2004,** *39*, 1197–1206.

Loor J.J.; X. Lin; J.H. Herbein. Dietary *trans*-vaccenic Acid (*trans*-11-18:1) Increases Concentration of cis9,trans11-conjugated Linoleic Acid (rumenic acid) in Tissues of Lactating Mice and Suckling Pups. *Reprod Nutr Dev.* **2002,** *42*, 85–99.

McGuire M.A.; D.E. Bauman. Milk Biosynthesis and Secretion. In *Encyclopedia of Dairy Sciences*. Roginski H.; J.W. Furquay; P.E. Fox, Eds. Elsevier Sciences: London, 2002; pp. 1828–1834.

Mensink R.P.; M.B. Katan. Effects of Dietary *trans* Fatty Acids on High-density and Low-density Lipoprotein Cholesterol Levels in Healthy Subjects. *N Eng J Med* **1990,** *323*, 439–445.

Mensink R.P.; P.L. Zock; A.D.M. Kester; et al. Effects of Dietary Fatty Acids and Carbohydrates on the Ratio of Serum Total to HDL Cholesterol and on Serum Lipids and Apolipoproteins: a Meta-analysis of 60 Controlled Trials. *Am J Clin*

Nutr **2003,** *77*, 1146–1155.

Molkentin J.; Precht D. unpublished results, 1995.

Mougios V.; A. Matsakas; A. Petridou; et al. Effect of Supplementation with Conjugated Linoleic Acid on Human Serum Lipids and Body Fat. *J Nutr Biochem* **2001,** *12*, 585–594.

Nestel P.M.; M. Noakes; M. Belling; et al. Plasma Lipoprotein Lipid and Lp[a] Changes with Substitution of Elaidic Acid for Oleic Acid in the Diet. *J. Lipid Res.* **1992,** *33*, 1029–1036.

Noone E.J.; H.M. Roche; A.P. Nugent; et al. The Effect of Dietary Supplementation Using Isomeric Blends of Conjugated Linoleic Acid on Lipid Metabolism in Healthy Human Subjects. *Br J Nutr* **2002,** *88*, 243–251.

Oomen C.M.; M.C. Ocke; E.J.M. Feskens; et al. Association Between *trans* Fatty Acid Intake and 10-year Risk of Coronary Heart Disease in the Zutphen Elderly Study: a Prospective Population-based Study. *Lancet* **2001,** *357*, 746–751.

Palmquist D.L.; A.L. Lock; K.J. Shingfield; et al. Biosynthesis of Conjugated Linoleic Acid in Ruminants and Humans. In *Advances in Food and Nutrition Research.* Taylor L.L Ed. Elsevier: San Diego, CA; 2004.

Parodi P.W. Distribution of Isomeric Octadecenoic Fatty Acids in Milk Fats. *J Dairy Sci* **1976a,** *59*, 1870-1873.

Parodi P.W. Composition and Structure of Some Consumer-available Edible Fats. *J Am Oil Chem Soc* **1976b,** *53*, 530–534.

Parodi P.W. Conjugated linoleic acid: the early years. In *Advances in Conjugated Linoleic Acid Research Vol. 1.* Yurawecz M.P.; M.M. Mossoba; J.K.G. Kramer; M.W. Pariza; G.J. Nelson Eds. AOCS Press: Champaign, IL, 1999; pp 1–11.

Parodi P.W. Conjugated linoleic acid. In *Encyclopedia of Dairy Sciences. Vol. 3.* Roginski H.; J.W. Fuquay; P.F. Fox Eds. Academic Press: London, 2002; pp 1587–1594.

Parodi P.W. Conjugated linoleic acid in food. In *Advances in Conjugated Linoleic Acid Research, Volume 2.* Scbcdio J.-L.; W.W. Christie; R.O. Adlof, Eds. AOCS Press: Champaign, IL, 2003; pp. 101–122.

Parodi P.W. Milk Fat in Human Nutrition. *Aust J Dairy Technol* **2004,** *59*, 3–58.

Pfalzgraf A.; M. Timm; H. Steinhart. Gehalte von *trans*-fettsauren in Lebensmitteln. *Z Ernahrungswiss* **1993,** *33*, 24–43.

Pietinen P.; A. Ascherio; P. Korhonen; et al. Intake of Fatty Acids and Risk of Coronary Heart Disease in a Cohort of Finnish Men. The alpha-tocopherol, beta-carotene Cancer Prevention Study. *Am J Epiddemiol* **1997,** *145*, 876–887.

Piperova L.S.; J. Sampugna; B.B. Teter; et al. Duodenal and Milk *trans* Octadecenoic Acid and Conjugated Linoleic Acid (CLA) Isomers Indicate that Post-absorptive Synthesis is the Predominant Source of *cis*-9-containing CLA in Lactating Dairy Cows. *J Nutr* **2002,** *132*, 1235–1241.

Ponnampalam E.N.; N.J. Mann; A.J. Sinclair. Effect of Feeding Systems on omega-3 Acids, Conjugated Linoleic Acid and *trans* Fatty Acids in Australian Beef Cuts: Potential Impact on Human Health. *Asia Pac J Clin Nutr* **2006,** *15* (1), 21–29.

Precht D.; J. Molkentin. *Trans* Fatty Acids: Implications for Health, Analytical Methods, Incidence in Edible Fats and Intake. *Die Nahrung* **1995,** *5* (6), 343–374.

Precht D.; J. Molkentin. Rapid Analysis of the Isomers of *trans*-octodecenoic Acid in Milk Fat. *Int Dairy J* **1996,** *6*, 791–809.

Precht D.; J. Molkentin. Effect of Feeding on *trans* Positional Isomers of Octadecenoic Acid in Milk Fats. *Milchwissenschaft* **1997a,** *52* (10), 564–568.

Precht D.; J. Molkentin. Effect of Feeding on Conjugated *cis* Δ9, *trans* Δ11-octadecadienoic Acid and Other Isomers of Linoleic Acid in Bovine Milk Fats. *Nahrung* **1997b,** *41*, 330–335.

Precht D.; J. Molkentin. *Trans*-geometrical and Positional Isomers of Linoleic Acid Including Conjugated Linoleic Acid (CLA) in German Milk and Vegetable Fats. *Fett-Lipid* **1997c,** *99*, 319–326.

Precht D.; J. Molkentin. *Trans* Unsaturated Fatty Acids in Bovine Milk Fat and Dairy Products. *Eur J Lipid Sci Technol* **2000a,** *102*, 635–639.

Precht D.; J. Molkentin. Frequency Distributions of Conjugated Linoleic Acid and *trans*-fatty Acid Contents in European Bovine Milk Fats. *Milchwissenschaft* **2000b,** *55*, 687–691.

Renner E.; Y.C. Yoon. Untersuchungen Uber Isomere Formen Ungesattigter Fettsauren in Nahrungsfetten. 1. Isomere der Octadecensaure. *Milchwissenschaft* **1982,** *37*, 264–266.

Rickert R.; H. Steinhart; J. Fritsche; et al. Enhanced Resolution of Conjugated Linoleic Acid Isomers by Tandem-column Silver-ion High Performance Liquid Chromatography, *J. High Resol Chromatogr* **1999,** *22*, 144–148.

Riserus U.; P. Arner; K. Brismar; et al. Conjugated Linoleic Acid (CLA) Reduced Abdominal Adipose Tissue in Obese Middle-aged Men with Sign of the Metabolic Syndrome: a Randomized Controlled Trial. *Int J Obes Relat Metab Disord* **2001,** *25*, 1129–1135.

Riserus U.; P. Arner; K. Brismar; et al. Treatment with Dietary *tran*10*cis*12 Conjugated Linoelic Acid Causes Isomer-specific Insulin Resistance in Obese mMen with the Metabolic Syndrome. *Diabetes Care* **2002,** *25*, 1516–1521.

Ritzenthaler K.L.; M.K. McGuire; R. Falen; et al. Estimation of Conjugated Linoleic Acid Intake by Written Dietary Assessment Methodologies Underestimates Actual Intake Evaluated by Food Duplicate Methodology. *J Nutr* **2001,** *131*, 1548–1554.

Salminen I.; M. Mutanen; A. Aro. Dietary *trans* Fatty Acids Increase Conjugated Linoleic Acid Levels in Human Serum. *J Nutr Biochem* **1998,** *9*, 93–98.

Santora JE, Palmquist DL, Roehrig KL, *Trans*-vaccenic acid is desaturated to conjugated linoleic acid in mice. J Nutr 130:208-215, 2000.

Smedman A, Vessby B. Conjugated linoleic acid supplementation in humans—metabolic effects. *Lipids* 36:773-778, 2001.

Tavani A, Negri E, D'Avanzo BD, et al. Margarine intake and risk of nonfatal acute myocardial infarction in Italian women. Eur J Clin Nutr 32:40-57, 1997.

Terpstra A. Effect of conjugated linoleic acid on body composition and plasma lipids in humans: an overview of the literature. *Am J Clin Nutr* 79:352-361, 2004.

Tricon S.; G.C. Burdge; S. Kew; et al. Opposing Effects of *cis*-9, *trans*-11 and *trans*-10,*cis*-12 Conjugated Linoleic Acid on Blood Lipids in Healthy Humans. *Am J Clin Nutr* **2004,** *80*, 614–620.

Tricon S, Burdge GC, Jones EL, et al. Effects of dairy products naturally enriched with *cis*-9, *trans*-11 conjugated linoleic acid on the blood lipid profile in healthy middle-aged men. *Am J Clin Nutr* 83:744-753, 2006.

Turpeinen A.M.; M. Mutanen; A. Aro; et al. Bioconversion of Vaccenic Acid to Conjugated Linoleic Acid in Humans. *Am J Clin Nutr* **2002,** *76*, 504–510.

Tzonou A.; A. Kalandidi; C. Hsieh; et al. Diet and Coronary Heart Disease: a Case-control Study in Athens, Greece. *Epidemiology* **1993,** *4*, 511–516.

Ulberth F.; M. Henninger. Quantitation of *trans* Fatty Acids in Milk Fat Using Spectroscopic and Chromatographic Methods. *J Dairy Res* **1994,** *61*, 517–527.

van de Vijver L.; A. Kardinaal; C. Couet; et al. Association Between *trans* Fatty Acid Intake and Cardiovascular Risk Factors in Europe: the TRANSFAIR Study. *Eur J Clin Nutr* **2000,** *54*, 126–135.

US Department of Health and Human Services

US Department of Agriculture, Dietary Guidelines for Americas, 2005. Available at www.healthierus.gov/dietaryguidelines

Watkins B.A.; Y. Li; D.R. Romsos; et al. CLA and Bone Modeling in Rats. In *Advances in Conjugated Linoleic Acid Research, Volume 2*. Sebedio J.-L.; W.W. Christie; R.O. Adlof Eds. AOCS Press: Champaign, IL, 2003; pp. 218–250.

Weggemans R.M.; M. Rudrum; E.A. Trautwein. Intake of Ruminant versus Industrial *trans* Fatty Acids and Risk of Coronary Heart Diease – What is the Evidence? *Eur J Lipid Sci Technol* **2004,** *106*, 390–397.

Willet W.C.; M.J. Stampfer; J.E. Manson; et al. Intake of *trans* Fatty Acids and Risk of Coronary Heart Disease Among Women. *Lancet* **1993,** *341*, 581–585.

Wolff R.L. Contribution of *trans*-C18:1 Acids From Dairy Fat to European Diets. *J. Am Oil Chem Soc* **1994,** *71*, 277–283.

Wolff R.L.; C.C. Bayard. Improvement in the Resolution of Individual *trans*-18:1 Isomers by Capillary Gas-liquid Chromatography-use of a 100m CP-Sil 88 column. *J Am Oil Chem Soc.* **1995,** *72*, 1197–1201.

Wolff R.L. Content and Distribution of *trans*-18:1 Acids in Ruminant Milk and Meat Fats. Their Importance in European Diets and Their Effect on Human Milk. *J Am Oil Chem Soc.* **1995,** *72*, 259–272.

Zock P.L.; M.B. Katan. Hydrogenation Alternatives: Effects of *trans* Fatty Acids and Stearic Acid versus Linoleic Acid on Serum Lipids and Lipoproteins in Humans. *J Lipid Res* **1992,** *33*, 399–410.

Chapter 6

Consumption of *Trans* Fatty Acids in North America

Margaret C. Craig-Schmidt

Department of Nutrition and Food Science, Auburn University, Auburn, AL 36849 USA

Introduction

Dietary *trans* fatty acids originate from two sources: biohydrogenation in the rumen of animals and commercial hydrogenation or industrial processing of vegetable and marine oils (Craig-Schmidt and Holzer, 2000). Meat and dairy products from ruminants contain 1–8% of the fat as *trans* fatty acids, whereas industrial processing of soybean and other oils results in products such as margarine, shortening, and frying fats that contain *trans* fatty acids up to 40–50% of total fatty acids. Deep-fried fast foods, as well as bakery goods, snack foods and other commercial products made with industrially hydrogenated fat, are major sources of dietary *trans* fatty acids (Craig-Schmidt and Holzer, 2000). Thus, populations consuming relatively large amounts of commercially hydrogenated fat have a greater dietary *trans* fatty acid content than do populations whose dietary fat is primarily derived from ruminant sources. Populations who use olive oil and other nonhydrogenated vegetable oils as the primary dietary fat have relatively minor amounts of *trans* fatty acids in the diet.

The purpose of this review is to summarize data on the consumption of *trans* fatty acids in the United States and Canada from an historical perspective. Earlier reviews or government reports which summarize estimates of *trans* fatty acids in the North American diet include a report by the Life Sciences Research Office for the U.S. Food and Drug Administration compiled by Senti (1985), the International Life Sciences Institute Expert Panel on *Trans* Fatty Acids and Coronary Heart Disease (ILSI, 1995), the American Society for Clinical Nutrition/American Institute of Nutrition Task Force on *Trans* Fatty Acids (ASCN/AIN, 1996), and the International Life Sciences Institute Expert Panel on *Trans* Fatty Acids and Early Development (ILSI, 1997). A comprehensive review of world-wide consumption of *trans* fatty acids through the late 1990s was published by Craig-Schmidt (1998). In addition, a brief update to this review has been published recently (Craig-Schmidt, 2006).

Methods Used to Estimate *Trans* Fatty Acid Consumption

Methods used to estimate *trans* fatty acids in the diet include: (1) estimates based on “food disappearance” or market share data, (2) analysis of dietary consumption data of a representative population, (3) laboratory analysis of duplicate portion or composite diets, and (4) biomarkers. Even though each method has inherent advantages and disadvantages, collectively they can give reasonable estimates of *trans* fatty acid consumption in a population. Because of variability in nutrient databases with respect to *trans* fatty acids and the inherent disadvantages in assessing the diet of individuals, the use of biomarkers rather than the content of the diet has become increasingly popular in studies designed to relate consumption of *trans* fatty acids to health risk for various diseases.

Estimates Based on "Food Disappearance" or Market Share Data

The amount of a commodity "consumed" in grams per day per person can be calculated using government statistics for the amounts of commodities that are available for human consumption in a given year, in combination with the population of the country. This method which is based on food disappearance or market share data has the advantage in that it accounts for the availability of fat for the total population. On the other hand, the estimates are dependent on pairing the market share data for a given commodity with the "average" or "typical" *trans* fatty acid content for that commodity. The *trans* fatty acid content of oils and fats varies widely with brand names and changes in methods of processing, and therefore, values for the *trans* fatty acid content of various commodities are at best rough estimates. Wastage, i.e., oil that is discarded or fat that is purchased but not consumed, must also be considered, and these "wastage" factors are also rough approximations (Hunter and Applewhite, 1993).

Estimates of *trans* fatty acid consumption based on food disappearance or market share data represent availability in the food supply rather than individual consumption and thus tend to overestimate *trans* fatty acid consumption. In general, this method for estimating *trans* fatty acid consumption results in greater values than those obtained by other methods.

Analysis of Dietary Consumption Data of a Representative Population

The diets of individuals can be assessed by using diet recalls, diet records, and food frequency questionnaires. The amounts of various food items consumed can then be combined with data from food composition tables to give the average daily consumption of a nutrient or other component in the diet. The advantage of this method is that consumption by the individual rather than availability in the population is assessed. However, food composition tables for the *trans* fatty acid content of food are subject to inaccuracies because of changes in food processing, the wide variability in *trans* fatty acid content of similar types of foods, and differences in the analytical methods used to determine the *trans* fatty acid content of each food item. The inherent problems with assessing the diet of an individual, as well as the limitations of food composition databases for the *trans* fatty acid values, tend to result in underestimations of the *trans* fatty acid content of the diet compared to values obtained by other methods.

The problems with dietary methodology in assessing the intake of *trans* fatty acids have been illustrated by Innis et al. (1999a). These investigators analyzed over 200 food items, clearly demonstrating the variability in *trans* fatty acid content among food items or brands within a product category. For example, in 17 brands of crackers analyzed, *trans* fatty acid content ranged from 23 to 51% of total fatty acids or differences of 1 to 13 g *trans* fatty acids per 100 g of crackers. The significance of this variability to the estimation of *trans* fatty acid consumption from dietary intake data was illustrated by analyses of a single diet taken from a diet record. In this analysis, each food was calculated using the minimum and maximum values for *trans* fatty acids within a given category. For this single diet, *trans* fatty acid intake ranged from a low of 1.4 to 25.4 g/day. Thus, Innis et al. (1999a) illustrated that the wide variability in *trans* fatty acid content of different food items within a category may result in large errors in the estimation of *trans* fatty acid intake of individuals and groups.

Laboratory Analysis of Duplicate Portion or Composite Diets

In the duplicate portion or composite diet technique, diets are collected from a representative population or formulated based on a diet composition considered to be "typical" of a population. The diets are then analyzed in the laboratory. The advantage of this method is that dietary *trans* fatty acids can be determined very accurately, especially if sophisticated analytical methodology is used. However, laboratory analyses are very expensive, and thus, it is possible to analyze only a limited number of diets. Although the *trans* fatty acid content of the diet can be determined with a high degree of accuracy, this method is useful only as the subset of the population from which diet is collected is representative or only as the composite diets to be analyzed are reflective of the diet of the population as a whole.

Biomarkers of Trans Fatty Acid Consumption

The *trans* fatty acid content of human milk reflects the *trans* fatty acid content of the mother's diet on the previous day (Aitchison et al., 1977; Craig-Schmidt et al., 1984); adipose tissue, on the other hand, is believed to reflect long term consumption patterns (Beynen et al., 1980; Hirsch, 1960). Both of these tissues in combination with an appropriate equation have been used to estimate the consumption of *trans* fatty acids. In addition, the *trans* fatty acid content of adipose tissue and other tissues such as red blood cells is used as a relative biomarker of consumption even without extrapolating to concentrations in the diet itself.

Human Milk

Because the *trans* fatty acid content of human milk is reflective of the maternal diet, relative estimates of the amounts of *trans* fatty acids being consumed at the time the milk samples were collected can be obtained. Also, distribution of positional *trans* isomers of the fatty acids in breast milk appears to reflect the predominant source of *trans* fatty acids in the maternal diet. The presence of relatively large amounts of vaccinic acid (*trans*-Δ11 18:1) and the relative amount of *trans*-Δ16 18:1 indicate that milk and other dairy products are the major sources of *trans* fatty acids in the maternal diet (Chardigny et al., 1995; Wolff, 1995). However, a more equal distribution of *trans*-Δ11 18:1 and *trans*-Δ10 18:1 indicates that the *trans* fatty acids in the maternal diet are derived primarily from commercially hydrogenated fats (Chen et al., 1995a).

In interpreting *trans* fatty acid values in human milk, one should be cognizant of the analytical methodology used. In analyzing the fatty acid content of human milk, Ratnayake and colleagues (Chen et al., 1995a) used a combination of capillary gas chromatography and silver nitrate thin-layer chromatography, a method in which excellent separation of geometric and positional isomers of fatty acids is achieved. Thus, these investigators (Chen et al., 1995a) obtained a relatively high value for the 18:1*t* content of Canadian human milk samples (Chen et al., 1995a) due in part to sophisticated analytical methodology compared to direct gas chromatographic techniques used often by other investigators. Wolff (1995) has suggested that values of 18:1*t* obtained by direct gas chromatography should be multiplied by a factor of 1.25 to make them comparable to those obtained by the method of Ratnayake and colleagues (Chen et al., 1995a). Thus, in comparing *trans* fatty acid values for human milk, it is important to use data obtained by similar analytical methods whenever possible or to make corrections such as those suggested by Wolff (1995) for differences in methodology.

Both Craig-Schmidt et al. (1984) and Wolff (1995) have formalized the relationship between *trans* fatty acids in the milk and *trans* fatty acids in the maternal diet. In 1984, Craig-Schmidt et al. (1984) quantified the relationship between the amount of 18:1*t* in the maternal diet and the amount of 18:1*t* in human milk samples obtained the following day. This linear relationship can be expressed by the equation:

$$y = 1.49 + 0.42x$$

where y=18:1*t* in human milk, expressed as percent of total fatty acids and x=18:1*t* in the maternal diet of the previous day, expressed as % of total fatty acids. The 18:1*t* in the milk was highly correlated ($r=0.909$) with 18:1*t* in the diet. The equation was based on analysis of 80 milk samples from eight United States lactating women who were 2 months postpartum. The women consumed diets of known *trans* fatty acid content during two 5-day experimental periods in which milk samples were collected daily. Use of this equation in combination with analysis of baseline milk samples enabled Craig-Schmidt et al. (1984) to estimate the 18:1*t* content of the women's self-chosen diets to be 7.8% of the total fatty acids.

In a later study, Wolff (1995) plotted 18:1*t* values for human milk from four countries against the estimated daily per capita 18:1*t* intake for each of these countries to obtain the following relationship:

$$y = 0.20 + 0.76x$$

where y = 18:1*t* in human milk, expressed as percent of total fatty acids and x = daily per capita consumption of 18:1*t*, expressed as g/person/day. In deriving this equation, Wolff (1995) used *trans* fatty acid values for human milk based on his own data for France, the data of Boatella et al. (1993) for Spain, the data of Koletzko (1986) for Germany, and the data of Craig-Schmidt (1984) for the United States, standardized to account for differences in methodology. One limitation of the Wolff equation is that the dietary data and the human milk data were not obtained from the same subjects, as was the case in the study of Craig-Schmidt et al. (1984).

The International Life Sciences Institute Expert Panel on *trans* Fatty Acids and Early Development (ILSI, 1997) used the regression equation derived by Craig-Schmidt et al. (1984) to calculate the *trans* fatty acids in the diet of lactating women in the United States. The value they obtained was 4.2 g/day or 5.3% of total dietary fat. The human milk data used in this calculation were based on the *trans* fatty acid content (18:1*t*) of 57 human milk samples collected from American women by Finley et al. (1985).

Similarly, Chen et al. (1995a) used the equation of Craig-Schmidt et al. (1984) to estimate the *trans* fatty acid content of the Canadian diet. These investigators analyzed 198 human milk samples collected in 1992 from nine provinces in Canada to calculate the total *trans* fatty acids in the Canadian diet as 10.6 g/person/day or 4.0% of energy. Chen et al. (1995a) based their estimate on the total *trans* fatty acids in the milk and not on the 18:1*t* content of the milk on which the original equation was derived. Moreover, the equation of Craig-Schmidt (1984) was based on women consuming experimental maternal diets formulated to be adequate in all nutrients, including energy, and may not be applicable to milk samples obtained from women who are consuming inadequate calories (Chappell et al., 1985).

Human Adipose Tissue

Human adipose tissue is also believed to reflect the amount of *trans* fatty acids in the habitual diet (Hirsch et al., 1960; Beynen et al., 1980), and analysis of this tissue can be used as an indicator of the relative consumption of *trans* fatty acids. Also, the distribution of isomeric *trans* fatty acids in adipose tissue may be used as an indication of the type of fat being consumed. For example, Chen et al. (1995b) found that the *trans* isomeric distribution of 18:1 in adipose tissue from Canadian subjects was similar to the distribution of these isomers in partially hydrogenated vegetable oils, rather than butter. These investigators interpreted these data to be an indication that industrially processed fat was the predominant source of *trans* fatty acids in the Canadian diet. Thus, adipose tissue can serve as a valuable biomarker of both the amount and type of *trans* fatty acids in the habitual diet.

The relationship between *trans* fatty acids in adipose tissue and the diet has been quantified by several investigators. Hudgins et al. (1991) assumed a 1:1 correlation between *trans* fatty acids in adipose tissue and diet; other investigators (ILSI, 1995) have assumed a 1:2 relationship between the *trans* fatty acid content of adipose tissue and that in the diet. Still other investigators have derived equations describing this relationship in more detail. For example, Enig et al. (1990) derived the following equation based on studies in animals:

$$y = 0.97 + 0.44x$$

where y=% adipose tissue 18:1*t* and x=% dietary 18:1*t*. Others (ILSI, 1997) have used *trans* fatty acid concentrations in adipose tissue from human subjects to derive an equation for the relationship between *trans* fatty acid content in adipose tissue and the *trans* fatty acid content of the diet routinely consumed:

$$y = 1.52 + 2.40x$$

where y=*trans* fatty acids in the diet in g/day and x=% 18:1*t* in adipose tissue.

Estimation of *trans* fatty acid intake on the basis of the *trans* fatty acid content of adipose tissue is subject to several limitations. First, use of different analytical methods may make data difficult to compare among different laboratories. Second, the quantity of *trans* fatty acids stored versus that metabolized may be dependent on dietary factors such as caloric intake. Third, the *trans* fatty acid content may differ for various sampling sites, e.g., abdominal versus lateral thigh (Chen et al., 1995b).

In spite of these limitations, adipose tissue, as a biomarker of *trans* fatty acid intake has been validated by a number of investigators, including Garland et al. (1998) who demonstrated a correlation (r=0.40) between dietary intake and adipose tissue concentration of *trans* fatty acids. Lemaitre et al. (1998) found an even greater correlation (r=0.67) between dietary intake of *trans* fatty acids and concentrations in adipose tissue. When adjustments were made for energy intake, age, and body mass index, the correlation coefficients were 0.76 among men and 0.52 among women. Hu, et al. (2000) reported a similar correlation (r=0.51) between adipose tissue concentrations and intake of *trans* fatty acids.

Red Blood Cells and Other Tissues

The concentration of *trans* fatty acids in various fractions of the blood has been used increasingly as a biomarker to relate the incidence of disease to *trans* fatty acid consumption. For example, this method has been used by to relate *trans* fatty acids in the red blood cell membrane to risk of primary cardiac arrest (Lemaitre et al., 2002) and to systemic inflammation in heart failure (Mozaffarian et al., 2004).

King et al. (2005) used serum *trans* fatty acids as a surrogate to demonstrate a relationship between *trans* fatty acid intake and the risk of prostate cancer. *Trans* fatty acids in plasma phospholipids also have been used by van de Vijver et al. (1996) as a biomarker for dietary *trans* fatty acid consumption .

Estimates of *Trans* Fatty Acids in the North American Diet

All of the methodologies described previously have been used to estimate the content of *trans* fatty acids in diets consumed in Canada and the United States. In interpreting these data, one should consider the date of collection and method of *trans* fatty acid analysis because the food supply is constantly changing and analytical methodology is becoming more sophisticated.

In addition, there is considerable individual variation in estimates of *trans* fatty acid intake. Even in a selected, highly homogeneous population, individual variation in *trans* fatty acid consumption is great. For example, van den Reek et al. (1986a) used 7-day diet records to assess the diets of eight, apparently healthy, U.S. adolescent Caucasian girls, aged 12 to 15 years. The mean 18:1*t* content of the diet was 5.3% of total fatty acids, with a range of 3.0 to 9.1%. Also, the variation was wide for a single subject. For example, over the 7-day period, the amount of total *trans* fatty acids in the daily diet of one subject ranged from 2 to 12% of total fatty acids (van den Reek et al., 1986b). Because of this variation among individuals and the day-to-day variation within the diet of a single individual in *trans* fatty acid intake, many investigators are reporting the range of values of consumption as well as the mean. Others represent the variation by dividing the population into quintiles of *trans* fatty acid intake. Median values, such as those reported by Allison et al. (1999), are useful also in this regard.

Given in Tables 6.1 and 6.2 are the values for *trans* fatty acid consumption for the United States and Canada. Values are listed chronologically by the date of publication.

United States

Trans *Fatty Acid Consumption Reported in the 1970s*

In 1975, Kummerow (1975) was one of the first investigators to estimate *trans* fatty acids in the diet of people in the United States. His estimate of 8% of fatty acids in the diet was based on disappearance data taken from USDA Household Consumption of 1971 (USDA, 1971) and typical *trans* fatty acid content of visible fats. Stick margarine was assumed to contain 25–35% of the fatty acids as *trans* fatty acids; tub margarines, 15–25%; shortenings, 20–30%; and salad oils, 0–15%. Using a similar approach, Enig et al. (1978) also estimated that 8% of fat consumed in the United States was *trans* fat. This would be equivalent to an average value of 12.1 g/day/person.

Aitchison et al. (1977) using the duplicate portion technique estimated that the diet of lactating women contained an average of 5% *trans* fatty acids, a value somewhat less than the estimates based on disappearance data. In addition, these authors documented the wide range of *trans* fat consumption in individual diets with minimum values ranging from 0–4.4 g/d and maximal values ranging from 2.3–9.9 g/d for 3-day diet collections. Overall, the range for dietary *trans* fat was 1.3–8.3% of total fatty acids.

Trans *Fatty Acid Consumption Reported in the 1980s*

Estimates of *trans* fat in the US diet based on disappearance data were 6.8, 8.3–10.2, and 7.6 g/d per person for the estimates of Emken (1981); Senti (1985); and Hunter and Applewhite (1986), respectively. Senti (1985) included a correction for discarded fat in his estimates and reported a value of 8.3 g/d or 5.9% of fat in the diet as *trans* fatty acids, compared to 10.2 g/d or 6.1% of fat without the correction.

Craig-Schmidt and colleagues (Craig-Schmidt et al., 1984; van den Reek et al., 1986a, 1986b) collected detailed data on *trans* fatty acid consumption of individuals from selected populations, rather than on the general population. Based on laboratory analysis of duplicate portion diets of adolescent females, a value of 6.5% of total fatty acids was obtained (van den Reek et al., 1986b). Using diet records of the same subjects and the food composition database compiled by Enig et al. (1983), the *trans* fatty acid content of the diet was estimated to be 5.3% of total fat (van den Reek et al., 1986a), a value only slightly less than the value obtained by laboratory analysis. For self-selected diets of lactating women, the 18:1*t* content of the diet was estimated to be 7.8% of total fatty acids. This value was obtained based on an equation derived by Craig-Schmidt et al. (1984), describing the relationship of 18:1*t* in the maternal diet and human milk of the following day, and was the first example of using a biomarker to estimate the *trans* fatty acid content of the diet. Also, possible extremes in home-cooked meals were identified in this study (Craig-Schmidt et al., 1984) as 0.9 g/day for a 5-day menu made with no commercially hydrogenated fat to 9.7 g/day for the same 5-day menu made with commercially hydrogenated fat sources purchased from local grocery stores.

Trans *Fatty Acid Consumption Reported in the 1990s*

The value Hunter and Applewhite (1991) reported has been quoted widely as the *trans* fat content of the American diet. This value of 8.1 g/day/person was based on the availability of various foods in the U.S. food supply in 1989. Estimates of availability were based on industry market size and share data; the *trans* fatty acid content of food items was based on the Institute of Shortening and Edible Oils data for typical products.

In 1990, Enig et al. (1990) reported values for *trans* fatty acid consumption that were greater than previously reported values. Based on availability of fats in the population of the United States and weighted averages of *trans* content in categories of fats and oils, Enig et al. (1990) estimated daily *trans* fatty acid consumption to be 12.5–15.2 g/person or 5–15% of total fatty acids. In addition, Enig et al. (1990) combined a value of 8% of total fat as *trans* fatty acids with data from the Lipid Research Clinics on fat intakes of men and women, aged 20–65 years. For men with fat intakes of 40–259 grams per day, these investigators reported *trans* fatty intakes of 2.4–20.6 g/day; similarly, for women with fat intakes of 31–179 g/day, values of 1.9–14.3 g/day were reported. In interpretations of these data, the upper extremes were often emphasized. Median values for this population were not given.

During this time, estimates of *trans* fatty acid content of the diet were also based on the *trans* fatty acid content of adipose tissue as a biomarker. The study by Hudgins et al. (1991) underscored individual variation in the *trans* fatty acid consumption with values ranging from 1.3–7.6 g/day for men aged 23–78 years of age. These investigators assumed a one-to-one relationship between amounts in the diet and in the adipose tissue. Similarly, Hunter et al. (1992) used fat aspirates to estimate *trans* fatty acid intake as an average 4.2% of total fatty acids for a similar group of men. This value was somewhat less than the values of 4.7 and 5.4% of total fatty acids obtained using a semiquantitative food frequency questionnaire administered to the same population. Correlation between food frequency data and adipose tissue *trans* fatty acids was reported by Hu et al. (1997) to be 0.51 and by Garland et al. (1998) to be 0.40. Lemaitre et al. (1998) found greater correlations (0.52–0.58 for women and 0.67–0.76 for men) between the *trans* fatty acid intake obtained by the NCI/Block Food Frequency Questionnaire and adipose tissue *trans* fatty acids.

In the 1990s, a number of investigators began to combine data gathered earlier using the Willett semiquantitative food frequency questionnaire with *trans* fatty acid data in the USDA nutrient composition tables. Mean consumption values of approximately 3–4 g of *trans* fat /d were reported for 115 Boston-area post-menopausal women (London et al., 1991), 118 Boston-area men greater than 40 years of age (Hunter et al., 1992), 748 U.S. men aged 43–85 years (Troisi et al., 1992), 85,095 women in the Nurses' Health Study (Hu et al., 1997; Willett, 1993) and a subset of this population (Garland et al., 1998), as well as Boston area men and women (Ascherio et al., 1994).

In 1998, Lemaitre et al. (1998) reported the *trans* fatty acid intake of subjects in the Seattle, Washington area. These investigators used the NCI/Block food frequency questionnaire in combination with a revised USDA database. Men consumed an average of 2.57 g/day of *trans* fatty acids, whereas women consumed 1.99 g/day. The individual variation reported was great, with men consuming 0.7 to 8.6 g/day of *trans* fatty acids and women consuming as little as 0.3 g/day and as much as 22.8 g/day! The average value for consumption expressed as percent of total fatty acids was approximately 5% for both men and women.

Use of the food frequency questionnaire method has the advantage of allowing the diet of a large number of subjects to be assessed but generally suffers from incomplete or inaccurate *trans* fatty acid values in the food composition database used. The values for *trans* fatty acid consumption obtained using food frequency questionnaires (Hunter et al., 1992; London et al., 1991; Troisi et al., 1992; Willett et al., 1993; Ascherio et al., 1994) are generally less than those obtained from availability data for the United States (Senti, 1985; Enig et al., 1978, 1990; Hunter and Applewhite, 1986, 1991).

Other techniques to estimate dietary *trans* fatty acids were used and reported in the 1990s. Kromhout et al., (1995) analyzed food composites obtained in 1987 as part of the Seven Countries Study. U.S. consumption of *trans* fatty acids (18:1*t*) was estimated to be 3.9 g/day, a value that was less than that of the Netherlands and Finland, but greater than that of the former Yugoslavia, Italy, Greece or Japan.

Allison et al. (1999) reported data on *trans* fatty acid intake from a large number of subjects (n=11,258) divided into subgroups by age and gender. The study was based on food intake data assessed by 24-hour recall and 2-day diet records from the 1989–1991 Continuing Survey of Food Intakes by Individuals (CSFII) in combination with the USDA database. Total *trans* fatty acid intake in the US population was reported to be 5.3 g/day. The mean in the adult population ranged from 4.2 g/day for women, aged 50–69 years, to 6.6 g/day for men, aged 20–49 years. Teenage boys had the greatest consumption at 7.1g/day, with 2.8% of energy derived from *trans* fatty acids. Consumption of *trans* fatty acids by teenage girls was greater than the values reported for women (5.1 g/day for teenage girls compared to 4.2–4.6 g/day for women). Surprisingly high values of 4.1 and 5.3 g/day were reported for children 3–5 years of age and 6–11 years of age, respectively. Individual values for intake of dietary total *trans* fatty acids ranged from 1.6 g/day in women to 11.6 g/day in young men.

Trans *Fatty Acid Consumption Reported in 2000 or Later*

Recent studies have attempted to document a decline in *trans* fatty acid consumption in the United States. Using the Willett semiquantitative food frequency questionnaire, Hu et al. (2000) analyzed the habitual diet of women from the Nurses' Health Study from 1980–1990. These investigators reported a decrease in *trans* fatty acid consumption from 2.2–1.52% of energy during this time period. Intermediate values of 1.9% and 1.68% were reported for 1984 and 1986, respectively. Lopez-Garcia et al. (2005) analyzed a subset of the Nurses' Health Study and reported consumption of *trans* fatty acids to be 2.4 g/person/day or 1.4% of energy or 4.3% of total fatty acids for the mean of questionnaires administered in 1986 and 1990. These values are less than those reported in 1993 by Willett et al. (1993) for the 1980 data from the Nurse's Health Study: 4.0 g/person/day or 5.8% of total fatty acids. Harnack et al. (2003) reported a decline in *trans* fatty acid consumption from 1980–1997 in the adult population of Minneapolis-St. Paul, Minnesota. Values for *trans* fatty acid consumption were given for women/men as g/person/day during four periods: 5.4/8.4 for 1980–1982, 4.9/8.0 for 1985–1987, 4.5/7.0 for 1990–1992, and 4.7/6.4 for 1995–1997. Thus, a general decrease in *trans* fatty intake over this seventeen year period can be noted.

Canada

Trans *Fatty Acid Consumption Reported in the 1970s and 1980s*

In 1981, Brisson (1981) was one of the first to estimate *trans* fatty acid intake in the Canadian population. A typical Canadian person was believed to consume an average of 9.1 g/day, with a maximal value of 17.5 g/day. These estimates, based on the average *trans* fatty acid content of selected food items and Canadian government consumption data from 1977, were in the same range as early estimates of *trans* fatty acid consumption by the population of the United States.

Trans *Fatty Acid Consumption Reported in the 1990s*

Using the equation of Craig-Schmidt et al. (1984) in combination with human breast milk analysis of 198 samples collected in 1992, Ratnayake and colleagues (Chen et al., 1995a) calculated that *trans* fatty acid intake by lactating women ranged from 3.0 g/day to 20.3 g/day, with the majority of women consuming about 10 g/day. Estimates by Ratnayke and Chen (1995) also were determined on the basis of fat and calorie intakes reported in the Nova Scotia Dietary Survey along with the assumption that *trans* fatty acids were 10.4% of the total dietary fat. For this population, estimates ranged from 5.2 g/day for elderly women to 12.5 g/day for young men.

Even more extreme ranges in *trans* fatty acid consumption were reported by Innis et al. (1999a) who calculated values of 1.4–25.4 g/person/day for a sample Canadian diet. This wide variability in intake depended on the brand of food item or database chosen. These investigators used a database of food items purchased in Vancouver, British Columbia, Canada, and analyzed in Innis' laboratory or the FOOD PROCESSOR database (ESHA Research, Salem, Oregon). Using a combination of these databases and three-day diet records, Innis and King (1999b) reported the average *trans* fatty acid intake by Canadian lactating women (n=21) to be 6.87±1.07 g/person/day.

Trans *Fatty Acid Consumption Reported in 2000 or Later*

In a cross-sectional prospective study of healthy, pregnant women in Vancouver, Canada, Elias and Innis (2002) reported values for *trans* fatty acid intake that were somewhat less than the earlier estimates for the Canadian diet. The values of 3.8±0.3 g/day for women in the second trimester of pregnancy and 3.4±0.3 g/day for women in the third trimester of pregnancy were based on a food intake questionnaire designed specifically to estimate *trans* fatty acid intakes from a database of food items analyzed in the investigator's own laboratory. Using a similar approach, Innis et al. (2004) also reported values for the *trans* fatty acid intake of Canadian children aged 18–60 months. Intake at 4.81±3.12 g/day for this vulnerable population was even greater than that reported for the pregnant women. In a recent publication, however, Friesen and Innis (2006) reported the *trans* fatty acid content of human milk samples in Canada has declined since the introduction of *trans* fat food labeling, implying that *trans* fatty acid consumption has also decreased recently.

Comparison of *Trans* Fatty Acid Intake in North America with Other Regions of the Word

Historically, the *trans* fatty acid intake in North America has been one of the greatest in the world (Craig-Schmidt, 1998). The United States and Canada have lagged behind Europe in making changes in the food supply that would eliminate or reduce industrially produced *trans* fatty acids. A recent study by Stender and colleagues (Stender et al., 2006) has compared the *trans* fatty acid content of fast food in 20 countries. French fries and chicken nuggets were purchased from McDonald's or Kentucky Fried Chicken establishments between November, 2004, and September, 2005, and analyzed by capillary gas chromatography in a single laboratory using a standard method (Leth et al., 2003) accredited by the International Organization for Standardization. The results document that the United States is among the countries with the greatest *trans* fatty acid content in fast foods. For fast food from McDonald's restaurants, the United States was the greatest in *trans* fatty acids (10 grams or 23% of total fat in French fries and 11% in chicken nuggets) with Denmark being the least (less than 1 g or 1% of total fat in both products). Outlets in Peru, United Kingdom, South Africa, and Poland were also near the top. Obviously, the study does not give a complete picture because foods from conventional restaurants or home-cooked meals were not analyzed. In addition, consumption rates from these two fast food chains were not factored into the study design. The report, however, does provide relative amounts of *trans* fatty acids at two major fast food chains in a cross-cultural population and is useful as long as one recognizes the limitations of the study.

Future Trends in *Trans* Fatty Acid Consumption

Two legislative approaches have been taken in an effort to eliminate industrially produced *trans* fatty acids from the food supply (Ratnayake and Zehaluk, 2005). Industry was targeted directly in the approach taken by Denmark. Legislation, effective Jan. 1, 2004, restricted the use of industrially produced *trans* fatty acids to a maximum of 2% of the fat in any food product in Denmark (Stender and Dyerberg, 2003). This legislation has effectively eliminated industrially produced *trans* fatty acids from the food supply in Denmark (Stender et al., 2006). In contrast, the United States (Department of Health and Human Services, FDA, 2003) and Canada (Regulations Amending the Food and Drug Regulations, 2003) have targeted the consumer through food-labeling regulations. In both of these North American countries, amendments to food-labeling regulations were passed in 2003, requiring mandatory declaration of *trans* fat on nutrition labels. The effect of this legislation on *trans* fatty acid consumption has not been

TABLE 6.1
Estimates of *Trans* Fatty Acid Consumption in the United States

Reference	Population, Diet Assessment Method and Dietary Data Source	*trans* Fatty Acid Intake		
		g /person /day	energy %	% Total Fatty Acids
Kummerow, 1975	Fat disappearance of general U.S. population. Household Consumption Data (USDA, 1971) and typical *trans* content of visible fats; assumed 25–35% *trans* content of stick margarines, 15–25% for tub margarines, 20–30% for shortenings and 0–15% for salad oils.			8.0
Aitchison et al., 1977	Duplicate portion diets of lactating U.S. women (n=11). Laboratory analysis of three daily diet collections from each subject.	min.= 0.0–4.4 max.= 2.3–9.9		5.0 (range= 1.3–8.3)
Enig et al., 1978	Amount available in U.S. food supply of general U.S. population. Estimates based on government fats and oils availability data and known *trans* fatty acid content of foods.	12.1		8.0
Emken, 1981	Fat disappearance data of general U.S. population. Estimates based on consumption of hydrogenated soybean oil containing 20% average *trans* fatty acid.	6.8 (18:1*t*)		
Craig-Schmidt et al., 1984	Self-selected diets of lactating U.S. women (n=8). Estimates based on equation indicating relationship of 18:1*t* in diet and human milk of the following day.			7.8
Craig- Schmidt et al., 1984	Diets adequate for lactating women formulated using normal foods. Laboratory analysis (18:1*t*) of experimental diets made with and without commercial sources of hydrogenated fats. • With hydrogenated fat (n=5) • Without hydrogenated fat (n=5)	 9.7 0.9		 11.8 1.0

TABLE 6.1, CONT
Estimates of *Trans* Fatty Acid Consumption in the United States

Reference	Population, Diet Assessment Method and Dietary Data Source	*trans* Fatty Acid Intake		
		g /person /day	energy %	% Total Fatty Acids
Senti, 1985	Food disappearance data (1983) of general U.S. population. Estimates based on availability of fat in the U.S. food supply as indicated by U.S. production and sales data. • With correction for discarded fat • Without correction of discarded fat	 8.3 10.2		 5.9 6.1
van den Reek et al., 1986b	Laboratory analysis of duplicate portion diets (7 daily diets per subject) for Caucasian adolescent females, 12–15 years of age (n=8)	3.14 ± 0.26 (18:1*t*)		6.5±0.42 (18:1*t*)
van den Reek et al., 1986a	Calculations from 7-day weighed diet records of Caucasian adolescent females, 12–15 years of age, and 18:1*t* content of foods from Enig et al. (1983) (n=8)	2.8±0.26		5.3±0.44
Hunter and Applewhite, 1986	Fat availability in U.S. food supply of general U.S. population (1980, 1984). Estimates based on retail sales for household use, amount of fat used by food service industry and amount of fats used by bakeries and other food processors and typical values for *trans* fatty acid content of foods.	7.6		
Enig et al., 1990	Amount available for total US population. Estimates based on government (USDA-ERS) fats and oils data and weighted averages for *trans* content of categories of fats and oils	13.3 (range= 12.5–15.2)		range=5–15
Enig et al., 1990	Amount available for adults, age 20–65 years, Lipid Research Clinics. Estimates based on reported fat intake values and 8% of total fat as *trans* fatty acids. • men (with fat intakes of 40–258 g fat/day) • women (with fat intakes of 31–179 g fat/day)	 2.4–20.6 1.9–14.3		

TABLE 6.1, CONT
Estimates of *Trans* Fatty Acid Consumption in the United States

Reference	Population, Diet Assessment Method and Dietary Data Source	*trans* Fatty Acid Intake		
		g /person /day	energy %	% Total Fatty Acids
Hunter and Applewhite 1991	Food availability in general U.S. food supply (1989). Estimates based on industry market size and share data combined with ISEO (Institute of Shortening and Edible Oils) data for typical amounts of *trans* fatty acids in products	8.1		
Hudgins et al., 1991	Habitual diet of adult Caucasian males, aged 23–78 (n=76) (1981). Values based on one-to-one relationship between *trans* content of adipose tissue and diet assuming a 3000 calorie diet with 38% of calories as fat.	range = 1.3 – 7.6		
London et al., 1991	Habitual diet of Boston-area post-menopausal U.S. women free from cancer, aged 36–81 years (*n*=115) (1986–1987). Willett semiquantitative food frequency questionnaire containing 116 food items combined with USDA nutrient composition data and other food composition data	3.44 ± 1.84 (total) 2.83 ± 1.63 (18:1*t*)		5.83±1.68 (total) 4.77±1.56 (18:1*t*)
Hunter et al., 1992	Habitual diet of Boston area men, aged 40–75 years (n=118) (1986-1987). Willett semiquantitative food frequency questionnaire containing 131 food items (administered twice) combined with USDA nutrient composition data and other food composition data. • initial questionnaire • final questionnaire	 3.0±1.6 3.4±1.6		 4.7±1.5 5.4±1.5
Hunter et al., 1992	Habitual diet of Boston area men, aged 40–75 years (n=118) (1986–1987). Estimates based on fat aspirates			4.2±0.9
Troisi et al., 1992	Habitual diet of U.S. men, aged 43–85 (n=748) (1987–1990). Willett semiquantitative food frequency questionnaire containing 116 food items combined with USDA nutrient composition data and other food composition data	3.4±1.2	1.6 (range= 0.3 – 3.8)	5.5

TABLE 6.1, CONT
Estimates of *Trans* Fatty Acid Consumption in the United States

Reference	Population, diet assessment method and dietary data source	*trans* fatty acid intake		
		g/person/day	energy %	% total fatty acids
Willett et al., 1993	Habitual diet of women from Nurses' Health Study (n=85,095) (1980). Willett semiquantitative food frequency questionnaire containing 61 food items combined with USDA nutrient composition data and other food composition data.	4.0±1.9 (range= 2.4 – 5.7)		5.8±1.8
Ascherio, et al., 1994	Habitual diet of Boston area white men and women, <76 years of age (1982–1983). Willett semi-quantitative food frequency questionnaire containing 116 food items combined with USDA nutrient composition data and other food composition data.			
	• men	4.36 ± 2.33	1.5	4.3
	• women	3.61 ± 2.24	1.7	4.8
	energy adjusted mean			
	• lowest quintile	3.0 – 3.1		
	• highest quintile.	6.7 – 6.8		
Kromhout et al., 1995	Food composites (1987) of typical intake in U.S. Railroad cohort (*n*=30) as assessed in the Seven Countries Study (1958–1964). Laboratory analysis of food composites.	3.9 (18:1*t*)		
Hu et al., 1997	Habitual diet of women from Nurses' Health Study (*n*=85,095) as assessed by Willett semiquantitative food frequency questionnaire (1980 plus follow-up in 1984, 1986, and 1990).Correlation between FFQ dietary data and *trans* fatty acid content of adipose tissue= 0.51		2.2	
	• lowest quintile	1.3		
	• highest quintile.	2.9		

TABLE 6.1, CONT
Estimates of *Trans* Fatty Acid Consumption in the United States

Reference	Population, diet assessment method and dietary data source	*trans* fatty acid intake		
		g/person/day	energy %	% total fatty acids
Garland et al., 1998	Habitual diet of women from Nurses' Health Study (validation population, n=140) as assessed by Willett semiquantitative food frequency questionnaire (FFQ), 18 months apart (1986–1987). Correlation between FFQ dietary data and *trans*-fatty acid content of adipose tissue= 0.40			
	FFQ-1			
	total *trans*	2.8±1.3		4.7±1.4
	18:1 *trans*	2.2±1.1		3.7±1.3
	18:2 *trans*	0.4±0.2		0.7±0.2
	FFQ-2			
	total *trans*	2.8±1.4		4.8±1.5
	18:1 *trans*	2.3±1.2		3.8±1.4
	18:2 *trans*.	0.4±0.2		0.7±0.2

TABLE 6.1, CONT
Estimates of *Trans* Fatty Acid Consumption in the United States

Reference	Population, diet assessment method and dietary data source	*trans* fatty acid intake		
		g/person/day	energy %	% total fatty acids
Lemaitre et al., 1998	Self-administered food frequency questionnaire modified from NCI/Block FFQ combined with revised US Department of Agriculture data; subjects (n=51 aged 51–78 years) from greater Seattle, Washington (1996). Correlation between FFQ dietary data and *trans*-fatty acid content of adipose tissue=0.67 for men and 0.58 for women; adjusted for energy 0.76 and 0.52, respectively; mean±SD (range)			
	Men (n=24)			
	total *trans*	2.57±0.69 (0.7–8.6)		5.46±1.68 (3.5–10.6)
	18:1 *trans*	2.24±0.60 (0.6–7.8)		
	18:2 *trans*	0.29±0.08 (0.09–0.83)		
	16:1 *trans*	0.016±0.002 (0–0.020)		
	Women (n=27)			
	total *trans*	1.99±0.88 (0.3–22.8)		4.98±2.39 (1.8-10.7)
	18:1 *trans*	1.78±0.78 (0.2–20.6)		
	18:2 *trans*	0.24±0.09 (0.04-2.33)		
	16:1 *trans*	0.014±0.002 (0-0.20)		

TABLE 6.1, CONT
Estimates of *Trans* Fatty Acid Consumption in the United States

Reference	Population, diet assessment method and dietary data source	*trans* fatty acid intake		
		g/person/day	energy %	% total fatty acids
Allison et al., 1999	Food Intake data from the 1989 – 1991 Continuing Survey of Food Intakes by Individuals (CSFII) using 24-hour recall and 2-day diet record methodology combined with US Department of Agriculture trans fatty acid database (*n*=11,258); total *trans* given unless otherwise stated; mean ± SE (range given as 10th percentile to 90th percentile)			
	Average intake for US population	5.3	2.6	7.4
	Boys and girls, aged 3–5 years of age (n=615)	4.1±0.12 (2.1 – 6.6)	2.6±0.05	7.5±0.13
	Boys and girls, aged 6–11 years of age (n=1,172)	5.3±0.13 (2.5 – 8.5)	2.7±0.05	7.7±0.13
	Boys, aged 12–19 years of age (n=618)	7.1±0.37 (3.1 – 11.4)	2.8±0.06	7.6±0.14
	Girls, aged 12–19 years of age (n=672)	5.1±0.16 (2.1 – 8.6)	2.7±0.06	7.6±0.13
	Men, aged 20–49 years of age (n=2,076)	6.6±0.16 (2.4 – 11.6)	2.6±0.03	7.1±0.08
	Women, aged 20–49 years of age (n=2,787)	4.6±0.10 (1.6–8.0)	2.6±0.03	7.4±0.09
	Men, aged 50–69 years of age (n=879)	5.8±0.19 (2.1–10.2)	2.6±0.06	7.3±0.14
	Women, aged 50–69 years of age (n=1,226)	4.2±0.13 (1.6–7.4)	2.6±0.05	7.5±0.13
	Men and Women, aged 70 years and older (n=1,213)	4.5±0.12 (1.8–8.1)	2.6±0.05	7.9±0.12
Hu et al., 2000	Habitual diet of Women from Nurses' Health Study (*n*=85,941) as assessed by Willett semiquantitative food frequency questionnaire (1980, 1984, 1986, and 1990). Correlation between FFQ dietary data and *trans* fatty acid content of adipose tissue= 0.51			
	"*trans* fat"			
	1980		2.20	
	1984		1.90	
	1986		1.68	
	1990		1.52	

TABLE 6.1, CONT
Estimates of *Trans* Fatty Acid Consumption in the United States

Reference	Population, diet assessment method and dietary data source	*trans* fatty acid intake		
		g/person/day	energy %	% total fatty acids
Harnack et al., 2003	Minnesota Heart Survey; adults aged 25 – 74 years of age in Minneapolis-St. Paul, Minnesota, metropolitan area; 24-hour diet recall combined with the University of Minnesota Nutrition Coordinating Center Food and Nutrient Database (mean intake value for women/men)			
	1980–1982			
	total *trans*	5.4/8.4	2.8/3.0	
	18:1 *trans*	4.7/7.3	2.4/2.6	
	18:2 *trans*	0.63/1.03	0.34/0.36	
	16:1 *trans*	0.08/0.12	0.04/0.04	
	1985-1987			
	total *trans*	4.9/8.0	2.6/2.8	
	18:1 *trans*	4.1/6.8	2.2/2.4	
	18:2 *trans*	0.64/1.01	0.34/0.35	
	16:1 *trans*	0.07/0.11	0.04/0.04	
	1990–1992			
	total *trans*	4.5/7.0	2.4/2.5	
	18:1 *trans*	3.9/6.0	2.0/2.1	
	18:2 *trans*	0.56/0.86	0.30/0.30	
	16:1 *trans*	0.06/0.08	0.03/0.03	
	1995–2007			
	total *trans*	4.7/6.4	2.2/2.2	
	18:1 *trans*	4.0/5.4	1.8/1.8	
	18:2 *trans*	0.64/0.85	0.29/0.29	
	16:1 *trans*	0.06/0.09	0.03/0.03	

TABLE 6.1, CONCLUDED
Estimates of *Trans* Fatty Acid Consumption in the United States

Reference	Population, diet assessment method and dietary data source	*trans* fatty acid intake		
		g/person/day	energy %	% total fatty acids
Lopez-Garcia et al., 2005	Habitual diet of women from Nurses' Health Study (n=730) as assessed by Willett semiquantitative food frequency questionnaire (FFQ), (mean of 1986 and 1990). mean ± SD of 3rd quintile (range = mean of 1st quintile to mean of 5th quintile			
	total *trans*	2.4±0.1 (1.5–3.7)	1.4±0.1 (0.9–2.1)	4.3±0.6 (3.1–5.9)
	18:1 *trans*	1.9±0.1 (1.1–3.0)		
	18:2 *trans* (*tt*)	0.1±0.02 (0.1–0.2)		
	18:2 *trans* (*tc,ct*)	0.2±0.1 (0.2–0.3)		
	16:1 *trans*	0.2±0.04 (0.1–0.2)		

TABLE 6.2
Estimates of *Trans* Fatty Acid Consumption in Canada

Reference	Population, Diet Assessment Method and Dietary Data Source	*trans* Fatty Acid Intake		
		g/d /person	energy %	% TFA
Brisson, 1981	Typical Canadian diet (1977) for • General Canadian population • Hypothetical Man, aged 20–39 Calculated from average *trans* fatty acid content of selected foods and Canadian government consumption data, 1977.	 9.1 (avg.) 17.5 (max) 11.1 (avg) 17.1 (max)		 9.5
Chen et al., 1995a	Usual diet for •Lactating Women (n=198) • Low-*trans* milk (n=18) • Medium-*trans* milk (n=155) • High-*trans* milk (n= 25) (1992) Calculated from human breast milk analysis and equation of Craig-Schmidt, et al. (1984) for the relationship of milk *trans* 18:1 to dietary *trans*-18:1.	 10.6 3.0 10.1 20.3	 4.0 1.1 3.9 7.7	 – 3.7 13.2 26.0
Ratnayake and Chen, 1995	Diet based on government survey Male, age 18–34 age 35–49 age 50–64 age 65–74 Female, age 18–34 age 35–49 age 50–64 age 65–74 Average Estimates based on fat and calorie intakes reported in the Nova Scotia Dietary Survey and assuming 10.4% of the total fat as *trans*-18:1 (range 0.6–32.3%), which was calculated using the Craig-Schmidt, et al. (1984) equation and *trans*-18:1 levels found in Canadian human milk	 12.5 9.6 9.1 7.8 7.0 6.4 5.6 5.2 8.4 (range= 0.5–26.1)	 3.7 3.7 3.7 3.5 3.7 3.7 3.4 3.4 3.7 (range= 0.2 – 11.3)	

TABLE 6.2, CONT
Estimates of *Trans* Fatty Acid Consumption in Canada

Reference	Population, Diet Assessment Method and Dietary Data Source	*trans* Fatty Acid Intake		
		g/d/ person	energy %	% TFA
Innis et al.1999a	Sample Canadian diet used to show variability in estimated *trans*-fatty acid intake depending on brand or database chosen; database of 200 food items purchased in Vancouver, British Columbia, Canada, analyzed in Innis' laboratory or FOOD PROCESSOR database (ESHA Research, Salem, Oregon)	1.4 – 25.4		
Innis and King, 1999b	Three-day diet records of Canadian lactating women (n=21); analyzed using FOOD PROCESSOR database (ESHA Research, Salem, Oregon) modified to include 300 brand name food products analyzed by the laboratory of Innis	6.87 ± 1.07	2.46 ± 0.38	
Elias and Innis, 2002	Cross-sectional prospective study using a food intake questionnaire designed specifically to estimate *trans*-fatty acid intakes calculated from a *trans*-fatty acid food database developed by food analysis; healthy, pregnant women (n=60) recruited in Vancouver, British Columbia, Canada; mean ± SE			
	second trimester of pregnancy	3.8±0.3	1.3±0.1	
	third trimester of pregnancy	3.4±0.3	1.3±0.1	
Innis et al., 2004	Validated 178-item food frequency questionnaire designed to assess patterns of fat and fatty acid intakes in a detailed, in-person interview with the primary caregiver; FOOD PROCESSOR food composition database (ESHA Research, Salem, Oregon) and the Canadian nutrient file modified to include 500 food products analyzed by the laboratory of Innis; Canadian children aged 18–60 months (n=84)			
	All children	4.81±3.12		
	18–24 mo old	3.47 ±1.86		
	25–36 mo old	5.27±3.86		
	37–90 mo old	4.98±3.05		

documented. However, many major food manufacturers are introducing "*trans* free" products (Tarrago-Trani et al., 2006). Although the food-labeling laws do not address food eaten in restaurants, recent lawsuits against fast food chains have brought this issue into the public arena. Because of increased consumer awareness about the detrimental health effects of *trans* fatty acids, it is expected that industrially produced *trans* fatty acids will be eliminated or greatly reduced in the food supply of North America within the next few years.

References

Aitchison J.M.; W.L. Dunkley; N.L. Canolty; L.M. Smith. Influence of Diet on trans Fatty Acids in Human Milk. *Am J Clin Nutr* **1977,** *30*, 2006–2015.

Allison D.B.; K. Egan; L.M. Barraj; C. Caughman; M. Infante; J.T. Heimbach. Estimated Intakes of trans Fatty and Other Fatty Acids in the US Population. *J Am Diet Assoc.* **1999,** *99*, 166–174.

American Society for Clinical Nutrition /American Institute for Nutrition Task Force on Trans Fatty Acids, Position paper on trans fatty acids. *Am J Clin Nutr* **1996,** *63*, 663–670.

Ascherio A.; C.H. Hennekens; J.E. Buring; C. Master; M.J. Stampfer; W.C. Willett. Trans-fatty Acids Intake and Risk of Myocardial Infarction. *Circulation* **1994,** *89*, 94–101.

Beynen A.C.; R.J.J. Hermus; J.G.A.J. Hautvast. A Mathematical Relationship Between the Fatty Acid Composition of the Diet and That of the Adipose Tissue in Man. *Am J Clin Nutr* **1980,** *33*, 81–85.

Boatella J.; M. Rafecas; R. Codony; A. Gibert; M. Rivero; R. Tormo; D. Infante; F. Sanchez-Valverde. Trans Fatty Acid Content of Human Milk in Spain. *J Pediatr Gastroenterol Nutr* **1993,** *16*, 432–434.

Brisson J. Lipids in Human Nutrition: An Appraisal of Some Dietary Concepts. Jack K. Burgess, Inc.: Eaglewood, New Jersey, 1981; pp 41–71.

Chappell J.E.; M.T. Clandinin; C. Kearney-Volpe. Trans Fatty Acids in Human Milk Lipids: Influence of Maternal Diet and Weight Loss. *Am J Clin Nutr* **1985,** *42*, 49–56.

Chardigny J.M.; R.L. Wolff; E. Mager; J.-L. Sebedio; L. Martine; P. Juaneda. Trans mono- and Polyunsaturated Fatty Acids in Human Milk. *Eur J Clin Nutr* **1995,** *49*, 523–531.

Chen Z.Y.; G. Pelletier; R. Hollywood; W.M.N. Ratanayke. Trans Fatty Acid Isomers in Canadian Human Milk. *Lipids* **1995a,** *30*, 15–21.

Chen Z.Y.; W.M.N. Ratanayke; L. Fortier; R. Ross; S.C. Cunnane. Similar Distribution of trans-fatty Acid Isomers in Partially Hydrogenated Vegetable Oils and Adipose Tissue of Canadians. *Can J Physiol Pharmacol* **1995b,** *73*, 718–723.

Craig-Schmidt M.C.; J.D. Weete: S.A. Faircloth; M.A. Wickwire; E.J. Livant. The Effect of Hydogenated Fat in the Diet of Nursing Mothers on Lipid Composition and Prostaglandin Content of Human Milk. *Am J Clin Nutr* **1984,** *39*, 778–786.

Craig-Schmidt M.C. Worldwide consumption of trans fatty acids. In *Trans-Fatty Acids in Human Nutrition*. Sebedio J.-L.; W.W. Christie Ed. The Oily Press: Dundee, Scotland, 1998; pp 59–113.

Craig-Schmidt M.C.; B.M. Holzer. Fatty acid isomers in foods. In *Fatty Acids In Foods and Their Health Implications. 2nd edition*. Chow CK, Ed. Marcel Dekker, Inc.:

New York, 2000; pp 307–356.

Craig-Schmidt M.C. World-wide Consumption of trans Fatty Acids. *Atherosclerosis Supplements.* **2006,** *7*, 1–4.

Department of Health and Human Services, Food and Drug Administration. 21 CFR Part 101 (Docket No. 94P-0036). Food labeling: Trans fatty acids in nutrition labeling, nutrient content claims and health claims. Washington, July 2003, p. 254.

Elias S.L.; S.M. Innis. Bakery Foods are the Major Dietary Source of trans-fatty Acids Among Pregnant Women with Diets Providing 30 Percent Energy From Fat. *J Am Diet Assoc* **2002,** *102*, 46–51.

Emken E.A. Metabolic Aspects of Positonal Monounsaturated Fatty Acids Isomers. *J Am Oil Chem Soc* **1981,** *58*, 278–283.

Enig M.G.; R.J. Munn; M. Keeney. Dietary Fat and Cancer Trends - a Critique. *Fed Proc* **1978,** *37*, 2215–2220.

Enig M.G.; L.A. Pallansch; J. Sampugna; M. Keeney. Fatty Acid Composition of the Fat in Selected Food Items with Emphasis on trans Components. *J Am Oil Chem Soc* **1983,** *60*, 1788–1795.

Enig M.G.; S. Atal; M. Keeney; J. Sampugna. Isomeric trans Fatty Acids in the U.S. Diet. *J Am Coll Nutr* **1990,** *9*, 471–486.

Finley D.A.; B. Lonnerdal; K.G. Dewey; L.E. Grivetti. Breast Milk Composition: Fat Content and Fatty Acid Composition in Vegetarians and Non-vegetarians. *Am J Clin Nutr* **1985,** *41*, 787–800.

Friesen R.; S.M. Innis. Trans Fatty Acids in Human Milk in Canada Declined with the Introduction of trans Fat Food Labeling. *J Nutr* **2006,** *136*, 2559–2561.

Garland M.; F.M. Sacks; G.A. Colditz; E.B. Rimm; L.A. Sampson; W.C. Willett; D.J. Hunter. The Relation Between Dietary Intake and Adipose Composition of Selected Fatty Acids in US Women. *Am J Clin Nutr* **1998,** *67*, 25–30.

Harnack L.; S. Lee; S.F. Schakel; S. Duval; R.V. Luepker; D.K. Arnett. Trends in the trans-fatty Acid Composition of the Diet in a Metropolitan Area: The Minnesota Heart Survey. *J Am Diet Assoc* **2003,** *103*, 1160–1166.

Hirsch J.; J.W. Farquhar; E.H. Ahrens Jr; M.L. Peterson; W. Stoffel. Studies of Adipose Tissue in Man: A Microtechnic for Sampling and Analysis. *Am J Clin Nutr* **1960,** *8*, 499–511.

Hu F.B.; M.D. Stampler; J.E. Manson; E. Rimm; G.A. Colditz; B.A. Rosner; C.H. Hennekens; W.C. Willett. Dietary Fat Intake and the Risk of Coronary Heart Disease in Women. *N Eng J Med* **1997,** *337*, 1491–1499.

Hu F.B.; M.J. Stampfer; J.E. Manson; F. Grodstein; G.A. Colditz; F.E. Speizer; W.C. Willett. Trends in the Incidence of Coronary Heart Disease and Changes in Diet and Lifestyle in Women. *N Engl J Med* **2000,** *343*, 530–537.

Hudgins L.C.; J. Hirsch; E.A. Emken. Correlation of Isomeric Fatty Acids in Human Adipose Tissue with Clinical Risk Factors for Cardiovascular Disease. *Am J Clin Nutr* **1991,** *53*, 474–482.

Hunter D.J.; H.B. Rimm; F.M. Sacks; M.J. Stampfer; G.A. Colditz; L.B. Litin; W.C. Willett. Comparison of Measures of Fatty Acid Intake by Subcutaneous Fat Aspirate, Food Frequency Questionnaire, and Diet Records in a Free-living Population of US Men. *Am J Epidemiol* **1992,** *135*, 418–427.

Hunter J.E.; T.H. Applewhite. Isomeric Fatty Acids in the US Diet: Levels and Health Perspectives. *Am J Clin Nutr* **1986,** *44*, 707–717.

Hunter J.E.; T.H. Applewhite. Reassessment of trans Fatty Acid Availability in the US Diet. *Am J Clin Nutr* **1991,** *54*, 363–369.

Hunter J.E.; T.H. Applewhite. Correction of Dietary Fat Availability Estimates for Wastage of Food Service Deep-frying Fats. *J Am Oil Chem Soc* **1993,** *70*, 613–617.

Innis S.M.; T.J. Green; T.K. Halsey. Variability in the trans Fatty Acid Content of Foods Within a Food Category: Implications for Estimation of Dietary trans Fatty Acid Intake. *J Am Coll Nutr.* **1999a,** *18*, 255–260.

Innis S.M.; D.J. King. Trans Fatty Acids in Human Milk are Inversely Associated with Concentrations of Essential all-cis n-6 and n-3 Fatty Acids and Determine trans, but not n-6 and n-3, Fatty Acids in Plasma Lipids of Breast-fed Infants. *Am J Clin Nutr* **1999b,** *70*, 383–390.

Innis S.M.; Z. Vaghri; D.J. King. n-6 Docosapentaenoic Acid is not a Predictor of Low Docosahexaenoic Acid Status in Canadian Preschool Children. *Am J Clin Nutr* **2004,** *80*, 768–773.

International Life Sciences Institute Expert Panel on Trans Fatty Acids and Coronary Heart Disease. Trans Fatty Acids and Coronary Heart Disease Risk. *Am J Clin Nutr* **1995,** *62*, 655S–708S.

International Life Sciences Institute Expert Panel on Trans Fatty Acids and Early Development. Trans Fatty Acids: Infant and Fetal Development. *Am J Clin Nutr* **1997,** *66*, 715S–736S.

King I.B.; A.R. Kristal; S. Schaffer; M. Thornquist; G.E. Goodman. Serum trans-fatty Acids are Associated with Risk of Prostate Cancer in $-carotene and Retinol Efficacy Trial. *Cancer Epidemiol Biomarkers Prev* **2005,** *14*, 988–992.

Koletzko B.; M. Mrotzek; H.J. Bremer. Fat Content and cis- and trans-isomeric Fatty Acids in Mature Human fore- and Hindmilk. In *Human Lactation 2, Maternal and Environmental Factors*. Hamosh M.; A.S. Goldman, Ed. Plenum Press: New York, 1986; pp 589–594.

Kromhout D.; A. Menotti; B. Bloemberg; C. Aravanis; H. Blackburn; R. Buzina; A.S. Dontas; F. Fidanza; S. Giampaoli; A. Jansen; et al. Dietary Saturated and trans Fatty Acids and Cholesterol and 25-year Mortality From Coronary Heart Disease: The Seven Countries Study. *Preventive Medicine* **1995,** *24*, 308–315.

Kummerow F.A. Lipids in Atherosclerosis. *J Food Sci* **1975,** *40*, 12–17.

Lemaitre R.N.; I.B. King; R.E. Patterson; B.M. Psaty; M. Kestin; S.R. Heckbert. Assessment of trans-fatty Acid Intake with a Food Frequency Questionnaire and Validation with Adipose Tissue Levels of trans-fatty Acids. *Am J Epidemiol* **1998,** *148*, 1085–1093.

Lemaitre R.N.; I.B. King; T.E. Raghunathan; R.M. Pearce; S. Weinmann; R.H. Knopp; M.K. Copass; L.A. Cobb; D.S. Siscovick. Cell Membrane trans-fatty Acids and the Risk of Primary Cardiac Arrest. *Circulation* **2002,** *105*, 697–701.

Leth T.; A. Bysted; K. Hansen; L. Ovesen. Trans Fatty Acid Content in Danish Margarines and Shortenings. *J Am Oil Chem Soc* **2003,** *80*, 475–478.

London S.J.; F.M. Sacks; J. Caesar; M.J. Stampfer; E. Siguel; W.C. Willett. Fatty Acid Composition of Subcutaneous Adipose Tissue and Diet in Postmenopausal US Women. *Am J Clin Nutr.* **1991,** *54*, 340–345.

Lopez-Garcia E.; M.B. Schulze; J.B. Meigs; J.E. Manson; N. Rifai; M.J. Stampfer; W.C. Willett; F.B. Hu. Consumption of trans Fatty Acids is Related to Plasma Biomarkers of Inflammation and Endothelial Dysfunction. *J Nutr* **2005,** *135*, 562–566 .

Mozaffarian D.; E.B. Rimm; I.B. King; R.L. Lawler; G.B. McDonald; W.C. Levy. Trans Fatty Acids and Systemic Inflammation in Heart Failure. *Am J Clin Nutr* **2004,** *80*, 1521–1525.

Ratnayake W.M.N.; Z.Y. Chen. Trans Fatty Acids in Canadian Breast Milk and Diet. In *Development and Processing of Vegetable Oils for Human Nutrition*. Przybylski R.; B.E. McDonald, Eds. American Oil Chemists' Society: Champaign, Illinois, 1995; pp 20–35.

Ratnayake W.M.N.; C. Zehaluk. Trans Fatty Acids in Foods and Their Labeling Regulations. In *Healthful Lipids*. Akoh C.C.; O.-M. Lai Ed. AOCS: Champaign, Illinois, 2005; pp 1–32.

Regulations Amending the Food and Drug Regulations (Nutrition Labelling, Nutrient Content Claims and Health Claims) Department of Health, Canada Gazette, Part 11, January 1, 2003.

Senti F.R., Ed. Health Aspects of Dietary Trans Fatty Acids. Bethesda, MD: Life Science Research Office, Federation of American Societies for Experimental Biology, 1985.

Stender S.; J. Dyerberg. The Influence of trans Fatty Acids on Health. A Report From the Danish Nutrition Council, 4th edition. Publication no. 34, 2003.

Stender S.; J. Dyerberg; A. Astrup. High Levels of Industrially Produced trans Fat in Popular Fast Foods. *N Eng J Med* **2006,** *354*, 1650–1652.

Tarrago-Trani M.T.; K.M. Phillips; L.E. Lemar; J.M. Holden. New and Existing Oils and Fats Used in Products with Reduced trans-Fatty Acid Content. *J Am Diet Assoc* **2006,** *106*, 867–880.

Troisi R.; W.C. Willett; S.T. Weiss. Trans-Fatty Acid Intake in Relation to Serum Lipid Concentrations in Adult Men. *Am J Clin Nutr* **1992,** *56*, 1019–1024.

United States Department of Agriculture (USDA). Civilian consumption of visible and invisible food fats per person, 1959-71. ERS/USDA Statistical Bulletin 489, U.S. Fats and Oils Statistics, 1950-1971. USDA, 1971, p 102.

van de Vijver L.P.L.; G. van Poppel; A. van Houwelingen; D.A.C.M. Kruyssen; G. Hornstra. Trans Unsaturated Fatty Acids in Plasma Phospholipids and Coronary Heart Disease: a Case-control Study. *Atherosclerosis* **1996,** *126*, 155–161.

van den Reek M.M.; M.C. Craig-Schmidt; A.J. Clark. Use of Published Analyses of Food Items to Determine Dietary trans Octadecenoic Acid. *J Am Diet Assoc* **1986a,** *86*, 1391–1394.

van den Reek M.M.; M.C. Craig-Schmidt; J.D. Weete; A.J. Clark. Fat in the Diets of Adolescent Girls with Emphasis on Isomeric Fatty Acids. *Am J Clin Nutr* **1986b,** *43*, 530–537.

Willett W.C.; M.J. Stampfer; J.E. Manson; G.A. Colditz; F.E. Speizer; B.A. Rosner; L.A. Sampson; C.H. Hennekens. Intake of trans Fatty Acids and Risk of Coronary Heart Disease Among Women. *Lancet* **1993,** *341*, 581–585.

Wolff R.L. Content and Distribution of trans-18:1 Acids in Ruminant Milk and Meat Fats. Their Importance in European Diets and Their Effect on Human Milk. *J Am Oil Chem Soc* **1995,** *72*, 259–272.

Chapter 7

Zero/Low *Trans* Margarine, Spreads, and Shortening

Gaey R. List[1] and Ted Pelloso[2]

[1]Food and Industrial Oil Research, National Center for Agricultural Utilization Research, Agricultural Research Service, U.S. Department of Agriculture, Peoria, IL 61604

[2]ConAgra Foods, Indianapolis, IN 46268

Names are necessary to report factually on available data: the USDA neither guarantees nor warrants the standard of the product, and the use of the name USDA implies no approval of the product to the exclusion of others that may also be suitable.

Introduction

The reduction and/or elimination and content of *trans* fatty acids in the food supply has attracted world wide interest (Hunter, 2004; List, 2004a, 2004b; Alonzo et al., 2000; Oveson et al., 1998; Ratnayake et al., 1998; Bayard and Wolff, 1995; Tekin et al., 2002; Tsanev, 1998; List et al., 2000a; Aro et al., 1998; Satchithanandam et al., 2004; van Erp-Baart et al., 1998; van Poppel et al., 1998; Tarrago-Trani, 2006; Matsuzaki et al., 2002). Europe, Canada, and the United States have enacted legislation regulating the levels of *trans* fatty acids in foods. Major sources of *trans* fatty acids in the U.S. diet are shown in

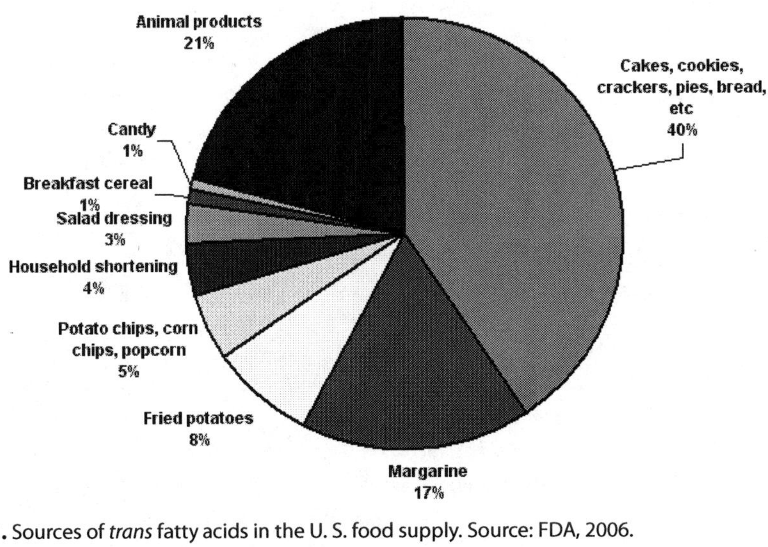

Fig. 7.1. Sources of *trans* fatty acids in the U. S. food supply. Source: FDA, 2006.

Fig. 7.1. Major sources include margarine (17%), animal products (21%) and baked goods (40%). Household shortenings account for 4%. Thus, over 60% of the *trans* fatty acids consumed in the United States come from margarine/spreads and baking shortenings. The reliability of the margarine data has been questioned, claiming that it is based on government data taken from 1994–1996 and that, since then, the average *trans* fat content of margarine has been reduced by more than 75% to less than 0.5 g/serving and the saturated fat content reduced by 1/3 to about one g/serving (Cristol, 2006). A detailed review of trends in *trans* fatty acid consumption in North America can be found in Chapter 8 and the literature up to 1998 has been reviewed by the same author (Craig-Schmidt, 1998).

Extensive discussions of various oil processing technologies to reduce *trans* fatty acids in food oil are beyond the scope of this chapter. The reader is directed to a number of reviews on chemical/enzymatic interesterification (Rozendall and Macrae, 1997; Huyghebaert et al., 1994; Erickson, 1995), fractionation (Krishnamurthy and Kellens, 1996; Gibon, 2006; Timms, 1997), modified hydrogenation (Ariaansz and Okonek, 1998; Beers and Mangus, 2004; Beers et al., 2004a, 2004b; Berben et al., 1994), structurally modified oilseeds (Gunstone, 2001) and combinations thereof (Rajah, 1994). This chapter will focus on the use of these technologies to formulate low/zero *trans* margarine/spreads and shortenings.

General Consideration

General considerations for using *trans*-free fats in foods were recently reviewed by Floter and Dujin (2006). Partial hydrogenation has been the technology of choice prior to *trans* fat labeling because of the high stability toward oxidation, the ease by which melting profiles can be attained and hydrogenated fats have favorable crystallization properties because they do so quickly and deliver structure to products. Factors to be considered for *trans*-free products include crystallization behavior, mouth feel and product stability. Other requirements for *trans*-free replacements include functional characteristics and the nutritional value of the product. Crystallization of *trans*-free fats should undergo polymorphic transitions quickly in order to maintain the sensory properties of the finished product. (Floter and Dujin, 2006)

Loh (2005, 2006) recently summarized some of the strategies used by the food industry to eliminate *trans* fats from their products. One approach involved changing the fat system. However, those who attempted to replace standard hydrogenated baking shortenings soon learned that few, if any, drop in solutions are available. By doing so, the product was different and other adjustments were required in the manufacturing and/or distribution system. For example, a dough made with *trans*-free shortening may require longer or shorter tempering time or frozen products may require longer freezing time. Other lessons learned from reformulation include the need for extensive testing of new products, as well as engineering changes including the need for additional storage tanks and piping.

Some products showed subtle changes after reformulation. For example, baked products may become more brittle and prone to breakage during distribution, while others may have a softer texture or less shelf life. As a result, packaging changes are needed to accommodate product changes resulting from reformulation.

Positioning of reformulated products is important. While many customers check the nutrition panel prior to purchasing a food item, nearly all look at the front of the package. Some leading brands tout their reformulation on the front as well as on the nutritional label.

Challenges to Trans *Reduction/Reformulation for Nutrition Labeling*

Aside from technical challenges to reformulate products with reduced or low *trans* fatty acid content, keeping customers happy is one of paramount importance. New products must be as close to the old one as possible. Appearance, taste and performance are primary considerations. If new products are different from the old one, the new product should offer something positive, i.e., "*trans* free." Cost is a major consideration in new product development. Customers may be reluctant to pay more for a product, even if it is advantageous for them to do so. Any new or reformulated product must be competitive with others sitting on the grocer's shelf. Prior to the development of a new product, parameters must be defined beforehand. Examples include fat content, flavorings, minor components and emulsifier systems. The fats and oils technologist may well be forced to educate management and marketing staffs on issues like technical feasibility, economics and preferences.

TABLE 7.1
Spreads Meeting FDA Labeling for Zero grams TFA/serving[a]

	% Fat Spread						
	80	70	60	50	40	35	30
	g Fat/Serving						
	11.2	9.8	8.4	7.0	5.6	4.9	4.2
% *Trans* in Oil	g TFA/Serving						
5	0.560	*0.490	0.420	0.350	0.280	0.245	0.210
6	0.670	0.590	*0.500	0.420	0.370	0.290	0.250
7	0.780	0.690	0.590	*0.490	0.390	0.340	0.290
8	0.896	0.780	0.670	0.560	*0.450	0.390	0.340
10	1.120	0.980	0.840	0.700	0.560	*0.490	0.420

[a]Serving size = 14 g.

TABLE 7.2
Evolution of Margarine Properties

			Solid Fat (°C)			
		Drop Pt. (°F)	10	21.1	33.3	% *Trans*
1965	Stick	96–98	30–35	18–22	3–5	25–30
	Soft	90–95	15–20	10–12	2–4	15
1980	Stick	94–97	24–29	14–18	1.5–3.5	20
	Soft	90–93	8–11	4–7	0.5–2.0	12
2005	Stick	96–100	15–17	6–9	2–4	10
	Soft	87–92	5–10	3–6	1–2	7

Reformulation of Spreads to Meet Labeling Requirements for Trans *Fatty Acids*

Meeting the Jan. 1, 2006, deadline for 0 g *trans* fat per serving involves a product containing less than 0.5 g TFA/serving. (Anon., 2003) For margarine/spreads and shortenings a serving size is set by the FDA at 14 g, or 0.5 oz for spreads and 12 g for shortenings. The relationships between the *trans* content of the oil component and the fat content of spreads is shown in Table 7.1.

Hydrogenated soybean oil IV65, 40% *trans*, is typically used to formulate spreads by blending with liquid soybean oil or other vegetable oils. Soft spreads typically are prepared by blending about 25% IV65 oil with 75% liquid soy and will contain about 10% *trans* fatty acids. (List and King, 2006a) As such, this spread should not exceed 35% fat in order to meet a 0 g TFA claim. Table 7.1 shows that spreads can be reformulated to meet the 0 g claim by using an IV 65 basestock manufactured commercially.

The historical evolution of stick and soft spread properties over the past 40 years are shown in Table 7.2. In 1965, solid fat contents and melting points of stick products were formulated to match the properties of butter, while soft spreads contained about ½ of solid fat and *trans* fatty acids. By 1980, a shift toward lower *trans* and solid fat occurred. Currently, targets for *trans* fatty acids are set at 7–10% for soft and stick products respectively. By the mid 1990s, the spread industry began to reduce *trans* levels and, by 2000, a 56% reduction had occurred in soft spreads (List et al., 2000).

A historical summary of developments in the margarine/spread industry is shown in Table 7.3. Of particular interest is the passage of the Nutrition Education Labeling Act of 1990 and the final rules for mandatory labeling of *trans* acids published by the FDA in July, 2003. Effective Jan. 1, 2006, *trans* fatty acid contents are to be listed as a separate line on the nutrition facts label and a zero g *trans* fat claim is permitted providing it does not exceed 0.5 g/serving. Further discussion of the Nutrition Education Labeling Act and *trans* fatty acids can be found in a recent review (Moss, 2006).

Although zero *trans* spreads have been available in Europe for many years under the label Becel (Niewenhuis, 2000), cholesterol lowering spreads are a relatively new phenomenon dating back to the mid to late 1990s. These products appear to occupy a niche market in the U. S. (Hollingsworth, 2004) In Europe, these products have had more success (Young, 2003).

Enzymatic Interesterification

Production of low-*trans* fats by enzymatic interesterification was introduced as a commercial process in 2003 (Anon., 2004b). In addition, the enzymatic process is continuous, whereas chemical interesterification is usually a batch process. The concept involves passing a mixture of liquid and completely hydrogenated soybean or cottonseed oil flakes through a series of rectors containing an immobilized enzyme bed. Four reactors containing from 100–400 kg enzyme/reactor constitute the heart of the process. Compared to chemical interesterification with sodium methoxide, this process has certain disadvantages including no side-products such as FFA, methyl esters, stops or partial glycerides are formed, color bodies are absent negating the need for downstream processing bleaching, no loss of neutral oil, and the enzyme is 1–3 specific, which preserves the natural sharply-melting properties of the liquid component. Because common liquid vegetable oils have triglyceride structures with unsaturated acids in the middle 2 position, functional triglycerides of the SUS type (where S=saturated, U=oleic, linoleic, linolenic) are known to melt more sharply than mixtures of SUS and SSU produced in random interesterification by chemical means. As an example, cocoa butter is composed of over 80% SIS, POP, and SOP. These symmetrical triglycerides impart a hard, brittle structure at room temperature, yet melt sharply at body temperature. On the other hand, sheep tallow, which has nearly the same fatty acid composition, but with significant amounts of SSU triglycerides, melts over a much wider range (List et al., 2003).

The enzyme used in the interesterification process is a triacylglycerol hydrolase lipozyme TLIM and is derived from the microorganism, *Thermomyces lanuginosus*, developed by Novozyme. The enzyme is heat stable at temperatures of 75–90°C needed to solublize the trisaturate in the feedstock and may be used repeatedly before activity declines. The enzyme is subject to loss of activity by traces of impurities in the feedstock including free fatty acids and soaps. Installation of a precolumn packed with silica prolongs the life and activity of the enzyme. (Lee and Sleeter, 2003; Lee, 2005). Peroxides, secondary oxidation products, and nickel are also catalyst poisons. Traces of nickel may be found in the soy flakes used to prepare the feedstock. The feedstock must be thoroughly degummed, bleached, and deodorized.

Table 7.3
Milestones in Margarine/Spread Development[a]

Year	Development
1869	Margarine developed in France
1873	Margarine patented in United States
1875	Margarine introduced in United States
1886	Legislation taxes margarine
1890	Cultured milk used for flavor
1917	Animal fats replaced by coconut oil
1923	Margarine fortified with vitamin E
1934	Coconut oil replaced by hydrogenated fats
1934	Less than 2 million pounds soybean oil usage
1938	Scraped surface heat exchangers (votator) introduced
1941	Margarine composition defined by Federal Standards of Identity
1942	Soybean oil usage passes 100 million pounds
1946	Soybean oil usage passes 200 million pounds
1947	Coal tar coloring agents replaced by β carotene
1950	Margarine Act passed by Congress abolishing taxes
1950	Soybean oil usage passes 300 million pounds
1952	Soft spreading margarine introduced
1955	Soybean oil usage passes 700 million pounds
1956	Dairy/vegetable oil blends marketed
1957	Passes butter in sales
1958	Soybean oil usage passes 1 billion pounds
1960	Becel - no *trans* margarine marketed by Unilever
1964	Soft tub products introduced
1966	Diet margarine introduced
1967	Fluid margarine introduced
1975	Low fat spreads introduced
1981	Butter/margarine blends introduced
1981–1997	Proliferation of lower fat spreads
1990	Nutrition Labeling and Education Act passed by Congress
1991	No-fat margarine appears
1992	Soybean oil usage peaks at 1.97 billion pounds
1995	Benecol introduced in Finland
1998	Cholesterol-lowering spreads introduced, zero *trans*/serving spreads

Table 7.3
Milestones in Margarine/Spread Development[a]

2000	Heart healthy claims allowed by FDA
2003	*Trans* acid content mandated on food labels, effective Jan. 1, 2006
2005	Soybean oil in spreads drops below 1 billion pounds
2006	Jan. 1, *trans* fat labeling becomes mandatory

[a]Adapted from Bumbalough (2000)

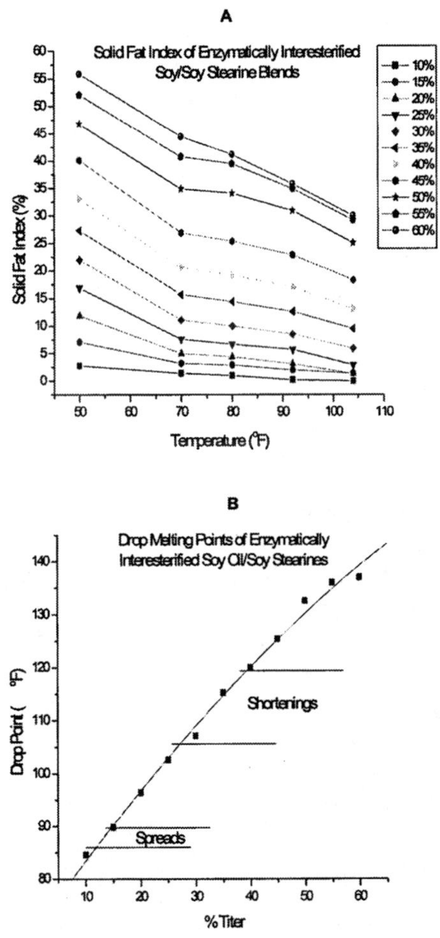

Fig. 7.2. Enzymatic interesterification. Source: ADM.

The enzyme is immobilized on the internal surface of macroporous silica granulates. Immobilization stabilizes the enzyme and is retained on the support such that it does not end up in the product. The enzyme may be reused, making the process more economical. Immobilization makes possible high reaction rates. A series of four reactors varying from 100 – 400 g enzyme/reactor makes it possible to produce 20–100 tons of product/day.

The process may be carried out either in batch or continuous mode. The batch process consists of mixing the enzyme with the oil in a tank until conversion is achieved. The enzyme is then recovered by sedimentation/filtration. The continuous process consists of the enzyme being held in a packed column reactor and the oil is pumped through at a controlled rate. The rate of conversion is dependent on the amount of time the feedstock remains in the reactor column. The continuous method is preferred over the batch method because it allows reuse of the enzyme and minimal cross contamination.

The continuous process may be carried out in single fixed bed or multiple fixed bed reactors connected in series. In single reactor systems, the activity of the enzyme decreases over time and the flow rate must be adjusted in order to achieve conversion. As such, the reaction must be monitored, usually by drop in melting point. In single fixed bed reactors the initial flow rate is quite high, 10 kg oil/kg enzyme/hr. Production declines steeply over time resulting in a low conversion rate at the end of the enzyme's lifetime. Use of multiple fixed bed reactors, each with different amounts of enzymatic activity, permits average production of 1–2 kg oil/kg enzyme/h. To ensure good performance in the process, the enzyme must be uniformly packed into the column because air trapped in the bed affects conversion when reduced.

Enzymatic interesterification qualifies as "green chemistry" since it reduces oil losses and waste products. Compared to chemical interesterification, the enzymatic process has the potential to save 180 million kg of soybean oil, 9 million kg of sodium methoxide, 53 million kg of soaps, 23 million kg of bleaching clay and 225 million liters of water. A further advantage is simplicity coupled with increased efficiency. Lower equipment investments and operational costs are further advantages compared to chemical interesterification.

The major disadvantage of the enzymatic process is the expense of the enzyme and its loss of activity with time. Silica pre-columns installed ahead of the reactors partially alleviate the loss of enzymatic activity (Lee et al., 2003; Lee, 2005). Enzymatic interesterification is a very versatile technology, allowing the formulation of a wide variety of fats with different melting points. By varying the titer or hard fat (i.e., hydrogenated soy or cottonseed flake) content of the feedstock, fats with increasing solid fat contents and melting points can be prepared with this technology (Fig. 7.2).

*Chemical Interesterification/Zero-*Trans *Fats*

Aside from catalytic hydrogenation, interesterification is the oldest fat

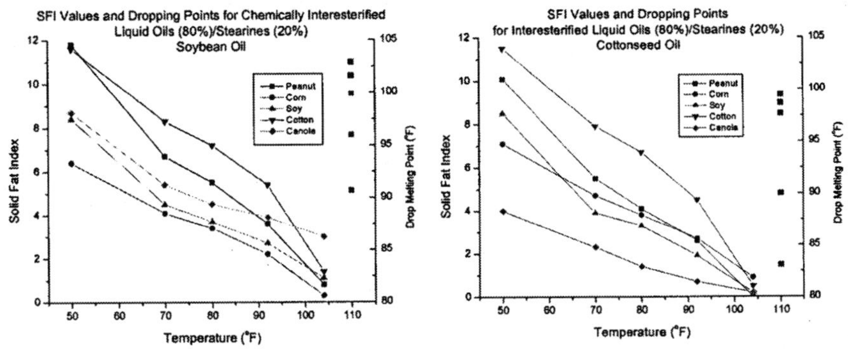

Fig. 7.3. Zero *trans* margarine/spread oils via chemical interesterification of liquid oils/stearines. Source: List 1995a.

modification technique available to the processing industry and dates back to the late 1920s (van Loon, 1930). Typically, the process involves treating a fat blend with sodium methoxide at temperatures ranging from 70–100°C, under a vacuum, with agitation. After the reaction is complete, the catalyst is inactivated with acid, the soaps formed during the reaction are washed out, and the fat is given a post bleach to remove color bodies formed during the reaction. Generally speaking, random interesterification of common vegetable oil raises the melting point, while animal fats remain essentially unchanged (Stirton, 1964). Interesterification of liquid oils with completely hydrogenated stearines, such as soybean or cottonseed, results in a marked lowering of the melting point. For example, a blend of 80% liquid soybean oil with 20% stearine melts at 62.8°C, while after random interesterification, the melting point is lowered to 36.8°C. Another advantage of random interesterification stems from the fact that many fats that crystallize in the *beta* form are changed to the more desirable *beta* prime after rearrangement leading to smoother textures in spreads and shortenings. *Beta* crystal formers tend to form "grainy" products.

Examples of random interesterification to the preparation of zero-*trans* fat, suitable for margarine and spreads, are shown in Fig. 7.3 where the solid fat profiles of interesterified liquid vegetable oil with completely hydrogenated soybean and cottonseed oils are shown. Although the solid fat profiles and melting points of these oils approximate those of traditional hydrogenated oils, limited data suggests that additional liquid oil may be required in the final blend (List et al., 1995a) to assure proper consistency and spreadability in the final product.

Shortenings have higher solid fat and melting point requirements and, to achieve this, the amount of stearine in the blend is increased to 50%. This basestock, when blended with additional liquid oil, allows formulation of all-purpose zero-*trans* shortening oils (Fig. 7.4) (List et al., 1995b).

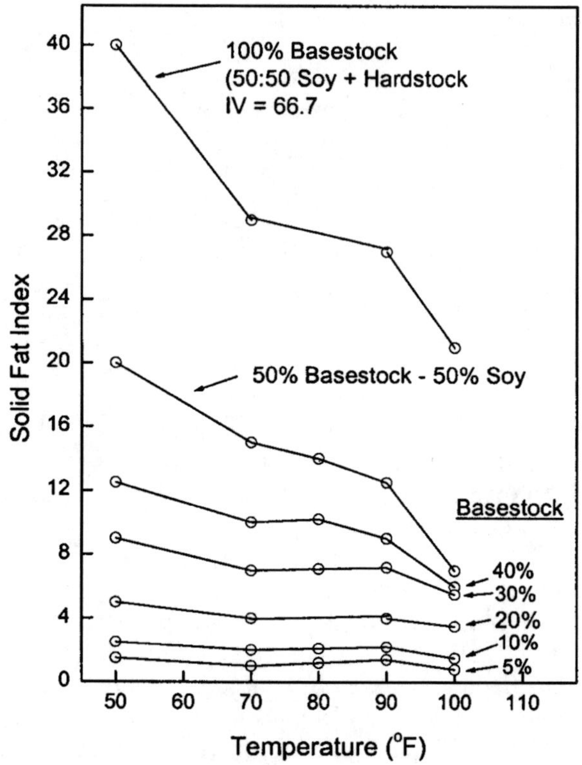

Fig. 7.4. Formulation of zero *trans* shortening oils by blending interesterified basestock with liquid soybean oil (List et al., 1995b).

Low/Zero Trans *Margarines/Spreads/Shortenings*

Zero/low *trans* margarines and spreads have been in the marketplace since the early 1960s (Niewenhuis, 2000) and zero *trans* shortenings date back to the 1880s when blends of lard, tallow and/or vegetable oils first appeared in grocery stores. Use of hydrogenated vegetable oils in margarines first appeared in 1934 and the original Crisco patent for hydrogenated all vegetable oil shortening dates to 1915.

Zero *trans* shortenings were reintroduced into the marketplace in 1998 under the name of "Transend." The product was prepared by the random interesterification of fully hydrogenated soybean oil and liquid canola oil and performed well in baking compared to traditionally formulated shortenings. Applications testing in cookies, biscuits, cakes, and icings deliver comparable performance relative to conventional all-purpose, cake and icing shortening. Other work compared non-*trans* baking shortenings in white cakes and similar flavor, texture, grain, and color were observed. Donuts made with non-*trans* shortening had similar spread, height, and weight and the flavor was judged to be the same.

Flake pie crusts were analyzed for texture and showed similar peak forces for penetration, although the standard crust had softer surfaces and exhibited more band compared to the non-*trans* crusts. The results demonstrated that a wide range of baked goods can be made with non-*trans* shortenings without requiring any changes in formulation and/or processing (Gilbertson et al., 2000).

Stick, pat, tub and liquid margarines can be prepared by blending liquid sunflower oil with interesterified babassu nut oil and saturated palm oil. Stick and pat products require 70–80% liquid oil while soft and liquid products require 80–95% sunflower oil in the blends (Ward, 1982a, 1982b).

Babassu oil is hydrogenated (at 300°F, 0.2% nickel) to an iodine value of less than 2 while palm oil is reduced to a similar IV. The melting points are 113°F (45°C) and 140°F (60°C), respectively. The hard stocks are blended in a proportion of 70:30 babassu:palm and rearranged at 250°F for 15 minutes. The interesterified fat has SFI solids of 36 at 92°F and a Wiley melting point of 109°F. No *trans* fatty acids were detected. Margarine oil prepared by blending 23% hard stock with 77% liquid sunflower had the following solid fat index profile: 13.9 at 50°F, 9.1 at 70°F and 2.4 at 92°F. Margarines prepared from the oil spread easily, were resistant to oil off, and had good mouth melt properties, penetration of 95.1 mm and ASTM grease cone at 45°F.

Soft tub products can be prepared by blending 16% of the hard stock with 84% liquid oil. SFI profile: 9.4 at 50°F, 6.2 at 70°F and 2.0 at 92°F, penetration 160. The soft product had good spreadablity, resistance to oil off and mouth melt.

The VandenBergh patents (Huizinga et al., 1999, 2000; Sassen, 1996) employ interesterification of dry, fractionated palm kernel stearines using a 1–3 specific lipase in a ratio of 55:45. The interesterified blend is then dry fractionated at 33–38°C. Margarine fat blends consist of 5–14% hard stock and 86–95% liquid oil.

Fully hardened high-lauric canola oil can be used to prepare *trans*-free spreads. Blends containing 8–10% of the high-lauric hard stock in sunflower oil have solid fat profiles suitable for soft spreads and have improved melt behavior compared to the use of interesterified hard stocks (Sassen, 1996).

Trans-free hard structural fats can be prepared from dry, fractionated palm and palm kernel stearines modified by interesterification. Margarine oils are prepared by blending 60–95% liquid with 5–40% of the hard, structural fat (Ullanoormadam, 2004).

A summary of patents relating to the preparation of low/zero *trans* fats is given in Table 7.4.

Modified Hydrogenation

Modified hydrogenation technology has shown promise in reducing the *trans* fatty acid content (Eller et al., 2005) of food oils. Among these include hydrogenation in critical fluids (King et al., 2001), pressure controlled hydrogenation (List et al.,

TABLE 7.4
U. S. Patents Relating to Interesterification, Partial Hydrogenation and Blending of Fats and Oils for Low/Zero *Trans* Acids Products

Subject	Oil Components	Inventor	Patent No.	Year
Margarine	Lauric acid oils	Lindsay	2,996,388	1961
Margarine	Lauric/non-lauric acid oils	Babayan	3,268,340	1966
Margarine	Long and short chain fatty acids	Sieden	3,353,964	1967
Margarine	Palm kernel oil	Galenkamp	3,210,197	1965
Margarine	Lauric acid oils	Sieden	3,592,661	1971
Margarine	High PUFA, low *trans*	Melnick	2,921,855	1960
Margarine	Cottonseed stearine	Gooding	3,099,564	1963
Margarine	Safflower, peanut oil	Schmidt	3,240,608	1966
Margarine	High PUFA	Westenberg	3,600,195	1971
Margarine	High PUFA, tropical oil	Fondu	3,634,100	1972
Margarine	High PUFA	McNaught	3,746,551	1973
Low calorie fats	Medium + long chain fatty acids	Sieden	5,288,512	1994
Shortening, filler fat	Hyd. oil, tallow, lard	Kidger	3,244,536	1966
Confectionary	Lauric acid oils, long chain alcohols	Brown	3,512,994	1970
Cake shortening	Lard, vegetable oil, itaconic acid	Gleason	2,966,412	1960
Cake shortening	Animal/vegetable fats, hydroxy carboxylic acids	Gleason	2,970,055	1961
Plastic shortening	Liquid oil, interesterified hardstock	Thompson	3,102,814	1963
All purpose shortening	Low *trans* hydrogenated oil, hardstock	Scavone	5,470,598	1995
Plastic shortening	Liquid oil, hyd. oil, esterified propoxylated glycerine	Mazurek	6,495,188	2002
All purpose shortening	Canola oil, hardstock	Robert	6,033,703	1994
Cake shortening	Animal fats	Steffen	2,855,310	1958
Plastic shortening	Lard, palm oil, vegetable oils	Nelson	2,855,311	1958
Margarine	Low *trans*, High in PUFA	Melnick	2,921,855	1960
Cholesterol lowering spread	Palm, soy, canola, olive	Sundram	5,578,334	1996
Cholesterol lowering spread	Canola, hyd. soy	Mittenen	5,502,045	1996
Sterol fortified food	Sunflower, rapeseed, palm, palm kernel	Amerongon	6,117,475	2000
Margarine hardstock	Soy, canola, sunflower, palm, palm kernel	Huizinga	6,156,370	2000

TABLE 7.5
Properties of Hydrogenated Soybean Oil Basestocks[a]

	Temp	Pressure	I.V.	Reaction Time	*Trans*	Selectivity		Melting Pt.	Fatty Acid Composition (%)					Solid Fat Content (°C)				
Oil	(°C)	(psi)		(min)	(%)	Ln	Lo	(°C)	C16	C18	C18:1	C18:2	C18:3	10	21.1	26.7	33.3	40
A	221	20	65	50	39.7	—	—	41.2	11.3	13.6	75.2	0	0	73.7	54.1	44.7	22.3	3.3
B	140	200	66.2	152	16.6	2.4	4.8	46.4	10.7	22.1	55.4	10.5	0.1	46.5	31.2	27	20.9	9.8
C	150	200	63.6	72	17.4	1.6	5.0	48.2	10.7	23.5	55.4	8.8	0.3	51.4	35	31.6	24.3	12.7
D	160	200	70.2	33	16.8	1.6	5.3	46.1	10.7	19.4	56.4	11.8	0.5	43.7	27.6	23.8	17.6	9.2
E	170	200	68.8	21	17.9	1.5	5.5	48.2	10.7	20.1	56.8	10.8	0.5	46.4	30.4	26.7	19.5	10.3

[a] Cannot be determined

[b] Nickel was used at 0.02% for all hydrogenations (Eller et al., 2005)

2000c), the use of Nobel metal catalysts (Ariaansz, 1998) and conditions known to suppress *trans* fatty acid formation including low temperatures, high pressures and increased catalyst loadings (Hasman, 1995). The following section deals with the use of higher pressures to prepare soybean oils suitable for margarine/spread formulation.

The compositional and physical properties of soybean oil hydrogenated at temperatures ranging from 140–170°C and a hydrogen pressure of 200 psi are shown in Table 7.5. For comparison, a commercially prepared margarine stock (IV 65) is included. Under commercial conditions of 221°C, 20 psi and 0.02% nickel catalyst concentration, extensive *trans* formation occurs. The experimental oils (B-E) show significant reduction in *trans* fatty content of the order of over 50%.

Previous publications from our laboratory showed that a temperature of 120°C and pressures up to 500 psi yields products with even lower TFA content (King et al., 2001). However, under these conditions, slow rates of reaction were observed. At a pressure of 200 psi and temperatures of 140–170°C, the time required to obtain iodine values of 65–70 are reduced to less than 45 minutes and are actually shorter than that observed in commercial practices. Under the conditions employed in the experimental oils (B-E), some selectivity is lost as evidenced by increased stearic acid content and by the fact that 8–10% linoleic acid remains at IVs of about 70. The commercially prepared margarine stock contains little if any linoleic acid at IV 65. Published data indicate that under selective conditions used for margarine, shortening and salad oil production, linolenate selectivity ratios of about 2 are observed (Ariaansz and Okonek, 1998). The experimental oil B-E shows linolenate selectivities of about 1.5–1.6. The solid fat content of commercial margarine stocks exhibit steep, sharply melting profiles with a melting point of about 41°C. The experimental oils (B-E) show melting points in the 46–48°C range largely because of the increased stearic acid content. The solid fat content of the experimental oils (B-E) are somewhat flatter than the commercial sample. While this is undesirable in spreads, would be an advantage in shortening formulations. None the less, these hydrogenated oils appear suitable as soft margarine stocks.

The properties of blends made from the hydrogenated oils (30%) and liquid soybean (70%) oil are shown in Table 7.6. The solid fat content and melting point of the blends qualify them as soft tub margarine and spreads. In addition, the blended oils contain only about 5–6% TFA which qualify them as 0 g *trans* fat/serving spreads at fat levels of 70% or below.

Thus with the acceptable reaction rates observed at higher pressures and lower temperatures along with greater than 56% reduction in TFA formation, it should be possible to prepare low *trans* spread oils via hydrogenation with commercially available nickel catalysts. The basestocks in Table 7.5 (17%) TFA should prove suitable for reduced *trans* shortening as well. Blends of the basestocks, liquid oils, and stearines will produce shortening fats below the usual

TABLE 7.6
Properties of Blended Oils

			Solid Fat Content (°C)					g TFA/a 4 g serving [b]			
Oil Blend[a]	*trans* (%)	Drop Point (°C)	10°C	21.1°C	26.7°C	33.3°C	40°C	80% fat	70% fat	60% fat	50% fat
A	11.9	31.4	17.2	8.6	5	1	0	1.33	1.16	1.00	0.83
B	5.0	37.7	11.1	6.3	5	3.1	1.3	0.56	0.49	0.42	0.35
C	5.2	39.4	11.8	7.9	6.1	3.6	1.8	0.58	0.51	0.47	0.36
D	5.8	36.9	9.5	6	4.4	2.2	0.6	0.65	0.51	0.44	0.41
E	5.4	37.8	10.3	6.6	5.1	2.6	1.5	0.60	0.53	0.45	0.38
Commercial Spread Oils[c]	5.3–14.5	31.6–33.1	10.0–18.0	4.4–8.0	2.8–4.7	0.9–1.7	0				

[a] Oil blends A–E were prepared by mixing 30% hydrogenated basestock with 70% liquid soybean oil.

[b] Fat content for blend shown as percent of blended oil in final product.

[c] Partially hydrogenated/liquid soybean oil based. (Eller et al., 2005)

levels of 20–25% TFA found in commercial products.

In commercial practice, hydrogenation converters can be operated at pressures up to about 50 psi and would require refitting to handle the higher pressures of 200 psi used in this study. However, rapid reaction rates, the benefit of TFA suppression and the ready availability of commercial catalysts suggest that the technology may be effective for producing reduced *trans* oils. A major deterrent is the higher pressures than normally employed in the industry.

*Zero-*Trans *Spreads—Structurally Modified Oilseeds High in Saturated Acids*

Soybean oil having elevated palmitic, stearic and high palmitic/high stearic acid contents were introduced in the 1990s through traditional plant breeding methods. (List et al., 2000b; List et al., 1996) Compared to standard cultivars containing about 15% saturated acids, the high saturate lines contain up to 40% saturates. A number of studies have shown that the high saturate soybean oils show promise as margarine, spread or shortening oils, but may require blending with harder components, such as palm, rearranged palm, or stearine. In their natural states, the high saturate oils have saturated acids esterified in the 1,3 positions leading to sharply melting triglyceride of the SUS configuration (where S=palmitic or stearic, U=unsaturated). As such, these oils possess insufficient solids over the 10–33.3°C temperature range and have melting points well below the 31–32°C ranges observed for hydrogenated oils. Random interesterification raises the melting point and the sold fat content to levels needed for soft margarine (Kok et al., 1999; List et al., 1997). Several laboratory studies have reported the properties of soft margarines made from interesterified, high stearic, and high palmitic/high stearic acid lines (Kok et al., 1999; List et al., 2001). Typically, these products tend to be somewhat harder and less spreadable compared to hydrogenated products, but otherwise, have more acceptable texture and mouth melt properties, resistance to oil/water loss and are resistant to the development of graininess during temperature recycling. Monsanto's Vistive™ line of high stearic soybean oils are expected to be commercialized for food use by the end of the decade and should be useful in spreads, shortenings, and frying applications.

Zero Trans *Spreads/Cholesterol Lowering*

Although zero *trans* spreads have been commercially available in Europe since the 1960s, introduction into the U.S. marketplace has been recent. Properties are shown in Table 7.7. Typically, these products are marketed as cholesterol lowering spreads, either through blending liquid oils and palm oil (Sundram et al., 1998) or from liquid and lightly hydrogenated oils fortified with plant sterols or stannols (Miettinen, 1996; van Amerongen and Cornells, 2000). Smart Balance™ and Smart Balance Omega Plus™ rely upon a patented blend of mono- and polyunsaturated fats including fish oils blended with palm oil to reduce serum cholesterol while Benecol™ and Take Control™ take advantage

TABLE 7.7
Composition and Properties of Cholesterol Lowering and *Trans* Free Spreads

		g Fat/	Fatty Acid Composition (%)[a]							Solid Fat Content (°C)[b]			Melting Point[c]	
		14g							%					Oil Off
Spread	% Fat	Serving	C16	C18	C18:1	C18:2	C18:3	C20:0	*trans*	10°	21.1°	33.3°	(°C)	(%)[d]
Take Control	35	5.0	8.6	6.3	40.8	37.3	0.4	5.3	0	8.0	2.3	0.6	24.7	3.1
Benecol	55	8.0	5.8	5.1	63.1	17.4	6.6	0.2	0	9.0	4.7	0.9	25.6	5.2
Smart Balance	67	9.0	30.3	3.8	35.6	23.9	3.6	0.2	0	9.3	4.0	0.1	25.6	3.8
Promise	60	8.0	6.9	13.0	40.8	49.5	0.6	1.6	0	11.3	4.3	0.2	28.4	6.2

[a]By GLC
[b]By pulsed NMR
[c]By Mettler drop point
[d]4 days at 70°F

of the cholesterol-lowering properties of plant sterols or stannols. Benecol™ was introduced in Finland in 1995 and was instantly successful, commanding a 160 million Euro/year market (Young, 2003) compared to a 60 million Euro/year market in the U.K. and France. Take Control™ and Benecol™ sales in the U.S. were about $60 million/year in the early stages of its debut in America and occupied about 1.5% of the market. Functional spread sales in Australia and New Zealand amounted to 140 million Australian with a 14% market share (Young, 2003).

Benecol (55% fat) is formulated from liquid canola oil, partially hydrogenated soybean oil and plant sterol esters and, when consumed according to the label, lowers LDL up to 14% (Daniels et al., 2001; U.S. Patent 5,502,045). Take Control (35% fat) is formulated from liquid canola and soybean oils, partially hydrogenated soybean oil and is fortified with plant sterol esters (U.S. Patent 6,156,370). Smart Balance has introduced two products (67% fat) into the market, Omega Plus and the Original. Omega Plus is formulated from palm, canola, menhaden, and olive oils and natural plant sterols. The Original is formulated from palm, soybean, canola, and olive oils and contains no plant sterols (U.S. Patents 5,843,497 and 5,578,334).

The Promise spreads (Light 35% fat) and Promise (60% fat) are 0 g *trans* fat and are formulated from liquid soybean, canola, sunflower, palm and palm kernel oils, and partially hydrogenated soybean oils (U.S. Patent 6,156,370). Interesterification technology is employed where interesterified palm and palm kernel stearines are used as the hardstock and blended with liquid sunflower or other vegetable oil to formulate the oil blend for Promise Light. The 60% spread is formulated from hydrogenated and liquid oil components.

Land o' Lakes has entered the zero *trans* market with soft baking and spreadable butter formulated from canola oil and cream.

Con-Agri Foods (Fleischmann's label) has introduced several no-*trans* spreads including an olive oil based (60% fat) and a light spread (31% fat). The former employs a blend of soybean and olive oils as the liquid component and partially hydrogenated soybean oil for the harder component. The light spread is formulated from liquid canola and partially hydrogenated corn oil.

Low Trans *Shortenings*

Low/zero *trans* shortenings (where low ≥ 5% TFA) were introduced in 1998 by Cargill under the trade name of Transend. Formulated from interesterified canola/soybean oils, these products were shown to perform equivalently to conventional shortenings in baking operations. However, increased cost was a major deterrent to their success in the marketplace. However, in view of the TFA labeling regulations, the need for such products is apparent.

In 2002, Bunge Foods introduced low *trans* baking margarine/shortenings into the marketplace. These products are formulated from partially hydrogenated soybean and cottonseed oils and typically contain 4% TFA and 23% saturated acids.

In 2003, Archer Daniels Midland introduced the Nova Lipid line of low *trans* shortenings based on both chemical and enzymatic interesterification of soybean and hydrogenated soybean and cottonseed oils. Products include an all purpose baking shortening, cake shortening, and a baking margarine. TFA contents of 9% permit a 1 g TFA/serving on nutritional labels. Other products in the ADM line include all purpose/pie shortenings, cake and icing shortenings, and all purpose vegetable shortenings formulated from interesterified soybean and cottonseed oils with TFA contents of 4.5–5.5%.

*Performance of Zero/Low-*Trans *Shortenings*

Prior to their introduction into the marketplace, several studies were conducted to compare the performance of low/zero-*trans* baking shortenings to products formulated from hydrogenated oils (Gilbertson et al., 2000). Cake and icing shortenings formulated with a proprietary technique contain 80% less *trans* fatty acids and, when tested in cookies, biscuits, cakes, and icings, delivered comparable performance relative to conventional all-purpose, cake and icing shortenings (Gilbertson et al., 2000). Other studies tested the functionality of non-*trans* shortenings in high ratio white cakes, cake donuts, pie crusts, and other baked goods by substituting the low *trans* shortenings for the hydrogenated products. White cake had similar volumes and symmetry, as well as flavor, texture, grain, and color. Donuts prepared with the non-*trans* shortening had similar spread, height and weight and the flavor, texture and mouth feel were judged to be the same when compared to the standard shortening. Pie crusts made with the two types of shortenings were compared for texture and the analysis showed similar peak forces but the standard crust had softer texture and exhibited more bend that the non-*trans* product (Daniels et al., 2001).

A number of palm oil based non-*trans* bakery shortenings are available commercially and limited published data indicate these products perform well in yellow cake and snap cookie applications. Both plastic and soft plastic formulations scored essentially the same in quality score sensory tests (Kodali and List, 2005).

The literature is replete with information on the performance of zero-*trans* shortenings made from lard and interesterified lard. Lard has long been prized as a shortening for pie crusts. The functional characteristics of pie shortening include plasticity at refrigerated temperatures, it must be loosely bound to provide lubricity, possess high solids for a flaky texture and must be sharply melting in the mouth. Lard, tallow, and blends made from them make them ideal for low-*trans* shortening applications (Kincs, 1985) despite their cholesterol content. In the early 1950s, Armour introduced interesterified lard to improve the shortening properties of lard (Slater, 1953). The standard for a good quality shortening is a volume of 250 cc / 100 g cake. Natural lard rarely gave volumes above 225 cc, with averages between 192–214 cc. Rearranged lard, when blended with a small amount of stearine, consistently yielded cake volumes above 260 cc.

References

Alonzo L.; M.J. Fraga; M. Juarez. Determination of trans Fatty Acids in Margarines Marketed in Spain. *J. Am. Oil Chem. Soc.* **2000,** *77*, 131–136.

Anon. Food labeling: Trans Fatty Acids in Nutrition Labeling, Final Rule. *Federal Register*. **2003,** *68*, 41433–41506.

Anon. ADM Chooses a Trouble-free Process for trans Free Fats. *Oil Mill Gazetteer*. **2004b,** *109*, 2–5.

Ariaansz R.F.; D.V. Okonek. Trans isomer control during edible oil processing. In *Proceedings of World Conference on Oilseed Processing, Vol. 1*. 1998.

Aro. A.; J. Van Amelsvoort; J. Becker; M.A. Van Erp-Baart; A. Kaftos; T. Leth; and G. Van Poppel. Trans fatty acids in dietary fats and oils from 14 European countries: The trans fair study. *J. Food Comp. and Analysis*. **1998,** *11* List, 2004a, 137–149.

Bayard, C.; R.L. Wolff. Trans 18:1 acids in French tub margarines and shortenings. *J. Am. Oil Chem. Soc.* **1995,** *72*, 1465–1489.

Beers, A.; G. Mangus. Hydrogenation of Edible Oils for Reduced trans Fatty Acid Content. *Inform*. **2004a,** *15* (7), 404–405.

Beers, A.; P. Berben; C. Groen; R. Jaqta; G. Lazar; G. Mangus. Lowering the trans content of edible oils. Englehard Technical Bulletin, Englehard Corporation, 2004b.

Berben, P.H.; F. Borninkhof; B.H. Recsink; E. Kuijpers. Production of low trans isomer containing products by hydrogenation. Abstact, American Oil Chemists' Society Meeting, Atlanta, GA. Paper available from Englehard Industries, Beachwood, OH 1994.

Bumbalough, J. Margarine types and preparation technology. In *Introduction to Fats and Oils Technology, 2nd edition*. Eds. O'Brien, R.D.; W. Farr; P.J. Wan. AOCS Press: Champaign, IL, 2000; pp 452–462.

Craig-Schmidt, M. C. Consumption of trans acids. In *trans Fatty Acids in Human Nutri-*

tion. Ed. Sebidio J.-L.; W.W. Christie. The Oily Press: Dundee, Scotland, 1998; pp. 59–113.

Cristol, R. Margarine products cut trans fat content 75%. *Wall Street Journal*, Dec. 16, p A11. 2006

Daniels, R.; T. Riley; N. Higgins. Low trans Shortenings in Bakery Applications. Abst. A53. American Oil Chemists' Society Meeting, Minneapolis, MN, 2001.

Eller, F.J.; G.R. List; J.A. Teel; K.R. Steidley; R.O. Adlof. Preparation of Spread Oils Meeting FDA Labeling Requirements for trans Fatty Acids. *J. Ag. Food Chem.* **2005,** *53*, 5982–5984.

Erickson, M.D. Interesterification. In *Practical Handbook of Soybean Processing and Utilization*. D.R. Erickson, Ed. AOCS Press: Champaign, IL, 1995; pp. 277–296.

Floter, E.; G. VanDujin. Trans free fats for use in food. In *Modifying Lipids for Use in Foods*. F.D. Gunstone, Ed., Woodhead Publishing Ltd.: Cambridge, England, 2006; pp. 492–443.

Gibon, V. Fractionation of lipids for use in foods. In *Modifying Lipids for Use in Foods*. F.D. Gunstone, Ed. Woodhead Publishing Ltd.: Cambridge, England, 2006; pp. 201–233.

Gilbertson, D; D. Lampert; W. Loh. Functional characteristics of non trans bakery shortening. Abst. A63. American Oil Chemists' Society Weekly, San Diego, CA, 2000.

Gunstone, F. D. Oilseed Crops with Modified Fatty Acid Composition. *J. Oleo. Sci.* **2001,** *50*, 269–279.

Hasman, J. Trans Suppression in Hydrogenated Oils. *Inform* **1995,** *6*, 1206–1213.

Hollingsworth, P. Margarine: The Over the Top Functional Food. *Food Technol.* **2004,** *55*, 59–62.

Huizinga, H.; C. Laurentius Sassen; L. Vermass; P. Schur. Process for making a margarine hardstock. U.S. Patent 5,858,445 (1999). Also 6,156,370 (2000) to VanDen-Bergh Foods, Conopco, Lisle, IL.

Hunter, J. E. Alternatives to trans Fatty Acids in Foods. *Inform*. **2004,** *15* (8) 510–512.

Huyghebaert, A.; D. Verhaeghe; H. DeMoor. Fat products using chemical and enzymatic interesterification in fats. In *Food Products*. D.P.J. Moran; K.K. Rajah, Eds. Blackie: London, 1994; pp. 319–345.

Kincs, F. Meat Fat Formulation. *J. Am. Oil Chem. Soc.* **1985,** *62*, 815–818.

King, J.W.; R.L. Holliday; G.R. List; and J.M. Snyder. Hydrogenation of Soybean Oil in Supercritical Carbon Dioxide and Hydrogen. *J. Am. Oil Chem. Soc.* **2001,** *78*, 107–113.

Kodali, D.; List, G.R. Trans fat alternatives. AOCS Press: Champaign, IL; 2005. pp 1–137.

Kok, L.L.; W.R. Fehr; E.G. Hammond; P.J. White. Trans Margarine From Highly Saturated Soybean Oil. *J. Am. Oil. Chem. Soc.* **1999,** *76*, 1175–1181.

Krishnamurthy, R.; M. Kellens. Fractionation and winterization. In *Bailey's Industrial Oil and Fat Products, 5th edition.* Y. H. Hui, Ed. John Wiley and Sons: New York, 1996; pp. 301–337.

Lee, Inmok; R. Sleeter. Method for producing fats or oils. U.S. Patent Appl. 2003/0054509A1. 2003.

Lee, Inmok. Method for producing fats or oils. U.S. Patent Appl. 2005/0014237A1. 2005.

List, G.R.; T. Pelloso; F. Orthoefer; M. Chrysam; T.L. Mounts. Preparation and Properties of Zero trans Soybean Oil Margarines. *J. Am. Oil Chem. Soc.* **1995a,** *72*, 383–384.

List, G. R.; T.L. Mounts; F. Orthoefer; and W.E. Neff. Margarine and Shortening Oils by Interesterification of Liquid and Trisaturated Triglycerides. *J. Am. Oil Chem. Soc.* **1995b,** *72*, 379–382.

List, G.R.; T.L. Mounts; F. Orthoefer; W.E. Neff. Potential Margarine Oils From Genetically Modified Soybeans. *J. Am. Oil Chem. Soc.* **1996,** *73*, 729–732.

List, G.R.; T.L. Mounts; F. Orthoefer; W.E. Neff. Effect of Interesterification on the Structure and Physical Properties of High-stearic Acid Soybean Oils. *J. Am. Oil. Chem. Soc.* **1997,** *74*, 327–329.

List, G.R.; K.R. Steidley; W.E. Neff. Commercial Spreads Formulation, Structures and Properties. *Inform* **2000a,** *11*, 980–986.

List, G.R.; F. Orthoefer; T. Pelloso; K. Warner; and W.E. Neff. Preparation and properties of low trans margarine and oils by interesterification, blending and genetic modification. In *Physical Properties of Fats, Oils and Emulsifiers with Application in Foods*. N. Widlak, Ed. AOCS Press: Champaign, IL., 2000b; pp. 226–237.

List, G.R.; W.E. Neff; R.L. Holliday; J.W. King; R. Holzer. Hydrogenation of Soybean Oil Triglycerides: Effect of Pressure on Selectivity. *J. Am. Oil Chem. Soc.* **2000c,** *77*, 311–314.

List, G.R.; T. Pelloso; F. Orthoefer; K. Warner; W.E. Neff. Soft Margarines From High Stearic Acid Soybean Oils. *J. Am. Oil Chem. Soc.* **2001,** *78*, 103–104.

List, G.R.; K.R. Steidley; W.E. Neff; G. Snowder. Physical Properties of Mutton Tallows. *Grasas y Aceities*. **2003,** *54* (List, 2004a), 81–83.

List, G.R. Decreasing trans and Saturated Fatty Acid Content in Food Oils. *Food Technol.* **2004a,** *58* (Hunter, 2004), 23–31.

List, G.R. Processing and Reformulation for Nutrition Labeling of trans Fatty Acids. *Lipid Technol.* **2004b,** *16* (8), 173–177.

List, G.R.; J.W. King. Hydrogenation of lipids for use in food. In *Modifying Lipids for Use in Foods*. F.D. Gunstone, Ed. Woodhead Publishing Ltd.: Cambridge, UK, 2006; pp. 173–200.

Loh, W. Trans fat reformulation is not a technical objective. In *Trans Fat Alternatives*. D. Kodali; G.R. List, Eds. AOCS Press: Champaign, IL, 2005; pp. 96–105.

Loh, W. Trans Fat Labeling: A Quiet Revolution in the Food Industry. *Inform*. **2006,** *17*, 283–284.

Matsuzaki, H.; T. Okamoto; M. Oyama; T. Mavruyama; I. Niiya; T. Yanagita; M. Sugano. Trans Fatty Acids Marketed in Eleven Countries. *J. Oleo Sci.* **2002,** *51*, 551–565.

Miettinen, T.; H. VanHanen; I. Wester. Use of a stannol fatty acid ester for reducing serum cholesterol. U.S. Patent 5,502,045, 1996, (Benecol).

Moss, J. Labeling of trans Fatty Content in Food, Regulations and Limits – The FDA View. *Artheroslerosis supp. 1*(2), **2006,** 57–59.

Niewenhuis, H. Putting science on the supermarket shelf. In *Scienta Gras, A Select History of Fat Science and Technology*. F.D. Gunstone; D. Firestone, Eds. AOCS Press: Champaign, IL, 2000, pp. 141–148.

Oveson, L.; T. Leth; K. Hansen. Fatty Acid Composition and Contents of trans Monounsaturated Fatty Acids in Frying Fats and Margarines and Shortenings Marketed in Denmark. *J. Am. Oil Chem. Soc.* **1998,** *75*, 1079–1083.

Rajah, K.K. Fat products using fractionation and hydrogenation. In *Fats in Food Products*. D.P.J. Moran; K.K. Rajah, Eds. Blackie Academic and Professional: London, 1994, pp. 277–313.

Ratnayake, W.M.N.; G. Pelletier; R. Hollywood; S. Bacler; D. Leyte. Trans Fatty Acids in Canadian Margarines: Recent Trends. *J. Am. Oil Chem. Soc.* **1998,** *75*, 1587–1594.

Rozendall, A.; A.R. Macrae. Interesterification of oils and fats. In *Lipid Technologies and Applications*. F. Gunstone, Ed. Marcel Dekker: New York, 1997, pp 223–263.

Sassen, C. Plastic fat spread comprising a hard stock. U.S. Patent 5,587,195. 1996.

Satchithanandam, S.; C.J. Oles; C.J. Spease; M.M. Brandt; M.P. Yurawecz; J.I. Rader. Trans, Saturated and Unsaturated Fat in Foods in the United States Prior to Mandatory trans-fat Labeling. *Lipids*. **2004,** *39* (Hunter, 2004), 11–18.

Slater, L.E. Lard Scores Again. *Food Eng*. **1953,** *100*, 72–73.

Stirton, A.J. Fat splitting, esterification and inter esterification. In *Bailey's Industrial Oil and Fat Products.* D. Swern, Ed. Wiley: New York, 1964, pp 931–972.

Sundram, K.; D. Perlman; K.C. Hayes. Increasing the HDL level and the HDL/LDL ratio in human serum by balancing saturated and polyunsaturated fatty acids. U.S. Patent 5,843,497. 1998 (Smart Balance)

Tarrago-Trani, M.T.; K.M. Phillips; L.E. Lamar; J.M. Holden. New and Existing Oils and Fats Used in Products With Reduced trans Fatty Acid Content. *J. Am. Dietetic Assoc*. **2006,** *106*, 867–880.

Tekin, A.; M. Cizmeci; H. Karabacak; M. Kayaban. Trans Fatty Acids and Solid Contents of Margarines Marketed in Turkey. *J. Am. Oil Chem. Soc.* **2002**, 79, 443–445.

Timms, R.E. Fractionation. In *Lipid Technologies and Applications* Gunstone F.. Ed., Marcel Dekker: New York, 1997; pp. 199–222.

Tsanev, R.; A. Russeva; T. Rizov; I.N. Dontcheva. Content of trans Fatty Acids in Edible Margarines. *J. Am. Oil Chem. Soc.* **1998,** *75*, 143–145.

Ullanoormadam, S.R. Trans-free hard structural fat for margarine blend and spreads. U.S. Patent 6,806,737. 2004. To premium vegetable oils berhad. Interesterified palm stearine/dry fractionated palm kernel oil. 60 – 95% liquid oil 5–40% structural fat.

van Amerongen M.P.; L.C. Cornells. Fat based food products comprising sterols. U.S. Patent 6,117,475. 2000. (Take Control)

van Erp-Baart, M.A.; C. Conet; C. Cuadrado; C. Kaftos; A. Stanley; G. Van Poppel. Trans Fatty Acids in Bakery Products From 14 European Countries: The trans Fair Study. *J. Food Comp. and Analysis.* **1998,** *11* (List, 2004a), 161–169.

van Loon, C.H. Process for the conversion of triglycerides with other esters. U.S. Patent 1,744,596. 1930.

van Poppel, G.; M.A. Van Erp-Baart; T. Leth; E. Gevers; J. Van Amelsvoort; D. Petithary-Lanzmann; A. Kaftos; A. Aro. Trans Fatty Acids in Foods in Europe: The trans Fair Study. *J. Food Comp. and Analysis.* **1998,** *11* (List, 2004a), 112–136.

Ward, J. Edible fat product I. U.S. Patent 4,341,812. 1982a.

Ward, J. Edible fat product II. U.S. Patent 4,341,813. 1982b.

Young, J. Introduction. In *Lipids for Functional Foods and Nutraceuticals*. F.D. Gunstone, Ed. Oily Press: Bridgewater, UK, 2003a; pp. 1–23.

Chapter 8

Labeling of *Trans* Fatty Acids

Julie Schrimpf-Moss, PhD, RD[1] and Virginia Wilkening, MS, RD[2]

[1]Food and Drug Administration, College Park, MD

[2]Food and Drug Administration, College Park, MD, Retired—2004

This chapter explains the United States (U.S.) Food and Drug Administration's (FDA or the agency) rationale for mandating the labeling of *trans* fatty acids (*trans* fat). This is the first change to the Nutrition Facts label since the Nutrition Labeling and Education Act of 1990 prompted mandatory nutrition labeling. The scientific basis that supports the labeling of *trans* fatty acids will also be explained. Lastly, this chapter will briefly summarize the *trans* fat regulatory considerations by other countries.

Background

The Nutrition Labeling and Education Act (NLEA) of 1990 amended the Federal Food, Drug, and Cosmetic Act (FFDCA) to provide, in part, that certain nutrients and food components be included in nutrition labeling. In the FFDCA, it states that the Secretary of the Department of Health and Human Services (and the FDA by delegation) can by regulation add or delete nutrients included in the food label or labeling if such action is found necessary to assist consumers in maintaining healthy dietary practices. In 1991, in response to this provision, the agency proposed (DHHS/FDA, 1991) that foods bear nutrition labeling listing certain nutrients and the amount of those nutrients in a serving of the food. Given the scientific knowledge about *trans* fatty acids at the time, FDA did not propose to require *trans* fatty acids to be listed.

In 1993, FDA issued a final rule (DHHS/FDA, 1993a) implementing the 1990 amendments that prescribed how nutrition labeling is to be provided on foods that are regulated by the agency. In that document, the agency required the declaration of total fat and saturated fat in the nutrition label with declaration of both monounsaturated fat and polyunsaturated fat (both defined as *cis* isomers only) required when claims are made about fatty acids and cholesterol. Based on its review of the comments, the agency stated that it was premature to include *trans* fatty acids in nutrition labeling because of a lack of agreement on the health implications of *trans* fatty acid intake. However, the agency acknowledged that it might be necessary to revisit the labeling of *trans* fatty acids in the future.

Trans Fat Labeling Proposed Rule

Nutrition Labeling

FDA received a citizen petition in 1994 (Docket No. 1994P-0036/CP1 and LET3) requesting that the agency initiate rulemaking to mandate the addition of *trans* fat to nutrition labeling. The petitioner stated that an increasing body of evidence suggests that dietary *trans* fatty acids raise blood cholesterol levels, thereby increasing the risk of coronary heart disease (CHD). In response to the petition and the growing science associating *trans* fat intake and risk of CHD, FDA issued a proposed rule in 1999 (the 1999 proposal) (DHHS/FDA, 1999). In the 1999 proposal, FDA proposed to amend its nutrition labeling regulation to require, in part, that the amount of *trans* fatty acids in a food, including dietary supplements, be included when the product contains 0.5 or more grams *trans* fatty acids per serving. Specifically, it proposed the amount of *trans* fat be included in the amount and percent Daily Value (%DV) declared for saturated fatty acids with a footnote indicating the amount of *trans* fatty acids in a serving of the product.

Subsequent to the 1999 proposal, the Institute of Medicine (IOM)/ National Academy of Sciences (NAS) issued a report (IOM/NAS report) (IOM/NAS, 2002) concluding that there is "a positive linear trend" between *trans* fatty acid intake and CHD risk, and recommending that "*trans* fat consumption be as low as possible while consuming a nutritionally adequate diet." The IOM/NAS report did not provide a Dietary Reference Intake (DRI) value for *trans* fat or information sufficient to support the agency establishing a Daily Reference Value (DRV) or other information on the label (i.e., %DV) for *trans* fat. In absence of a basis on which to establish a DV, FDA reopened the comment period of the 1999 proposal (DHHS/FDA, 2002a) to propose to require an asterisk (or other symbol) in the %DV column for *trans* fat that is tied to a similar symbol at the bottom of the Nutrition Facts box that is followed by the footnote statement "Intake of *trans* fat should be as low as possible." The agency stated that the statement was taken from the IOM/NAS report, was consistent with other recent scientific reports (USDA and DHHS, 2000; DHHS/NIH, 2002), and was for the purpose of assisting consumers in understanding the quantitative declaration of *trans* fat in the context of a total daily diet.

Nutrient Content and Health Claims

When FDA determined nutrient content claims for fat, fatty acids and cholesterol in 1993, FDA did not set a *trans* fat criterion for most claims because the evidence suggesting that *trans* fatty acids raise serum cholesterol was inconclusive at that time (DHHS/FDA, 1993b at p. 2332 and 2340). However, FDA set a *trans* fat criterion for the "saturated fat free" claim stating that "because of the uncertainty regarding this issue, the fact that consumers would expect a food bearing a 'saturated fat free' claim to be free of saturated fat and other components that significantly raise serum cholesterol, and the potential importance of a saturated fat free claim, the agency believes that it would be misleading for products that contain measurable amounts of *trans* fatty acids to bear a 'saturated fat free' claim" (DHHS/FDA, 1993b). FDA amended the definition of "saturated fat free" later that year based on comments to require that a food contain less than 0.5 g of *trans* fatty acids in addition to less than 0.5 g saturated fatty acids per reference amount customarily consumed and per labeled serving to be eligible to bear the claim (DHHS/FDA, 1993c).

In the 1999 proposal, FDA concluded that dietary *trans* fatty acids have adverse effects on blood cholesterol measures that are predictive of CHD risk (DHHS/FDA, 1999 at p. 62754). Consequently, to avoid misleading claims, the agency proposed that the amount of *trans* fatty acids be limited wherever saturated fat limits are placed on nutrient content claims, health claims, or disclosure and disqualifying levels. FDA also requested comment on whether "*trans* fat free" claims would help consumers maintain healthy dietary practices and whether they would provide incentive to the food industry to reduce the amount of *trans* fat in the food supply (DHHS/FDA, 1999 at p. 62759). FDA proposed a definition for the "*trans* fat free" claim. FDA concluded that there was no basis for defining the claim "low *trans* fat" without a quantitative recommendation for a daily intake of *trans* fat. Further, FDA did not define a "reduced *trans* fat" claim because it was concerned that a reduced *trans* fat claim would detract from educational messages that emphasize lower intakes of saturated fat. However, based on comments received about "reduced *trans* fat" claims, FDA reopened the comment period for the 1999 proposal on December 5, 2000 (DHHS/FDA, 2000) to consider comments that addressed "reduced *trans* fat" and "reduced saturated and *trans* fat" claims.

Review of the Science

Federal Government and the IOM/NAS

In the November 1999 proposal, FDA reviewed reports published by the U.S. Federal Government and the IOM/NAS. These reports, which were published between 1988 and 1995, showed that conclusions about the role of *trans* fat in raising low density lipoprotein (LDL) cholesterol, the primary risk factor for CHD, and dietary recommendations were evolving as results from new studies became available (NRC/NAS, 1989; DHHS, 1988; DHHS/NIH, 1993; USDA and DHHS, 1995).

Subsequent to the 1999 proposal, three expert panels convened to update, in light of new scientific evidence, the conclusions and recommendations in the reports previously mentioned. The Dietary Guidelines 2000 (USDA and DHHS, 2000) made the following statements regarding *trans* fatty acids and food sources of *trans* fat:

> Foods high in *trans* fatty acids tend to raise blood cholesterol. These foods include those high in partially hydrogenated vegetable oils,

> such as many hard margarines and shortenings. Foods with a high amount of these ingredients include some commercially fried foods and some bakery goods. (P. 28)
>
> Aim for a total fat intake of no more than 30 percent of calories, as recommended in previous editions of the Guidelines. If you need to reduce your fat intake to achieve this level, do so primarily by cutting back on saturated and *trans* fats. (P. 30)
>
> Limit use of solid fats, such as … hard margarines, …and partially hydrogenated shortenings. Use vegetable oil as a substitute. (P. 30)

In the report describing the basis for its recommendation, the Dietary Guidelines Advisory Committee 2000 (DGAC, 2000) suggested that information be provided to help the reader of the Dietary Guidelines 2000 distinguish among the different kinds of fats—saturated, *trans*, and unsaturated. The advisory committee summarized the scientific evidence on *trans* fatty acids as follows:

> *Trans* fatty acids are included because a definitive body of recent experimental evidence indicates that *trans* fatty acids raise the concentration of the most dangerous form of serum cholesterol (LDL-cholesterol). *Trans* fatty acids also tend to lower a protective form of serum cholesterol (HDL-cholesterol). Prospective epidemiological studies further note that higher intakes of *trans* fatty acids are associated with a higher incidence of coronary heart disease. (P. 37)

In 2002, the National Cholesterol Education Program (NCEP) issued a report on the detection, evaluation, and treatment of high blood cholesterol (DHHS/NIH, 2002). The 2002 NCEP report is an evidence-based report that extensively references the scientific literature. The expert panel concluded that:

> *Trans* fatty acids raise serum LDL-cholesterol levels. Through this mechanism, higher intakes of *trans* fatty acids thus should increase risk of CHD. Prospective studies support an association between higher intakes of *trans* fatty acids and CHD incidence (P. V-9)

Based on these conclusions, the Expert Panel recommended for individuals at increased risk for CHD that:

> Intakes of *trans* fatty acids should be kept low. The use of liquid vegetable oil, soft margarine, and *trans* fatty acid-free margarine are encouraged instead of butter, stick margarine, and shortening. (p. V–9)

Lastly, a 2002 report of the IOM/NAS found "a positive linear trend between *trans* fatty acid intake and LDL-cholesterol concentration, and therefore increased risk of CHD" (IOM/NAS, 2002). The IOM/NAS report summarized that this would suggest a Tolerable Upper Intake Level (UL) of zero, but because *trans* fats are unavoidable in ordinary diets and achieving such a UL would require extraordinary changes in dietary intake patterns that might introduce other undesirable effects and unknown health risks, a UL was not proposed. Instead, the report recommended "that *trans* fat consumption be as low as possible while consuming a nutritionally adequate diet."

In summary, the Dietary Guidelines 2000 (USDA and DHHS, 2000), NCEP (DHHS/NIH, 2002) and IOM/NAS (IOM/NAS, 2002) reports, based on current scientific evidence, consistently found that *trans* fatty acids are associated with increased LDL-cholesterol levels and that lower intakes of both saturated and *trans* fatty acids are important dietary factors in reducing the risk of CHD in the general population and for those at increased risk for CHD. Thus, there was strong agreement among the expert panels that the available evidence was sufficiently compelling to conclude that *trans* fat intakes increase CHD risk. Accordingly, these expert panels recommended, in addition to their longstanding recommendations that Americans consume diets limited in saturated fat, that consumers also select food products that are low in *trans* fat. These recommendations were made for the general population (IOM/NAS, 2002; USDA and DHHS, 2000) and persons at increased risk for CHD whose LDL-cholesterol is above goal levels (DHHS/NIH, 2002). Limiting *trans* fat intake along with saturated fat is also a recommendation from other public health sources, such as the World Health Organization, the American Heart Association and the Dietary Guidelines 2005.

Published Studies

To evaluate the evidence that dietary *trans* fat increases the risk of CHD, FDA reviewed the scientific evidence. FDA considered the findings from human studies to constitute evidence that is more directly relevant and persuasive than findings from animal studies. FDA gave greater weight to results from dietary intervention studies than to observational (epidemiological) studies because of an intervention study's ability to provide evidence of a cause-effect relationship. FDA regarded results from observational studies as indirect evidence for a relationship between *trans* fatty acid intake and CHD risk. FDA also reviewed estimates of dietary intake of *trans* fatty acids in the U.S. population.

In the 1999 proposal, FDA evaluated results of 12 dietary intervention studies (Almendingen et al., 1995; Aro et al., 1997; Judd et al., 1994, 1998; Lichenstein et al., 1993, 1999; Mensick & Katan, 1990; Nestel et al., 1992; Noakes and Clifton, 1998; Wood et al., 1993a, 1993b; Zock and Katan, 1992). FDA focused on the physiological measures of serum and plasma LDL-cholesterol concentrations to evaluate whether *trans* fatty acid intakes influence the risk of CHD because such measures are recognized as valid predictors of increased risk for CHD (DHHS/NIH, 1993). FDA concluded that controlled intervention studies in different population groups in the United States and other countries consistently indicate that consumption of diets containing *trans* fatty acids, like diets containing saturated fats, results in increase serum LDL-cholesterol (a major risk factor for CHD) compared with consumption of diets containing *cis*-monounsaturated or *cis*-polyunsaturated fat sources (DHHS/FDA, 1999). The agency also compiled reports of changes in serum total and high-density lipoprotein (HDL) cholesterol and serum lipoproteins to present a more complete picture of serum lipid changes (DHHS/FDA, 1999).

In the 1999 proposal, FDA also reviewed nine publications that examined associations between *trans* fatty acids, serum lipids and CHD endpoints (Aro et al., 1995; Ascherio et al., 1994, 1996; Hu et al., 1997; Kromhout et al., 1995; Pietenin et al., 1997; Roberts et al., 1995; Triosi et al., 1992; Willett et al., 1993). FDA stated that these epidemiological investigations of associations between dietary *trans* fatty acids and risk of CHD must be interpreted cautiously because no dose-response relationship was demonstrated in the studies (DHHS/FDA, 1999). FDA also stated that despite these generally recognized deficiencies in the observational studies, the repeated and consistent findings from these studies showed that consumption of *trans* fatty acids is associated with adverse effects on CHD risk in humans, which supports the findings from intervention studies (DHHS/FDA, 1999).

FDA acknowledged that estimates of dietary intakes of *trans* fat are subject to multiple sources of error. In the 1999 proposal, the agency reviewed intake estimates from three different types of data: (1) national food consumption data, (2) national disappearance data, and (3) observational studies done in U.S. population groups. By examining results from multiple methods of estimating intakes, the agency was able to assess some, but not all, of the uncertainties in current intake estimates. FDA noted the very limited composition data available for the *trans* fatty acid composition of foods and the difficulties in determining the accuracy of reported *trans* intakes with current knowledge and methods (DHHS/FDA, 1999). The available intake estimates suggested that average intakes of U.S. consumers probably fall within the range of 1.3 g–12.8 g/day (DHHS/FDA, 2003a). The agency noted that the scientific basis for and public health importance of *trans* fat labeling was based on the totality of available scientific evidence, including intervention and observational data as well as dietary intake data (DHHS/FDA, 2003a).

Subsequent to the 1999 proposal and before the issuance of the *trans* fat labeling final rule, additional studies on the topic of *trans* fatty acid intakes and CHD risk were published (Denke et al., 2000; de Roos et al., 2001a; de Roos et al., 2001b; Judd et al., 2002; Oomen et al., 2001). FDA reviewed the findings from these new studies to evaluate whether they differed significantly from the findings of studies included in the proposed rule. In general, the results from the more recent intervention and prospective studies were consistent with the results from the studies included in the November 1999 proposal in that they also found that diets containing *trans* fat increased LDL-cholesterol, and, therefore, CHD risk (Judd et al., 2002; de Roos et al., 2001a, 2001b; Denke et al., 2000) and that, in free-living populations, consumption of *trans* fat was associated with increased risk of heart attack and death from CHD (Oomen et al., 2001).

In summary, based on the consistent results across a number of the most persuasive types of study designs (i.e., intervention trials and prospective cohort studies) that were conducted using a range of test conditions and across different geographical regions and populations, the agency concluded that the available evidence for an adverse relationship between *trans* fat intakes and CHD risk was strong. FDA also found the results from the large prospective cohort studies among free-living U.S. population groups to be persuasive evidence that *trans* fat intakes associated with U.S. dietary patterns can have a significant adverse effect on CHD risk for U.S. consumers. Moreover, independent Federal Government expert panels reviewing the same scientific evidence as FDA consistently concluded that *trans* fat intakes are associated with increased CHD risk and recommend that U.S. consumers and those who need to lower their LDL-cholesterol level minimize their intakes of *trans* fat to reduce their risk of CHD. The scientific agreement for this relationship among the various expert groups provided further evidence of the strength of the science and the public health importance of lowering *trans* fat intakes for U.S. consumers. Therefore, FDA

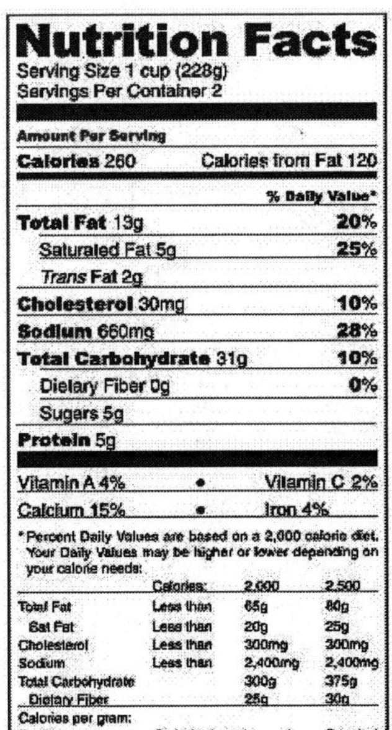

Nutrition Facts
Serving Size 1 cup (228g)
Servings Per Container 2

Amount Per Serving

Calories 260	Calories from Fat 120
	% Daily Value*
Total Fat 13g	**20%**
Saturated Fat 5g	**25%**
Trans Fat 2g	
Cholesterol 30mg	**10%**
Sodium 660mg	**28%**
Total Carbohydrate 31g	**10%**
Dietary Fiber 0g	**0%**
Sugars 5g	
Protein 5g	

Vitamin A 4% • Vitamin C 2%
Calcium 15% • Iron 4%

*Percent Daily Values are based on a 2,000 calorie diet. Your Daily Values may be higher or lower depending on your calorie needs:

	Calories:	2,000	2,500
Total Fat	Less than	65g	80g
Sat Fat	Less than	20g	25g
Cholesterol	Less than	300mg	300mg
Sodium	Less than	2,400mg	2,400mg
Total Carbohydrate		300g	375g
Dietary Fiber		25g	30g

Calories per gram:
Fat 9 • Carbohydrate 4 • Protein 4

Fig. 8.1. An example of a Nutrition Facts label that includes the declaration of *trans* fat.

found the overall weight of scientific evidence in support of this conclusion to be sufficiently compelling to warrant *trans* fat labeling so that consumers could know the *trans* fat content of foods and, thereby, make purchase decisions that take dietary recommendations into account.

Trans Fat Final Rule

On July 11, 2003, FDA issued a final rule requiring the mandatory declaration in the nutrition label of the amount of *trans* fat present in foods, including dietary supplements (DHHS/FDA, 2003a). As a result of the many comments to the 1999 proposal strongly opposing the inclusion of *trans* fat as part of the amount and %DV for saturated fat, the agency required that the declaration of *trans* fat be on a separate line immediately under the declaration for saturated fat. It was anticipated that the declaration of this nutrient on a separate line will help consumers understand that *trans* fat is chemically distinct from saturated fat and will assist them in making dietary choices that aid in maintaining healthy dietary practices. The declaration of *trans* fat is to be indented and expressed as grams per serving to the nearest 0.5 gram increment below 5 grams and to the nearest gram increment above 5 grams (see Fig. 8.1). Since there was no scientific basis for establishing a DV for *trans* fat, the final rule did not require the listing of a %DV as is required for some of the other mandatory nutrients, such as saturated fat.

For the purpose of nutrition labeling, *trans* fats are defined as the sum of all unsaturated fatty acids that contain one or more isolated (i.e., nonconjugated) double bonds in a *trans* configuration. This definition identifies *trans* fatty acids by their chemical structures regardless of origin of the *trans* fatty acid. Under FDA's definition, conjugated linoleic acid would be excluded from the definition of *trans* fat. The definition of *trans* fatty acids, excluding fatty acids with conjugated double bonds, is consistent with the way that *cis* isomers of polyunsaturated fatty acids are defined.

Based on evaluation of more than 1700 comments received, FDA withdrew sections of the 1999 proposal that pertained to the definition of nutrient content claims for "free" and "reduced" levels of *trans* fat and limits on the amounts of *trans* fat wherever saturated fat limits are placed on nutrient content claims, health claims, and disclosure and disqualifying levels. The agency also withdrew the proposed requirement (DHHS/FDA, 2002a) to include a footnote stating: "Intake of *trans* fat should be as low as possible."

The agency set the effective date at January 1, 2006, the next uniform effective date following publication of the final rule in 2003 (DHHS/FDA, 2002b). This is the date by which all foods entering interstate commerce must include *trans* fat in the Nutrition Facts label. This allowed firms more than two years to implement this final rule, providing some regulatory relief and economic savings for small businesses.

The following are some highlights of the *trans* fat final rule (DHHS/FDA,

2003a) that amended title 21 of the Code of Federal Regulations (CFR) part 101.9 Nutrition Labeling of Food:

- The rule states: "The word '*trans*' may be italicized to indicate its Latin origin." The word "may" is purposely used to allow manufacturers to either italicize the word *trans* or not.

- The rule states: "… that label declaration of *trans* fat is not required for products that contain less than 0.5 gram of total fat in a serving if no claims are made about fat, fatty acids or cholesterol. … If the serving contains less than 0.5 gram, the content when declared, shall be expressed as zero. …if a statement of the *trans* fat content is not required and, as a result, not declared, the statement "Not a significant source of *trans* fat" shall be placed at the bottom of the table of nutrient values." Therefore, when a product contains zero gram total fat, there are two options. Either label *trans* fat as zero or use the "Not a significant source…" statement. Similarly, the same options apply for saturated fat when a product contains zero gram total fat.

- The rule states: "(f) The declaration of nutrition information may be presented in the simplified format set forth herein when a food product contains insignificant amounts of eight or more of the following: Calories, total fat, saturated fat, *trans* fat, cholesterol…." FDA changed the rules regarding the simplified format in 21 CFR 101.9(f) to include *trans* fat in the list of mandatory nutrients that must be considered when determining whether a food qualifies for the simplified format. As a result, the number of mandatory nutrients that must be present at insignificant levels changed from seven to eight.

- The rule states: "(g)(5) A food with a label declaration of calories, sugars, total fat, saturated fat, *trans* fat, cholesterol, or sodium shall be deemed to be misbranded under section 403(a) of the act if the nutrient content of the composite is greater than 20 percent in excess of the value for that nutrient declared on the label." FDA changed the rules regarding misbranding in 21 CFR 101.9(g)(5) to include *trans* fat in the list of nutrients that are considered to be misbranded if under declared by more than 20 percent in the nutrition label.

Trans Fat Advance Notice of Proposed Rulemaking (ANPR)

Comments received in response to the 1999 proposal were very diverse. Comments relating to claims indicated strongly opposing views. Comments also raised concerns about the absence of consumer studies to determine how the proposed footnote and claims would be perceived. Hence, FDA concluded that based on information and arguments presented in the comments, it was premature to establish new or revised definitions for nutrient content claims or require the use of the proposed footnote statement in the nutrition label. Instead, FDA issued an ANPR on the same day that the *trans* fat final rule was published (Docket No. 2003N-0076) (DHHS/FDA, 2003b) to request comments and consumer research on: (1) an appropriate basis for establishing qualifying criteria for *trans* fat in *trans* fat nutrient content claims and current nutrient content claims for saturated fat and cholesterol, lean and extra lean claims, and health claims that contain a message about cholesterol-raising lipids as well as disclosure and disqualifying levels; (2) whether such claims mislead consumers about the total fatty acid profile if levels of all cholesterol-raising lipids are not addressed, and, if so, whether qualifiers or disclosure statements would remedy this problem; (3) the use of a footnote or disclosure statement about *trans* fat, either alone or in combination with saturated fat and cholesterol, to enhance consumer understanding; (4) the language that may be appropriate for use in such a footnote or disclosure statement; and (5) the impact of nutrient content or health claims or a footnote or disclosure statement on consumers' food selections.

The agency is now reviewing comments received as a result of the ANPR (DHHS/FDA, 2003b), evaluating data from consumer research studies pertaining to a possible footnote or disclosure statement as well as nutrient content and health claims, and reviewing the current scientific literature. Information obtained will be used by the agency to determine the need for, and content of, further rulemaking to establish criteria for certain nutrient content or health claims, or to require the use of a footnote or disclosure statement or other labeling approaches about one or more cholesterol-raising lipids in the Nutrition Facts label to assist consumers in maintaining healthy dietary practices.

Trans Fat Regulatory Considerations of Other Countries

The Codex Committee on Food Labeling agreed that the declaration of *trans* fat should be left to the discretion of national legislation (FAO/WHO, 2003). Whereas there are no defined nutrient content claims for *trans* fat, the Codex Guidelines for Use of Nutrition Claims state that in the case of the claim for "low in saturated fat", *trans* fatty acids should be taken into account where applicable. This provision consequentially applies to foods claimed to be "low in cholesterol" and "cholesterol free" (FAO/WHO, 2001).

In addition to the United States, other countries have considered *trans* fat labeling. Beginning in December 2005, most food products sold in Canada must include the absolute amount of *trans* fat in grams per serving of the food as well as the %DV for *trans* fat and saturated fat combined. The Canadian regulations also define nutrient content claims and a health claim for *trans* fat (Canadian Department of Health, 2003). As of 2006, the Mercosur countries (including Brazil, Argentina, Paraguay, and Uruguay) will require mandatory nutrition labeling that includes a *trans* fat declaration (ANVISA, 2003).

In the European Union, the Council Directives on nutrition labeling of food stuffs provides for voluntary nutrition labeling except where a nutrition claim is made (Council of European Communities, 1990). Currently, there is no provision for the declaration of *trans* fat, but *trans* fat intake and the effects on human health have been explored (European Food Safety Authority, 2004). The Food Standards Agency of the United Kingdom allows claims regarding *trans* fat, in which case their amount must be declared with nutrition labeling (FSA, 2003 and date unknown).

The Australia and New Zealand Food Standards Agency currently requires the declaration of *trans* fat when a claim is made for cholesterol, saturated fat, *trans* fat, polyunsaturated fat, or monounsaturated fat (Food Standards Australia New Zealand, 2002). And, South Africa has proposals that will require the mandatory declaration of *trans* fat, among other nutrients, whenever a nutrient or health claim is made, in addition to the declaration for the nutrient that is the subject of the claim (South Africa, 2002).

Besides labeling, another approach to decrease the intake of *trans* fat is a restriction on the level of *trans* fat in the food supply. Beginning in June 2003 under Order No. 160 of the Food Stuffs Act, Denmark limited the amount of *trans* fat in foods by prohibiting oils and fats containing >2% of *trans* fat/100 g. More recently, the Canadian House of Commons requested Health Canada and the Heart and Stoke Foundation of Canada to develop recommendations and strategies to eliminate or reduce processed *trans* fat to the lowest level possible. The final report recommends limiting the total *trans* fat amount to 2% of total fat content in vegetable oils and tub margarines and to 5% of total fat content in all other foods sold to consumers or used as an ingredient (*Trans* Fat Task Force, 2006). If these recommendations are implemented, Canada will be the second country to restrict *trans* fat content in the food supply.

References

Almendingen, K.; O. Jordal; P. Kierulf; B. Sandstad; J.I. Pedersen. Effects of Partially Hydrogenated Fish Oil, Partially Hydrogenated Soybean Oil, and Butter on Serum Lipoproteins and Lp(a) in Men. *Journal of Lipid Research*, **1995,** *36*, 1370–1384.

ANVISA (Agência Nacional de Vigilância Sanitária), National Agency for Sanitary Monitoring Brazil. Mercosul Regulamento Téchnico Sobre Rotulagem Nutricional De Alimentos Embalados. RDC N° 360, December 23, 2003.

Aro, A.; A.F.M. Kardinaal; I. Salminen; J.D. Kark; R.A. Riemersma; M. Delgado-Rodriguez; J. Gomez-Aracena; J.K. Huttunen; L. Kohlmeier; B.C. Martin; J.M Martin-Moreno; V.P. Mazaev; J. Ringstad; M. Thamin; P. van't Veer; F.J. Kok. Adipose Tissue Isomeric *trans* Fatty Acids and Risk of Myocardial Infarction in Nine Countries: The EURAMIC study. *Lancet* **1995,** *345* (8945), 273–278.

Aro, A.; M. Jauhiainen; R. Partanen; I. Salminen; M. Mutanen. Stearic Acid, *trans* Fatty Acids, and Dairy Fat: Effects on Serum and Lipoprotein Lipids, Apolipoproteins, Lipoprotein(a), and Lipid Transfer Proteins in Healthy Subjects. *American Journal of Clinical Nutrition*, **1997,** *65*, 1419–1426.

Ascherio, A.; C.H. Hennekens; J.E. Burling; C. Master; M.J. Stampfer; W.C. Willett. *Trans*-fatty Acids Intake and Risk of Myocardial Infarction. *Circulation* **1994,** *89*, 94–101.

Ascherio, A.; E.B. Rimm; E.L. Giovannucci; D. Spiegelman; M. Stampfer; W.C. Willett. Dietary Fat and Risk of Coronary Heart Disease in Men: Cohort Follow Up Study in the United States. *British Medical Journal* **1996,** *313*, 84–90.

Canadian Department of Health. Regulations Amending the Food and Drug Regulations (Nutrition Labelling, Nutrient Content Claims and Health Claims). *Canada Gazette, Part II*, Volume 137, Number 1, page 154, January 1, 2003.

Council of European Communities. Council Directives 90/496/EEC on Nutrition Labelling for Foodstuffs of 24 September 1990. *Official Journal L 276*, October 6. 1990.

de Roos, N.M.; M.L. Bots; M.B. Katan. Replacement of Dietary Saturated Fatty Acids by *trans* Fatty Acids Lowers Serum HDL Cholesterol and Impairs Endothelial Function in Healthy Men and Women. *Arteriosclerosis, Thrombosis, and Vascular Biology* **2001a,** *21*, 1233–7.

de Roos, N.M.; E.G. Schouten; M.B. Katan. Consumption of a Solid Fat Rich in Lauric Acid Results in a More Favorable Serum Lipid Profile in Healthy Men and Women Than Consumption of a Solid Fat Rich in *trans*-fatty Acids. *Journal of Nutrition*, **2001b,** *131*, 242–245.

Denke, M.A.; B. Adams-Huet; B.S. Nguyen. Individual Cholesterol Variation in Response to a Margarine- or Butter-based Diet—a Study in Families. *JAMA* **2000,** *284*, 2740–2747.

Department of Health and Human Services (DHHS). The Surgeon General's Report on Nutrition and Health. Washington, DC, 1988, p. 96. (Internet address: http://www.surgeongeneral.gov/library/reports.htm.)

Department of Health and Human Services (DHHS)/ Food and Drug Administration (FDA). Food Labeling; Reference daily intakes and daily reference values; mandatory status of nutrition labeling and nutrient content revision: Proposed Rule. Volume 56 Federal Register, p. 60366, November 27, 1991.

DHHS/FDA. Food Labeling: Nutrient content claims, general principles, petitions, definition of terms; definition of nutrient content claims for fat, fatty acid, and cholesterol content of food: Final Rule. Volume 58 Federal Register, p. 2302, January 6, 1993a.

DHHS/FDA. Food Labeling; Mandatory status of nutrition labeling and nutrient content revision, format for nutrition label: Final Rule. Volume 58 Federal Register, p. 2079, January 6, 1993b.

DHHS/FDA. Food Labeling: Nutrient content claims, general principles, petitions, definition of terms; definitions of nutrient content claims for the fat, fatty acid, and cholesterol content of foods; food standards: requirements for foods named by use of a nutrient content claim and a standardized term; technical amendment: Final Rule. Volume 58 Federal Register, p. 44020, August 18, 1993c.

DHHS/FDA. Food Labeling: *Trans* fatty acids in nutrition labeling, nutrient content claims, and health claims: Proposed Rule. Volume 64 Federal Register, p. 62746, November 17, 1999.

DHHS/FDA. Food Labeling: *Trans* fatty acids in nutrition labeling, nutrient content claims, and health claims: Proposed Rule, reopening of the comment period. Volume 65 Federal Register, p. 75887, December 5, 2000.

DHHS/FDA. Food Labeling: *Trans* fatty acids in nutrition labeling, nutrient content claims, and health claims: Proposed Rule, reopening of the comment period. Volume 67 Federal Register, p. 69171, November 15, 2002a.

DHHS/FDA. Uniform labeling compliance date for food labeling regulations: Final Rule. Volume 67 Federal Register, p. 79851, December 31, 2002b.

DHHS/FDA. Food Labeling: *Trans* fatty acids in nutrition labeling; nutrient content claims, and health claims: Final Rule. Volume 68 Federal Register, p. 41433, July 11, 2003a.

DHHS/FDA. Food Labeling: *Trans* fatty acids in nutrition labeling; consumer research to consider nutrient content and health claims and possible footnote or disclosure statements: Advance Notice of Proposed Rulemaking. Volume 68 Federal Register, p. 41507, July 11, 2003b.

DHHS/NIH. The second report of the expert panel on detection, evaluation and treatment of high blood cholesterol in adults (adult treatment panel II (ATPII)), National Cholesterol Education Program. Bethesda, MD, September 1993.

DHHS/NIH. Expert panel on detection, evaluation, and treatment of high blood cholesterol in adults, third report of the National Cholesterol Education Program (NCEP) expert panel on detection, evaluation, and treatment of high blood cholesterol in adults (adult treatment panel III), Chapter II, "Rationale for intervention" and Chapter V "Adopting healthful lifestyle habits to lower LDL cholesterol and reduce CHD risk." 2002 (Internet address: http://www.NHLBI.nih.gov).

Dietary Guidelines Advisory Committee (DGAC). Report of the dietary guidelines advisory committee on the dietary guidelines for Americans: To the secretary of health and human services and the secretary of agriculture. U.S. Department of Agriculture, Washington DC, 2000, p. 1–37. (Internet address: http://www.health.gov/dietaryguidelines/dgac/pdf/dgac_ful.pdf)

European Food Safety Authority. Opinion of the Scientific Panel on Dietetic Products, Nutrition and Allergies on a Request from the Commission Related to the Presence of *Trans* Fatty Acids in Foods and the Effect on Human Health of the Consumption of *Trans* Fatty Acids. Request No. EFSA-Q-2003-022. *The EFSA Journal* 81: 1-49, 2004.

FAO/WHO, Joint FAO/WHO Food Standards Programme, Codex Alimentarius. Codex Guidelines for Use of Nutrition Claims, CAC/GL 23-1997 (revised 2001). Food and Agriculture Organization of the United Nations and the World Health Organization. Rome, 2001. (Internet address: http://www.fao.org/DOCREP/005/Y2770E/y2770e00.htm)

FAO/WHO, Food and Agriculture Organization of the United Nations and the World Health Organization. Codex Alimentarius Commission. Report of the Thirty-First Session of the Codex Committee of Food Labelling. Alinorm 03/22A, 2003. (Internet address: ftp://ftp.fao.org/codex/alinorm03/al0322Ae.pdf)

Food Standards Australia New Zealand. Food Standards Code. Part 1.2 Labelling and Other Information Requirements. Standard 1.2.8 Nutrition Information Requirement, 2002. (Internet address: http://www.foodstandards.gov.au/foodstandardscode/)

FSA, United Kingdom Food Standards Agency. The Food Labelling Regulations 1996: Guidance Notes. 2003. (Internet address: http://www.food.gov.uk/foodindustry/guidancenotes/labelregsguidance/foodlabelregsguid)

FSA, United Kingdom Food Standards Agency. Fats Sugars Salts: Do *Trans* Fats Have to be Mentioned on Food Labels? Date unknown (Internet address: http://www.eatwell.gov.uk/asksam/healthydiet/fssq/#A218418)

Hu, F.B.; J. Meir; M.J. Stampfer; J.E. Manson; E. Rimm; G.A. Colditz; B.A. Rosner; C.H. Hennekens; W.C. Willett. Dietary Fat Intake and the Risk of Coronary Heart Disease in Women. *New England Journal of Medicine* **1997,** *337*, 1491–1499.

Institute of Medicine (IOM)/ National Academy of Sciences (NAS). Dietary reference intakes for energy, carbohydrate, fiber, fat, protein and amino acids (Macronutrients). Chapters 8 and 11. National Academy Press: Washington, DC, 2002; p. 335–432.

Judd, J.T.; B.A. Clevidence; R.A. Muesing; J. Wittes; M.E. Sunkin; J.J. Podczasy. Dietary *trans* Fatty Acids: Effects on Plasma Lipids and Lipoproteins of Healthy Men and Women. *American Journal of Clinical Nutrition*, **1994,** *59*, 861–868.

Judd, J.T.; D.J. Baer; B.A. Clevidence; R.A. Muesing; S.C. Chen; J.A. Weststrate; G.W. Meijer; J. Wittes; A.H. Lichenstein; M. Vilella-Bach; E.J. Schaefer. Effects of Margarine Compared With Those of Butter on Blood Lipid Profiles Related to Cardiovascular Disease Risk Factors in Normolipemic Adults Fed Controlled Diets. *American Journal of Clinical Nutrition*, **1998,** *68*, 768–777.

Judd, J.T.; D.J. Baer; B.A. Clevidence; P. Kris-Etherton; R.A. Muesing; M. Iwane. Dietary *cis* and *trans* Monounsaturated and Saturated Fatty Acids and Plasma Lipids and Lipoproteins in Men. *Lipids*, **2002,** *37*, 123–131.

Kromhout, D.; A. Menotti; B. Bloemberg; C. Aravanis; H. Blackburn; R. Buzina; A. Dontas; F. Fidanza; S. Giampaoli; A. Jansen; et al. Dietary Saturated and *trans* Fatty Acids and Cholesterol and 25-year Mortality From Coronary Heart Disease: The Seven Countries Study. *Preventive Medicine*, **1995,** *24*, 308–315.

Lichtenstein, A.H.; L.M. Ausman; W. Carrasco; J. Jenner; J.M. Ordovas; E.J. Schaefer. Hydrogenation Impairs the Hypolipidemic Effect of Corn Oil in Humans: Hydrogenation, *trans* Fatty Acids, and Plasma Lipids. *Arteriosclerosis and Thrombosis*, **1993,** *13*, 154–161.

Lichtenstein, A.H.; L.M. Ausman; S.M. Jalbert; E.J. Schaefer. Effects of Different Forms of Dietary Hydrogenated Fats on Serum Lipoprotein Cholesterol Levels. *New England Journal of Medicine*, **1999,** *340*, 1933–1940.

Mensink, R.P.; M.B. Katan. Effect of Dietary *trans* Fatty Acids on High-density and Low-density Lipoprotein Cholesterol Levels in Healthy Subjects. *New England Journal of Medicine*, **1990,** *323*, 439–445.

National Research Council (NRC)/National Academy of Sciences (NAS). *Diet and health, implications for reducing chronic disease risk.* National Academy Press: Washington, DC, 1989; p. 4, 8, 42, 193, 196, 213, and 657.

Nestel, P.J.; M. Noakes; G.B. Belling; R. MacArthur; P. Clifton; E. Janus; M. Abbey. Plasma Lipoprotein Lipid and Lp(a) Changes with Substitution of Elaidic Acid for Oleic Acid in the Diet. *Journal of Lipid Research*, **1992,** *33*, 1029–1036.

Noakes, M.; P.M. Clifton. Oil Blends Containing Partially Hydrogenated or Interesterified Fats: Differential Effects on Plasma Lipids. *American Journal of Clinical Nutrition*, **1998,** *68*, 242–247.

Oomen, C.M.; M.C. Ocke; E.J. Feskens; M.-A. van Erp-Baart; F.J. Kok; D. Kromhout. Association Between *trans* Fatty Acid Intake and 10-year Risk of Coronary Heart Disease in the Zutphen Elderly Study: A Prospective Population-based Study.

Lancet, **2001,** *357*, 746–51.

Pietenin, P.; A. Ascherio; P. Korhonen; A.M. Hartman; W.C. Willett; D. Algmanes; J. Virtamo. Intake of Fatty Acids and Risk of Coronary Heart Disease in a Cohort of Finnish Men: The alpha-tocopherol, beta-carotene Cancer Prevention Study. *American Journal of Epidemiology*, **1997,** *145*, 876–887.

Roberts, T.L.; D.A. Wood; R.A. Riemersma; P.J. Gallagher; F.C. Lampe. *Trans* Isomers of Oleic and Linoleic Acids in Adipose Tissue and Sudden Cardiac Death. *Lancet*, **1995,** *345* (8945), 278–282.

South Africa, Minister of Health, Draft Regulations Relating to Labelling and Advertising of Foodstuffs. No. R 1055, August 8, 2002. (Internet address: http://www.doh.gov.za/docs/regulations/2002/reg1055.html)

Trans Fat Task Force. *Trans*forming the Food Supply. June 2006. (Internet address: http://www.hc-sc.gc.ca/fn-an/nutrition/gras-*trans*-fats/tf-ge/tf-gt_rep-rap_e.html)

Troisi, R.; W.C. Willett; S.T. Weiss. *Trans* Fatty Acid Intake in Relation to Serum Lipid Concentrations in Adult Men. *American Journal of Clinical Nutrition*, **1992,** *56*, 1019–1024.

U.S. Department of Agriculture (USDA) and DHHS. Nutrition and Your Health: Dietary Guidelines for Americans. 4th ed. Washington DC, Home and Garden Bulletin No. 232, 1995. Internet address: http://www.health.gov/dietaryguidelines/dga95/default.htm)

USDA and DHHS. Nutrition and Your Health: Dietary Guidelines for Americans. 5th ed. Washington DC, Home and Garden Bulletin No. 232, 2000. (Internet address: http://www.health.gov/dietaryguidelines/)

Willett, W.C.; M.J. Stampfer; J.E. Manson; G.A. Colditz; F.E. Speizer; B.A. Rosner; L.A. Sampson; C.H. Hennekens. Intake of *trans* Fatty Acids and Risk of Coronary Heart Disease Among Women. *Lancet*, **1993,** *341* (8845), 581–585.

Wood, R.; K. Kubena; B. O'Brien; S. Tseng; G. Martin. Effect of Butter, mono- and Polyunsaturated Fatty Acid-enriched Butter, *trans* Fatty Acid Margarine, and Zero *trans* Fatty Acid Margarine on Serum Lipids and Lipoproteins in Healthy Men. *Journal of Lipid Research*, **1993a,** *34*, 1–11.

Wood, R.; K. Kubena; S. Tseng; G. Martin; R. Crook. Effect of Palm Oil, Margarine, Butter, and Sunflower Oil on the Serum Lipids and Lipoproteins of Normocholesterolemic Middle-aged Men. *Journal of Nutritional Biochemistry*, **1993b,** *4*, 286–297, .

Zock, P.L.; M.B. Katan. Hydrogenation Alternatives: Effects of *trans* Fatty Acids and Stearic Acid versus Linoleic Acid on Serum Lipids and Lipoproteins in Humans. *Journal of Lipid Research*, **1992,** *33*, 399–410.

Chapter 9

Analysis and Characterization of *Trans* Isomers by Silver-Ion HPLC

Richard Adlof and Gary R. List

Food and Industrial Oil Research, National Center for Agricultural Utilization Research, USDA, Agricultural Research Service, Peoria, IL 61604 USA

Introduction

Produced by the partial hydrogenation of vegetable oils and present in animal fats and selected seed oils, *trans* fatty acids [both configurational (*cis* versus *trans*) and positional (double bond location may vary from the $\Delta 8$ to the $\Delta 15$ position on the fatty acid carbon chain) isomers (Slover et al., 1985; Mossoba, et al., 1991; Ulberth and Henniger, 1992; Ratnayake, 1998) continue to be an integral, if somewhat controversial "bad" (Wood, 2004; Katan, 2000; de Roos et al., 2003; Dlouhy, et al., 2003; Lamaitre, et al., 2002), "not bad" (Clifton et al., 2004), and/or "inconclusive" (Dyerberg et al., 2004; Sanders et al., 2003) part of the fats and oils industry, and silver ion high performance liquid chromatography (Ag-HPLC) continues to be an important part in their analysis and/or isolation (Fritsche and Steinhart, 1997). Other silver-based applications include thin layer chromatography (TLC; Ratnayake, 2004a; Molkentin and Precht, 2004), reverse-phase HPLC (Juanéda, 2002; Dugo, et al., 2004), and silver ions attached to membranes (Ozawa et al., 2001; Kim et al., 2000). (For reviews utilizing Ag-HPLC for analysis of *trans* isomers from seed oils see Cruz-Hernandez et al., 2004, Nikolova-Damyanova, 2003, Ratnayake, 2004a and 2004b, Ledoux et al., 2000, and Steinhart et al., 2002.) Ag-HPLC, applied to the separation and isolation of *trans*-monoenoic fatty acid isomers (with the individual positional isomers of the isolated *cis* and *trans* fractions converted into suitable derivatives before further characterization by GC or GC/MS), has contributed significantly to better understanding the role of *trans* fatty acids in human nutrition and metabolism (Firestone and Shepard, 1992; Ratnayake, 2001). Due to the large number of isomers generated during the hydrogenation process (Ratnayake and Pelletier, 1992), attempts to determine the *trans* monoenoic (18:1) fatty acid isomer distribution in a given animal or vegetable oil formulation by analysis of the triacylglycerols (TAG) continues to present problems (Christie, 2003; Evershed, 1993). A review by Nikolova-Damyanova (2003) provides a comprehensive overview of silver ion chromatography, including LC, TLC, HPLC and theory. In this chapter we will review Ag-HPLC history, theory, current applications (including advantages and limitations) of Ag-HPLC for the analysis and/ or isolation of *trans* fatty acids, as acids, esters and triacylglycerols, as well as possible future applications for this powerful methodology.

History

Mikes and co-workers (1973) utilized a Corasil© silica gel column coated with 0.8% silver nitrate and 1.75% ethylene glycol to partially separate methyl elaidate from methyl oleate. One of the shortcomings of silver nitrate/ silica systems is the gradual loss of silver ions when polar solvents such as methanol or acetonitrile are used and often result in loss of column efficiency and detector sensitivity. (The latter due to slow Ag coating of the HPLC detector cells.) The primary advantage of silver ion chromatography, with the silver ions bound to a substrate, is retention of the substrate silver ions, even after months of service. Thus in the mid-1970s, Rohm & Haas of Philadelphia, PA, USA, introduced Amberlyst© XN1010 resin, an ion-exchange resin (a cross-linked, macroreticular, polystyrene-divinylbenzene co-polymer with sulfonic acid groups attached) which, after the addition of Ag^+ ions, could be applied to the analytical and/ or preparative separation of *cis* and *trans* FAME isomers (Adlof, et al., 1994a), TAG (Adlof and Emken,1981) and phospholipids (Adlof, 1991). The resin (received in 16–50 mesh/ sulfonic acid form) was first ground in a burr mill, then wet-sieved to obtain a useful (60–80 (180–250μ), 100–120 (120–145μ), or finer mesh) resin fraction, and the sulfonic acid protons were then completely (Emken et al., 1964) or partially [partial argentation resin chromatography (PARC); (Adlof et al., 1980a, 1980b, 1981)] exchanged with silver ions. The high silver loading (28% or so by weight of resin) allowed the separation/purification of 25–30gm batches of *cis* and *trans* fatty acid esters on a 5cm x 30 cm glass column (Adlof, et al., 1978). DeJarlais and coworkers (1983) suggested acetonitrile as solvent for silver resin chromatography. Unfortunately, XN1010 production ceased in the early 1980s. While bulk quantities of this resin are no longer available, alternate sources (Phenomenex, Torrance, CA, USA; Shodex, Colorado Springs, CO, USA; Agilent Technologies, Folsom, CA, USA) of cationic exchange resins exist (see Future).

In the mid-1980s, Christie (1987a) prepared an Ag-HPLC column by exchanging the H^+ protons of a commercially-available HPLC column packed with 5–10μ Nucleosil SA© (Machery-Nagel, Düren, Germany; sulfonic acid protons on a silica backbone) with silver (Ag^+) ions. A solvent system (developed by Christie,1987a) composed of dichloromethane, dichloroethane and small amounts (0.01%–0.025%) of acetonitrile (ACN) was used with both the Nucleosil® and Chrompack® HPLC columns. The use of UV detectors was precluded unless phenacyl derivatives of the fatty acids were used (Christie, 1987b; personal communication) and others (Ratnayake, 1998) noted a gradual loss in resolution sometimes after only 6 months of column use. Whether this loss of resolution was due to silver ion removal by the ACN (W.W. Christie and Breckenridge, 1989, and private conversations with W.W.C.), to the use of chlorinated solvents (silver salt formation?), or column contamination by sample or solvent impurities remains unclear.

Chrompack (now Varian-Chrompack), Middelburg, The Netherlands, in the late 1980s introduced an HPLC column (the ChromSpher™ Lipids HPLC column) already containing silver ions. Adlof and Lamm (1998) were one of the first groups to utilize these commercially available columns. In order to avoid the use of the more hazardous chlorinated solvents employed by Christie and other groups, Adlof utilized a simple binary mobile phase of acetonitrile in hexane. The number of solvents compatible with the silver ion columns and UV detectors (200–210 nanometers) is limited. Hexane, methanol, isopropyl alcohol, ethyl ether, and acetonitrile (ACN) solvent systems were studied, with the isocratic solvent system of ACN in hexane as the final choice. Sample retention and resolution could be readily controlled by careful adjustment of the percentage of ACN in the hexane. (Table 9.1). The volatility of the hexane tended to promote bubble formation in the HPLC pump(s), but baseline stability was achieved by initial degassing of the solvents in an ultrasonic bath, and the addition of a magnetic stirrer, a helium sparger, and a small fan, the latter to cool the solvent pump mixing solenoid.

The isocratic solvent system of ACN (0.3–1.5%) in hexane (Fig. 9.1) was utilized to separate a wide variety of fatty acids (as FAME) by:

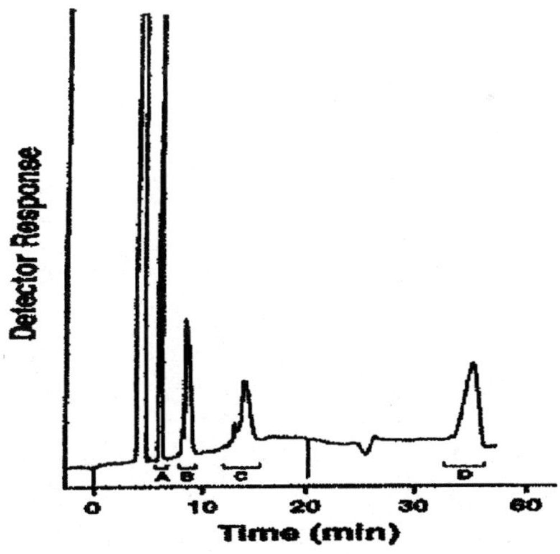

Fig. 9.1. Analysis of partially hydrogenated vegetable oil methyl esters by Ag-HPLC (Adlof, 1995). Single column, 20μg sample, 1.0 mL/ min 0.15% ACN in hexane, RI detector. Fractions: A, saturates; B, *trans*-18:1; C, *cis*-18:1; D, 18:2.

TABLE 9.1
%ACN versus Sample Retention
System: Dual-Column Ag-HPLC; 1.0 mL/ min: ACN in hexane Solvent
TAG std: OPO (O = *cis*-9-18:1; P = 16:0)

Sample	%ACN	Elution time (min.)
1	0.5	37
2	0.6	32
3	0.8	28
4	1.0	21
5	1.2	17

i. Geometry/ configuration (*cis* versus *trans*),

ii. The Location, and/or

iii. The number of double bonds.

Due to the low solubility of ACN in hexane (ca. 1.5% at ambient temperature (Adlof, 1994a), the procedure is less suitable for solvent gradients. To enable gradients and improve the resolution for fatty acid derivatives, heptane has been substituted for hexane (Elfman-Borjesson et al., 1997), or by the addition to hexane of dichloromethane (Momchilova et al., 1998), isopropanol [for triacylglycerols (Han, ct al., 1999)] or diethyl ether (EE; Delmonte et al., 2004b) as mediators. A relative comparison of sample retention (OPO TAG std.) with %ACN (isocratic solvent system) are tabularized in Table 9.1. (Please note: Elution times were collected over a 3-year period on a number of different Ag-HPLC columns, and are included for comparison purposes only.)

Theory

A number of hypotheses [see Nikolova-Damyanova and Momchilova, (2002), Nikolova-Damyanova (2003), and Frenking and Frolich (2000) and the references cited therein] have been applied to explain the separation of fatty acid isomer derivatives and TAG by Ag-HPLC. Most agree the separation is due to the formation of a 2-center bond with the pi-electrons of the FA unsaturation and the unpaired electrons of the carbonyl oxygen in the ester moiety. Some (Gunstone et al., 1967, Gunstone and Lie Ken Jie, 1970) suggest the two centers interact separately, the unsaturation with the Ag ion and the oxygen electrons of the esters with the unreacted polar sites of the Ag-HPLC silica substrate. Other data (Kasai et al., 1980) suggests that the silver ions may interact with two ethylenic molecules. The FA or TAG molecule must adopt a conformation which minimizes the distance between the reaction sites (the unsaturated bond(s) and the Ag^+ ion(s)) and the distance between the polar sites and the silver ions in the adsorbent.

Solvents like benzene, toluene and acetonitrile are assumed to interact predominantly with the silver ions, thus reducing their interaction with the double bonds of the substrate (Heath et al., 1977; Powell, 1982) and resulting in more rapid elution of the sample. Acetonitrile, which complexes strongly with silver ions (thus displacing unsaturated solutes), is of special value in Ag-HPLC on ion exchange media (DeJarlais et al., 1983) and may be the most- often used mobile phase modifier. In contrast, methanol, isopropanol and acetone may act by reducing the strength of the interaction between the solute and residual silanol groups and other polar moieties (Powell, 1981) on the silica-based substrate. While the addition of either acetonitrile or methanol/ isopropanol to hexane- or dichloromethane-based mobile phases for analysis (as phenacyl esters) of a series of fatty acid isomers resulted in similar separation/elution patterns, isopropanol was found to reverse the elution order of several of these difficult-to-resolve isomers (Momchilova and Nikolova-Damyanova, 2000b). These solvent/ solute/ substrate interactions can be manipulated to yield highly selective resolution of complex lipid species (FA, FA esters or TAG) differing only in the number, configuration and/or position of double bonds. (Solvents of high purity and HPLC grade should, of course, be used.) The presence of water can also significantly effect resolution, so solvents should be anhydrous or, under humid laboratory conditions, should be stored over drying agents such as magnesium or sodium sulfate.

Factors: Sample Retention, Ag-HPLC

Acyclic FA, FA Esters

i. *Cis* isomers are retained significantly more strongly than *trans* isomers. Retention is assumed to be (steric argument) directly related to the stability of the Ag/double bond complex formed (de Ligny, 1976), both isomers are retained more strongly than compounds with a triple bond.

ii. Location of the unsaturation on the carbon backbone of the fatty acid may also impact retention order, yielding the "sinusoidal curve" observed by Gunstone, et al. (1967) during analysis (as FAME) of the $\Delta 2-\Delta 18$ positional isomers of methyl *cis*- and *trans*-octadecenoates.

iii. Sample retention decreases as the number of substituents on the double bond increases (steric effects) in the order $R{\cdot}CH{=}CH_2 > R_2C{=}CH2 > R_2{\cdot}C{=}CH{\cdot}R > R_2C{=}CR_2$ (de Ligny, 1976; Nonadek, 1991) but, sample retention increases when the hydrogen atoms on the double bond(s) of the molecule ($R{\cdot}CH{=}CH{\cdot}R$) are replaced with deuterium or tritium atoms. The latter result may be due in part to the greater basicity of the molecule (Nonadek, 1991) or the decrease in steric hindrance due to the shorter bond length between the carbon and deuterium versus carbon and hydrogen atoms (Adlof, 1994a).

iv. In dienes, retention increases with increasing distance between the double bonds (with conjugated double bonds being held less strongly than methylene-interrupted double bonds). The most stable diene complexes are formed by 1,5-diene systems (Nikolova-Damyanova et al., 1992).

v. Methylene-interrupted trienes are retained more strongly than dienes, which are retained much more strongly than monoenes.

vi. Sample retention decreases with increasing chain-length, with longer-chain length FA having the same unsaturation being retained less-strongly and thus separable ["Normal-phase" effects (Adlof, 1997)].

For TAG: The three FA (Acyclic Compounds) per TAG Molecule are Assumed to Obey the Same Configuration/Retention Rules as Listed Above, Plus:

i. Fatty acids in the 1(3)-position of the TAG molecule are assumed to be equal and to exert stronger influence (steric effects) on retention than the FA in the 2-position (Fig. 9.2).

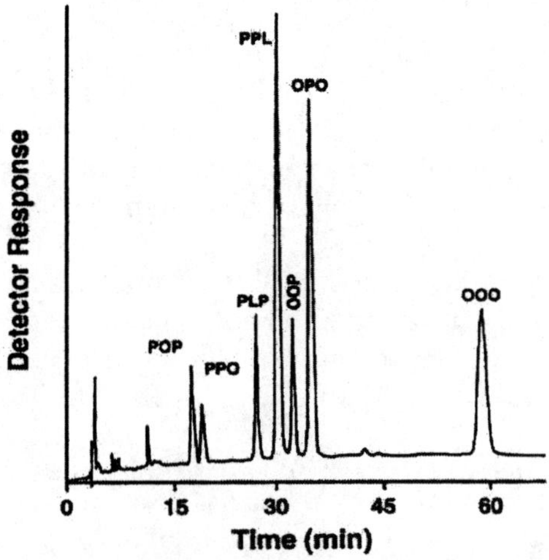

Fig. 9.2. Analysis of TAG positional isomer mix (Adlof, 1995) containing POP, PPO, OOP, OPO, PLP, PPL and OOO (12mg/ mL isooctane) by dual column Ag-HPLC. Sample size: 12 µg. Flow rate: 1.0 ml/min 0.5% ACN in hexane. UV detector at 206 nm. Chart speed: 10cm/ hour. (Reproduced by kind permission of Journal of High Resolution Chromatography and redrawn from the original).

ii. As noted for the acyclic compounds above, sample retention increases with increasing number of double bonds in TAG FA (where trilinolein = 6 double bonds), but the pattern is further complicated by contributing factor iii below.

iii. To a lesser extent, TAG retention decreases with increasing carbon number, again due to "Normal-phase" effects (Adlof, 1997), with SSS eluting more rapidly than AcAcAc (S = 18:0 and Ac = 2:0).

As noted above, the elution order of TAG composed of different FA (mixed TAG) is significantly more difficult to predict. The retention of individual, mixed FA TAGs is dependent not only on the number of double bonds per FA molecule, but also on the location of the FA molecule on the TAG glycerol backbone and, perhaps, even on the proximity of one unsaturated FA to another. As demonstrated previously (Adlof, 1995; Fevrier et al., 2001), the FA(s) located on the 1(3)-positions of a TAG exert a greater influence than the FA located on the 2-position.

Thus PPO is more strongly retained than POP (P = 16:0; O = *cis*-9-18:1). While this type of information should allow the prediction of TAG elution orders with Ag-HPLC, contributions by normal phase (Adlof, 1997) and temperature effects (Adlof and List, 2004), may significantly alter elution patterns. A complete understanding of the individual contributions and interactions of these complex factors has not yet been achieved.

Technology

Column(s)

While a single column has been used to fractionate fifteen of the sixteen *cis*/*trans* isomers of isomerized methyl arachidonate (*cis*-5,*cis*-8,*cis*-11,*cis*-14-20:4; Adlof, 1994a), improved resolution can usually be achieved by connecting two (or more) Ag-HPLC columns in series and, more importantly, by decreasing sample sizes to 0.5 μg or less (Adlof, 1995). A UV detector was required for the analysis of the positional 18:1 FAME isomers due to the UV detector's greater sensitivity when compared with refractive index or flame ionization detectors. Based on first principles, UV responses for monoenoic FAME were assumed (and later proven; Adlof 1994b) to be the same whether the double bond was located from the Δ8 to the Δ15 position on the carbon chain. The Ag-HPLC elution orders of the 18:1 positional isomers (*cis* and *trans*) were found to be the reverse of those obtained by SP2560 capillary GC (Adlof et al., 1995). As many as six Ag-HPLC columns linked in series (Sehat et al., 1999) have been utilized to maximize separation(s).

The Ag-HPLC method offers several other advantages, including rapid (15–20 min. run times) analysis and complete separation of the *cis* and *trans*

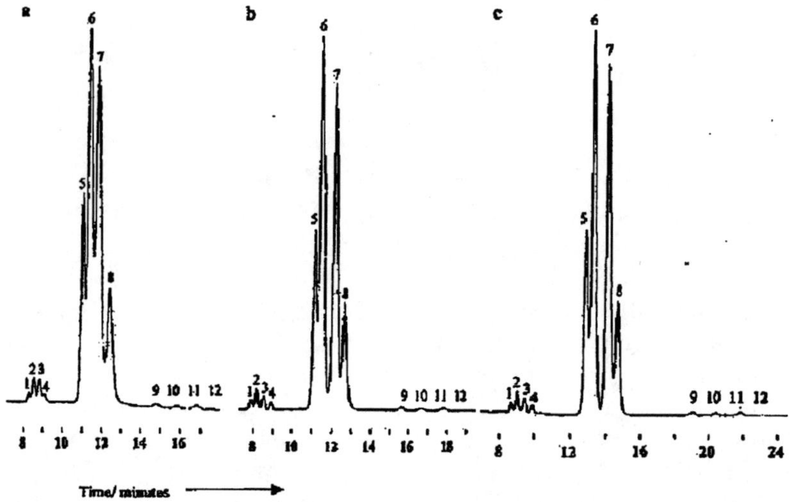

Fig. 9.3. Optimized silver ion chromatograms of standard CLA-based isomer mixtures (Cross, et.al., 2000) on single Lipids © Ag-HPLC column, obtained at 1.0 mL/ min, using a variety of hexane-based (isocratic) solvent systems, for analysis of a) CLA methyl esters, eluted with 0.1% ACN in hexane, b) Underivatized CLA, eluted with 0.025% ACN/ 2.5% acetic acid in hexane, and c) Underivatized CLA, eluted with 3.0% acetic acid in hexane. The individual isomers are designated: 1=11*t*,13*t*; 2=10*t*,12*t*; 3=9*t*,11*t*; 4=8*t*,10*t*; 5=11,13*c*,*t*/*t*,*c*; 6=10,12*c*,*t*/*t*,*c*; 7=9,11*c*,*t*/*t*,*c*; 8=8,10*c*,*t*/*t*,*c*; 9=11*c*,13*c*; 10=10*c*,12*c*; 11=9*c*,11*c*; 12=8*c*,10*c*. (Reproduced by kind permission of Journal of Chromatography A and redrawn from the original).

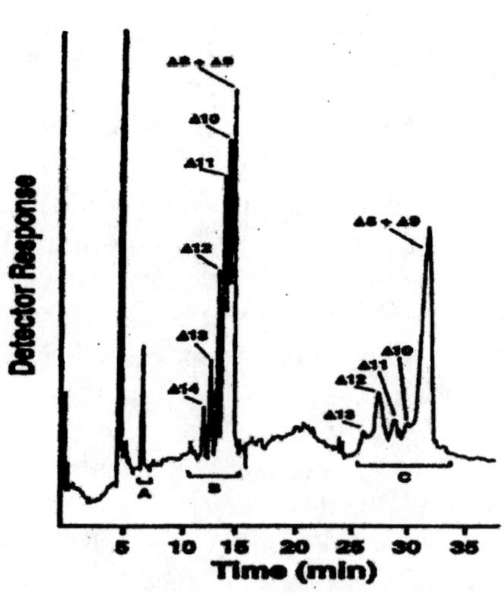

Fig. 9.4. Analysis of partially hydrogenated vegetable FAME by Ag-HPLC (Adlof, 1995). Sample size 0.4 µg; UV detection at 206 nm. Solvent flow rate: 1.0 mL/min 0.08% ACN in hexane. Fractions: A = Saturates, B = *trans*-18:1, C = *cis*-18:1.

18:1 FAME isomers with no *cis*/*trans* overlap. The complete separation of *cis* and *trans* isomers (Fig. 9.5) can be used to provide quantitative data for positional isomers not separated by GC. But Ag-HPLC is more limited than GC as a stand-alone method for the study of individual, positional FAME isomers. This technology is therefore a good alternative to silver nitrate TLC for preliminary isomer fractionation before GC analysis and as a method for isolation of most of the 18:1 positional isomers.

Specific Applications

Fatty Acids

Cross and co-workers (Ostrowska et al., 2000; Cross et al., 2000) recently utilized Ag-HPLC to analyze CLA as free fatty acids. The separation was performed again on ChromSpher Lipids column with a solvent system of 2.5% acetic acid and 0.025% acetonitrile in hexane and detection at 234 nm. This approach had the advantage that the intermediate steps of production and purification of derivatives (and potential sample losses, artifact formation and possible double bond migration related to these procedures) were omitted. And, unlike FA esters or TAG, the free fatty acids contributed yet another retention factor. As noted by the authors (Cross et al., 2000), "There can be little doubt that the additional strong CLA-stationary phase interactions not displayed by the CLAMEs (methyl esters) are carboxylic acid-silanol (residual/on the column) hydrogen bonding." Ostrowska and co-workers (2000) compared (Fig. 9.3) the quantification and elution patterns of CLA isomers, as FA versus FAME, and utilizing several solvent systems.

Fatty Acid Esters (Methyl, Phenacyl, Methoxyphenacyl, etc.)

The hexane/ACN solvent system offers several advantages over the dichloromethane/ dichloroethane/ACN solvent systems described by Christie and others for analytical, silver ion HPLC. The UV detector is the most common detector available in laboratory HPLC systems and the methyl ester form of fatty acids is the most common form for analysis by GC, GC/MS, and thin layer chromatography. The UV extinction coefficients for dichloromethane (233 nm) and dichloroethane (226 nm) require the use of mass, evaporative light scattering, or flame ionization detectors, or the conversion of the FA or FAME to phenacyl derivatives for analysis by UV. (Samples can then be converted to methyl esters for analysis by GC or GC/MS.) The disposal costs of chlorinated solvents are also higher than for hexane. The hexane/ACN solvent system could be used with UV detection at 206 or 210 nm, the desired fraction (*trans*, for instance) isolated and, after removal of the solvent(s), analysis (as FAME) by GC. Addition of an internal standard such as 17:0 FAME would allow quantitation of *trans* or other desired FA isomers (Ulberth and Buchgraber, 2003; Daun and Cantrill, 2003). As illustrated in Fig. 9.1, FA isomers can be rapidly fractionated by configuration (*cis*/*trans*) and/or number of double bonds. Retention of monoenoic fatty acid

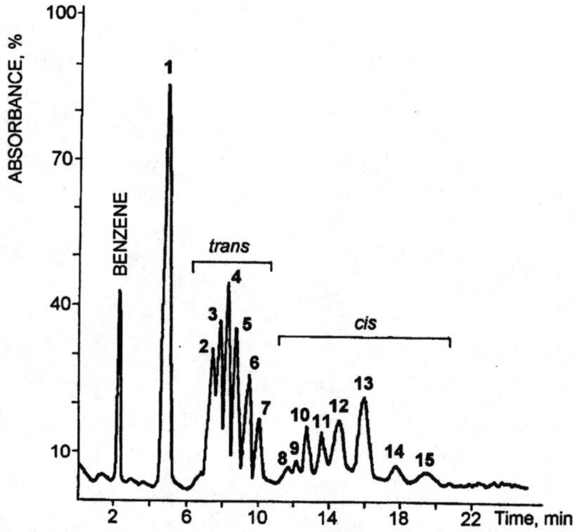

Fig. 9.5. Separation of isomeric fatty acids in partially hydrogenated sunflower oil as *p*-methoxyphenacyl esters on silver loaded Nucleosil™ 5SA column (Momchilova et.al., 1998). Isocratic elution, hexane-dichloromethane-acetonitrile, 60:40:0.2 (v/v/v); UV detection at 245 nm. 1 = saturated, 2 = unknown, 3 = 11-*trans*-18:1, 4 = unknown, 5 = 9-*trans*-18:1; 6 = unknown; 7 = 6-*trans*-18:1; 8 = 13-*cis*-18:1; 9 = 12-*cis*-18:1; 10 = 15-*cis*-18:1; 11 = 11-*cis*-18:1; 12 = unknown; 13 = *cis*-9-18:1; 14 = unknown; 15 = 6-*cis* and 7-*cis*-18:1. (Reproduced by kind permission of Journal of Chromatography A and redrawn from the original).

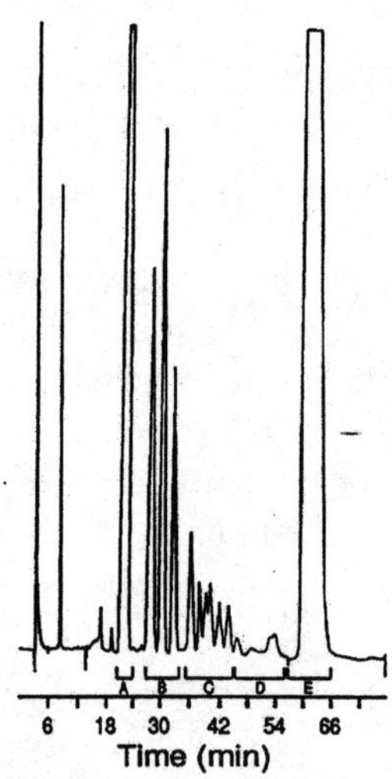

Fig. 9.6. Separation of isomerized methyl arachidonate. Sample size: 100 µg. Flow rate: 1.0 mL/min 0.125% ACN in hexane. UV detection at 210 nm. Group A = (4-*trans*); B = (3-*trans*, 1-*cis*); C = (2-*trans*, 2-*cis*); D = (1-*trans*, 3-*cis*); E = (4-*cis*). (Reproduced by kind permission of Journal of Chromatography A and redrawn from the original).

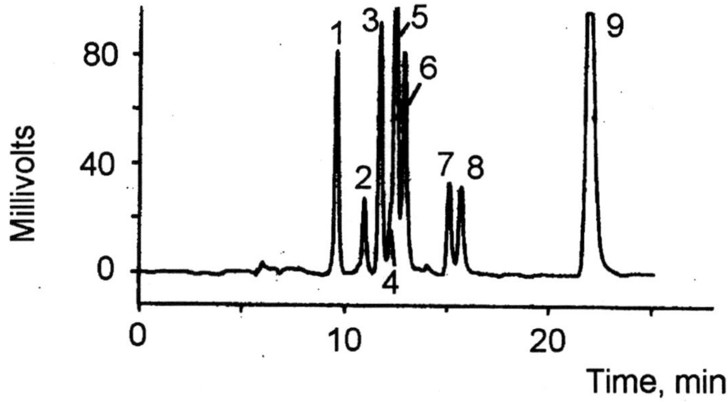

Fig. 9.7. Separation of a complex mixture of isomeric fatty acid methyl esters by silver ion HPLC on Chromspher Lipids column (Adlof and Lamm, 1998). Isocratic elution with 0.5% acetonitrile in hexane, detection at 206 nm. 1=9-*trans*-18:1; 2=9-*trans*, 11-*trans*-18:2; 3, 9-*trans*, 12-*trans*-18:2, 4=9-*trans*, 11-*cis*-18:2; 5=9-*cis*, 11-*trans*-18:2; 6=9-*cis*-18:1; 7=9-*trans*, 12-*cis*-18:2; 8=9-*cis*, 12-*trans*-18:2; 9=9-*cis*, 12-*cis*-18:2. (Reproduced by kind permission of the authors and Journal of Chromatography A and redrawn from the original).

esters is also dependent on the location/position of the FA double bond(s). By careful control of the ACN in hexane solvent system and injection of very small sample sizes, individual *cis*- and *trans*-positional FAME isomers could be separated on a commercially-available Ag-HPLC column without *cis-trans* overlap (Fig. 9.4; Adlof et al., 1995).

As noted before, the elution orders are the reverse of those obtained on GC. Positionally isomeric 18:1 fatty acids (as FAME) have also been resolved on a silver nitrate-impregnated column of silica gel (Bataglia and Frohlich, 1980), and the elution presented (according to increasing retention) as *trans* 13-, 12-, 11-, 10-, 9-, 8-, 7-18:1 and *cis* 12-, 11-, 10-, 9-, 8-18:1. In another example (Fig. 9.5), Momchilova and co-workers (1998) utilized a silver-loaded Nucleosil™ 5SA column to separate (as p-methoxyphenacyl esters) the isomeric fatty acids in partially hydrogenated sunflower oil.

The power of Ag-HPLC is further illustrated/exemplified by its utilization (a single Ag-HPLC column; isocratic ACN in hexane as solvent) for the analysis of isomerized methyl 20:4 (Adlof, 1994a). Resolution of 15 of the 16 possible *cis*/*trans* 20:4 isomers far exceeds the capabilities of current GC, HPLC or other analytical methodologies. Again, the eluted isomers are grouped by total *cis* and *trans* double bonds, with an elution pattern of A (4 *trans* isomers), B (3 *trans*, 1 *cis*), C (2 *trans*, 2 *cis*), D (1 *trans*, 3 *cis*) and E (4 *cis*). At 0.15% ACN in hexane, the 20:4 isomers could be separated within 25 minutes. With this solvent system, all four (3-*trans*, 1-*cis*) isomers could be separated, but one of the other isomers was unresolved. These results may be contrasted to earlier work by Lanser

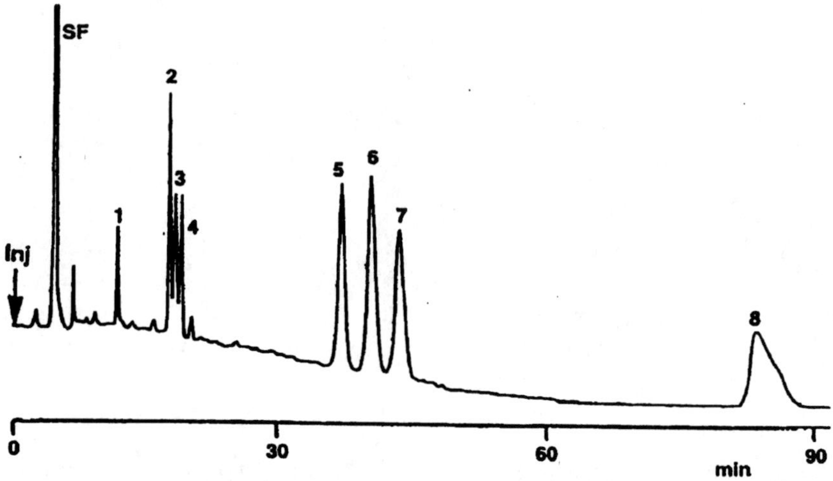

Fig. 9.8. Separation of all configurational isomers of 18:3 fatty acid as phenacyl esters by silver ion HPLC on silver loaded Nucleosil™ 5SA column (Juanéda et al., 1994). Gradient elution with dichloroethane (A) and methanol (B) starting with 95% A-5% B and changed to 75% A-25% B over 30 min, maintaining this for 15 min and then changing to 50% A-50% B over 5 min and maintaining this for additional 40 min. Detection at 238 nm. 1, *trans*-9, *trans*-12- , *trans*-15-18:3; 2, *trans*-9, *cis*-12, *trans*-15-18:3; 3, *cis*-9, *trans*-12, *trans*-15-18:3; 4, *trans*-9, *trans*-12, *cis*-5-18:3; 5, *cis*-9-*cis*, *trans*-12, *cis*-15-18:3; 6, *trans*-9, *cis*-12, *cis*-15-18:3; 7, *cis*-9, *cis*-12, *trans*-15-18:3; 8, *cis*-9, *cis*-12, *cis*-15-18:3. (Reproduced by kind permission of Journal of High Resoution Chromatography and redrawn from the original).

and Emken (1983), who separated isomerized methyl arachidonate into only 5 fractions on an XE254 silver ion column.

Relative Ag-HPLC retention times (two ChromSpher Lipids columns connected in series and an isocratic solvent system of 0.5% ACN in hexane) for a series of *trans* FAME is "summarized" in Fig. 9.7 (Adlof and Lamm, 1998), by resolution of a "*trans*" reference mixture of *trans*-9-18:1 > *trans*-9,*trans*-11-18:2 > *trans*-9,*trans*12-18:2 > *trans*-9,*cis*-11-18:2~*cis*-9,*trans*-11-18:2 > *cis*-9-18:1 >> *trans*-9,*cis*-12-18:2 > *cis*-9,*trans*-12-18:2 >>> *cis*-9,*cis*-12-18:2 (eluting in this order), baseline resolution achieved within 25 minutes.

While analysis of the fatty acids as methyl esters allows their analysis by Ag-HPLC and, analysis by GC after isolation, the utilization of FA in phenacyl or methoxyphenacyl forms often resulted in improved separation of *cis*/*trans* isomers (see Conjugated FA Esters for CLA) by Ag-HPLC (Momchilova et al., 1998, Momchilova and Nikolova-Damyanova, 2000a, 2000b, 2002a, 2002b). The strong UV absorption of the aromatic ester derivatives allows direct quantitation (UV detector) of the eluted components (Christie and Breckenridge, 1989), and the phenacyl or methoxyphenacyl esters could then be converted to FAME for analysis by GC or GC/MS.

Partial separation of linoleic and linolenic acid isomers as phenacyl esters was demonstrated by Christie and Breckenridge (1989) on a Nucleosil 5SA column

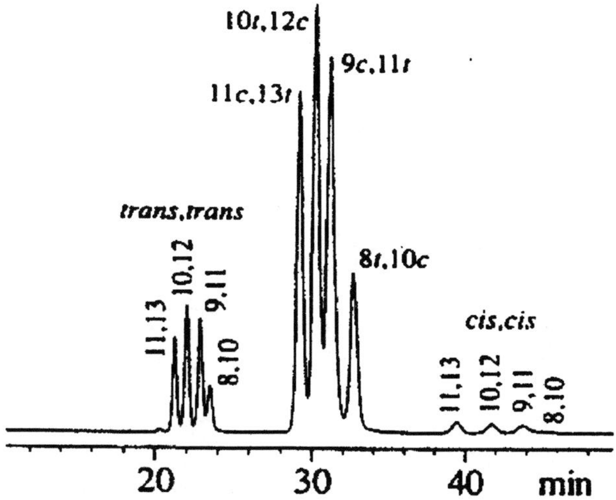

Fig. 9.9. Analysis of CLA isomers (as FAME) in a commercial CLA formulation by dual-column Ag-HPLC (Sehat et.al., 1999). Solvent: 0.1% ACN in hexane, photodiodarray detector.

loaded with silver ions by isocratic elution with dichloromethane-dichloroethane-acetonitrile, 49:75:0.5, v/v/v. Later, all eight *trans*-9, *trans*-12, *trans*-15-18:3 >> *trans*-9,*cis*-12,*trans*-15-18:3 > *cis*-9,*trans*-12,*trans*-15-18:3 > *trans*-9,*trans*-12,*cis*-15-18:3 >>> *cis*-9,*trans*-12,*cis*-15-18:3 > *trans*-9,*cis*-12,*cis*-15-18:3 > *cis*-9,*cis*-12,*trans*-15-18:3 >>>> *cis*-9,*cis*-12,*cis*-15-18:3 configurational isomers of linolenic acid (as phenacyl esters, eluting in this order) were fully resolved (Fig. 9.8) utilizing the same column with a linear gradient of 5–50% methanol in dichloromethane (*v*:*v*) over 30 min (Juanéda et al., 1994).

Conjugated FA Esters

Conjugated linoleic acid (CLA; *cis*-9,*trans*-11-octadecadienoic acid), present in dairy products and beef, has been shown in animals to have anticarcinogenic, growth-promoting, antiatherogenic, anti-diabetic and lean body mass-enhancing properties (Adlof, 2001 and references cited therein), with the *cis*-9, *trans*-11 isomer considered responsible for CLA's anti-carcinogenic properties (and an intermediate step in *cis*-9,*cis*-12-18:2 peroxidation (Smedman et al., 2004), and the *trans*-10,*cis*-12 isomer responsible for the observed weight loss/muscle-mass enhancement effects (Park et al., 1999; Gavino et al., 2000). Ag-HPLC has become a powerful technology for the analysis of the large number of conjugated *trans* isomers found in commercial CLA formulations and the analysis of CLA precursors such as vaccenic (*trans*-11-18:1) acid (Adlof et al., 2000). An example of CLA isomer analysis in a commercial CLA formulation (as FAME) by dual-column Ag-HPLC (Sehat et al., 1999) is illustrated in Fig. 9.9.

Since the positional isomers (as FAME) were only partially resolved on a single column, up to six Ag-HPLC columns were connected in series (Sehat et al.,

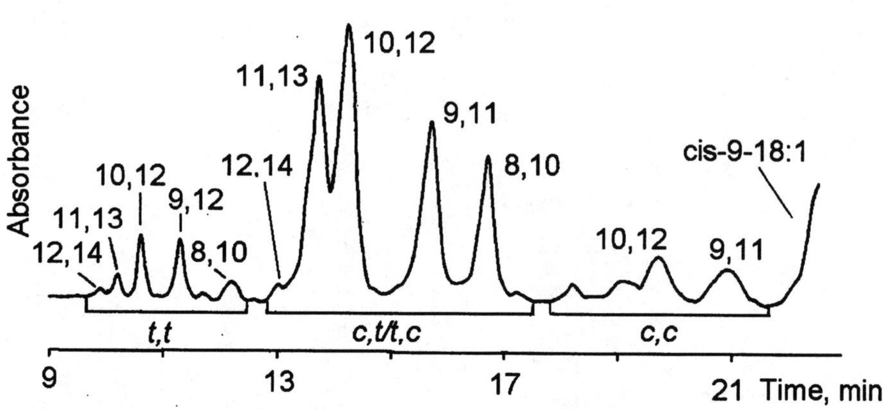

Fig. 9.10. Analysis of commercial sample of CLA as methoxyphenacyl esters by Ag-HPLC (Nikolova-Damyanova et al., 2000). Solvent: hexane-dichloromethane-acetonitrile (50:50:0.2; v/v/v; isocratic). UV detection at 270 nm. (Reproduced by kind permission of Journal of High Resolution Chromatography and redrawn from the original).

1999; Eulitz et al., 1999; Fritsche et al., 2000), resulting in improved separations, but also lengthy analysis times (increased to >120 min.)

Juanéda and Sébédio (1999) utilized a combination of RP- and Ag-HPLC to isolate the conjugated C_{20} metabolites from rats fed a diet of CLA isomers. The liver lipids were extracted, converted to CLAME and characterized by Ag-HPLC and GC-MS. Ag-HPLC allowed the isolation of such metabolites as *cis*-8,*trans*-12,*trans*-14-20:3, *cis*-5,*cis*-8,*trans*-12,*cis*-14-20:4 and *cis*-5,*cis*-8,*cis*-11,*trans*-13-20:4. Fritsche et al. (2000) used triple-column Ag-HPLC with UV detection (233 nm) to follow the influence of growth-promoting implants on composition of CLA isomers in beef fat. For quantitative determinations, a calibration plot of area under curves at 233 nm to mg CLA isomer was developed. The method was comparable to standard GC methods for quantitative determination of individual CLA isomers but, unlike GC, was considered more accurate due to lack of co-eluting FAME.

A combination of Ag-HPLC and GC was utilized by Philips and co-workers (1997) to directly determine *trans*-18:1 in diet composites, and by Kramer and co-workers (1998) to study the distribution of CLA isomers and metabolites in the tissue lipid classes of pigs fed a commercial CLA mixture. Steinhart and co-workers (2001) identified conjugated isomers of linolenic acid and arachidonic (20:4) acid in four groups of cheeses (Cheddar, Edam, Camembert and Cream cheese) by applying a combination of RP-HPLC, Ag-HPLC and GC. After extraction of the cheese fat and conversion to FAME, the FAME were prefractionated by preparative RP-HPLC and the 18:3/20:4 fraction further analyzed by Ag-HPLC (photodiode array detection to detect conjugation). Conversion to 4,4-dimethyloxazoline followed by GC-MS analysis allowed the

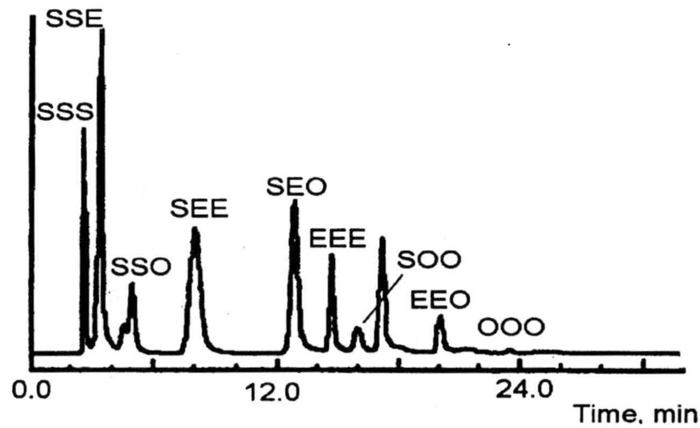

Fig. 9.11. Separation of triacylglycerols containing configurationally isomeric fatty acyl residues by silver ion HPLC on Chromspher Lipids column (Smith et al., 1994). Elution with (A), dichloromethane-dichloroethane, 1:1 (v/v); (B), acetone; (C), acetonitrile, and gradient from 98% A-2% B to 97% A-3% B over 5 min, to 40% A-60% B over 20 min and then to 80% B-30% C over further 8 min, evaporative light scattering detector. S, O, and E denoted saturated, oleic, and elaidic fatty acyl residues. (Reproduced by kind permission of the Journal of American Oil Chemists' Society and redrawn from the original).

identification of 13 conjugated isomers of arachidonic acid and 6 conjugated isomers of linolenic acid.

While most Ag-HPLC analyses have traditionally been done using CLAME, Nikolova-Damyanova et al. (2000) noted improved separation and quantization (single-column Ag-HPLC) of the *cis/trans* 8,10- through 11,13-18:2 isomers in a commercial sample of CLA, when phenacyl- or p-methoxyphenacyl-, rather than methyl ester derivatives, were analyzed (Fig. 9.10).

Resolution of the *cis/trans* isomers was similar to that obtained for CLAME using two or three Ag-HPLC columns connected in series (DelMonte et al., 2004a). A stepwise solvent gradient of 100% hexane-dichloromethane-acetonitrile (40:60:0.2; *v/v/v*; 30 min.) to 100% dichloromethane-acetonitrile (100:1) over 10 min and UV detection at 270 nm of the phenacyl esters yielded a semi-quantitative estimation of fatty acid composition comparable to, but more detailed than, GC analysis as CLAME. The method was also applied to a beef/pig fat sample, but pre-fractionation by RP-HPLC was required to concentrate the low levels of CLA isomers prior to analysis by Ag-HPLC.

The Ag-HPLC elution orders for the 18:2 conjugated and the 18:2 non-conjugated FAME are similar to the 9,12-18:2 GC patterns except that the *cis/trans*-18:2 elution orders are reversed (*trans/cis* elutes before *cis/trans*) and the *trans*-9,*trans*-11-18:2 and *cis*-9,*cis*-11-18:2 isomers can be readily separated from the *cis/trans* 9,11-isomer pair (Adlof and Lamm, 1998). While Ag-HPLC is capable of separating the non-conjugated *cis/trans* isomers of 9,12-18:2 FAME, it is less successful in separating the conjugated *trans*-9,*cis*-11 and *cis*-9,*trans*-

11-18:2 isomers, and also results in more overlap (FAME) between *cis*-9-18:1, and the *cis/trans* isomers of 9,11- and 9,12-18:2. This is in contrast to the GC elution order, where the majority of the *cis*-9,*cis*-11-18:2 isomers (as FAME) elute before the *cis*-9,*cis*-12-18:2 isomers. (See DelMonte et al., 2004b; and Adlof, 2003 for reviews of CLA analysis by Ag-HPLC.)

TAG

While Ag-HPLC has shown itself to be a very powerful technology for the separation of fatty acid isomers (as fatty acids and fatty acid esters) and symmetrical versus non-symmetrical TAG isomers (Fig. 9.2), its application to the analysis of more complex TAG formulations has been limited primarily to prefractionation of TAG mixtures for subsequent analysis by RP-HPLC, GC, FTIR, MS, or NMR (See review by Nikolova-Damyanova, 2003, and Limitations section, below.)

While Ag-HPLC has been applied to the separation of molecular species of *cis* and *trans*-containing TAGs in *trans*-hardened confectionary fats (Fig. 9.11; Smith et al., 1994) and partially hydrogenated vegetable oils (Macher and Holmquist, 2001), its application as a stand alone method may be applicable to the analysis of seed oil TAG samples containing only a limited number of FA isomers. Smith and co-workers (1994) presented an interesting comparison of the TAG separations achievable (saturated, *trans* and *cis* monoenoic FA mix)

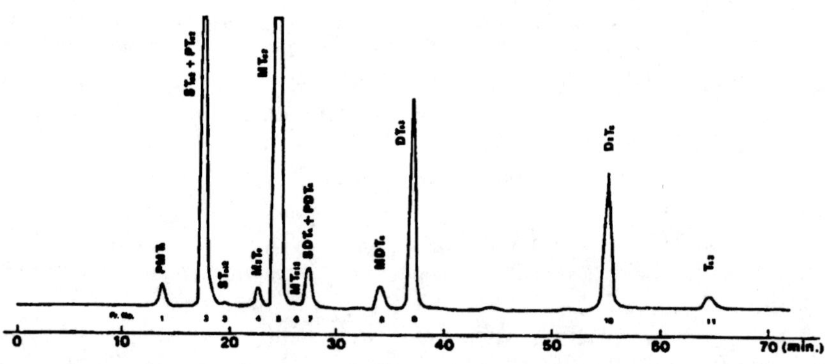

Fig. 9.12. Fractionation of TAG from seed oils of *Momordica charantia* by Ag-HPLC. [*P* = palmitic acid; S = stearic acid; M = monoenoic acids; D = dienoic acids; Tc = *cis*-9,*trans*-11,*trans*-13-18:3 ("-eleostearic acid); Tci = *cis*-9,*trans*-11,*cis*-13-18:3 (punicic acid) or *trans*-9, *trans*-11,*trans*-13-18:3??] Instrument: Hewlett-Packard HPLC Model 1050 w/quaternary delivery system. Column: Nucleosil 5 SA (250 x 4.6mm i.d.), impregnated with silver. Detector: Sedex 55 ELSD. Solvents: 1,2-dichloromethane/dichloroethane (1:1, v/v) A, acetone B and acetone/acetonitrile (9:1, v/v) C. Gradient: Linear gradient (1.0 mL/ min) from 100% A to 50% A/ 50% B over 5 min, then to 20% A/ 50% B/ 30% C over 60 min, to 50% B/ 50% C for 5 minutes more. Injection volume: 10 µL (1–2 mg/ mL in 1,2-dichloroethane). (Reproduced by kind permission of the Journal of Japan Oil Chemists' Society and redrawn from the original).

on the Chromspher Lipids Ag-HPLC (dichloroethane/dichloromethane/acetone/ ACN) column versus a silver nitrate on silica gel (hexane/ toluene/ ethyl acetate) column. They also note a gradual deterioration/decrease in sample retention times (number of samples not specified) over a several month period, but also noted that the Lipids columns could be regenerated 2–3 times to "as good as new" by flushing the Ag-HPLC column with volumes of ACN, methanol, HPLC-grade water, then with an aqueous solution of silver nitrate, followed (in reverse order) by water, methanol, and ACN. Christie (1988) utilized a 1:1 solution of methanol/ ACN and then silver nitrate in ACN to regenerate his columns.

Some success has recently been achieved utilizing Ag-HPLC for analysis of seed oils high in conjugated dienoic or trienoic fatty acids. Utilizing a Nucleosil (TM) 5 SA column (4.6 x 250 mm) saturated with silver ions and a complex solvent gradient of dichloromethane-dichloroethane (1:1) to acetone and acetone-acetonitrile (9:1)/programmed at 1 mL/min, Joh and Kin (1998) separated TAGs from the Chinese melon Fig. 9.12; Momordica charantia; and found 57.1 mol% conjugated trienoic FA, primarily "-eleostearic (*cis*-9,*trans*-11,*trans*-13-18:3) and punicic (*cis*-9,*trans*-11,*cis*-13-18:3), together with small amounts of the isomeric *cis*-9,*trans*-11,*trans*-13-18:3 FA.

TAG containing two conjugated trienoic and one dienoic fatty acid were found to elute much earlier than TAG containing two dienoic and one conjugated trienoic fatty acid, despite the former having the greater total number of double bonds. This elution order is similar to elution patterns noted for conjugated versus methylene-interrupted dienoic fatty esters (Adlof, 1994a), yet the limitations of the Ag-HPLC/isocratic solvent system for analysis of more complex TAG mixtures should also be recognized.

A similar problem exists in the analysis of isolated *trans*-containing TAG

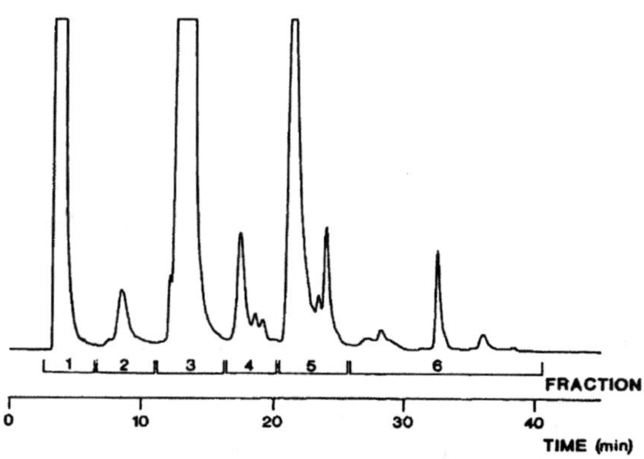

Fig. 9.13. Separation of butterfat triacylglycerols into six fractions according to unsaturation obtained by silver ion HPLC (Laakso and Kallio, 1993). Fractions/composition: 1 = SSS; 2 = SSMt ; 3 = SSMc ; 4 = SMcMt ; 5 = SMcMc ; 6 = More unsaturated molecules. [S = saturated; Mc = *cis*-monoenoic; Mt = *trans*-monoenoic fatty acyl residue.]

fractions from margarines or spreads. Reversed-phase HPLC, rather than Ag-HPLC, is often utilized to fractionate TAG by carbon number (FA chain length) and degree of unsaturation. The number of components is often so large that identification and quantification is difficult or even impossible. Ag-HPLC, which separates according to degree of unsaturation, produces much simpler and easier-to-interpret chromatograms than reverse-phase HPLC and thus provides unique information about such details (fingerprints) of the structure such as configuration and position of double bonds in a single acyl moiety or position of the fatty acid residues on the glycerol backbone (Laakso, 1996, Christie, 1995, Christie, 1997).

Laakso and co-workers (1992 and 1993) utilized a combination of preparative Ag-HPLC (Laakso and Kallio, 1993; Fig. 9.13) and RP-HPLC to the analysis of two solid and two liquid TAG fractions obtained from butterfat by an industrial melt process. Ag-HPLC was used to separate the TAG (by unsaturation/total # double bonds in TAG molecule(s)) into six subfractions, each of which was further fractionated (combined chain lengths) by RP-HPLC. Each fraction was then converted to FAME and analyzed by GC.

The combination of silver ion and reverse-phase chromatography may be utilized to provide information which cannot be achieved by these technologies alone. Ag-HPLC had been found comparable to the chemical (Nikolova-Damyanova et al., 1990) or enzymatic (Christie, 1987b) methods (selective removal of FA for further analysis by standard procedures such as GC or HPLC) used to determine the specific FA at the 1(3) or 2-position of a synthesized or isolated TAG (Adlof, 1995). Gas chromatography can be used to separate TAG positional isomers, but on-column injection and careful manipulation of

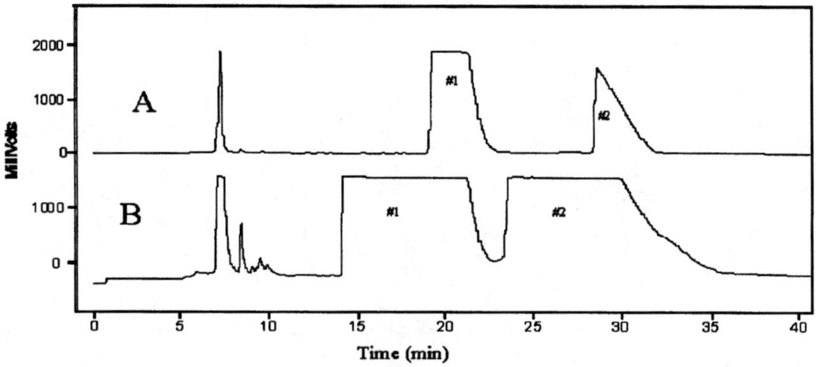

Fig. 9.14. Fractionation of CLA sample by dual-column Ag-HPLC system of a mixture of methyl 9,11-18:2 isomers by Ag-HPLC (Adlof, 2004). Sample sizes: 5 mg (A) and 10 mg (B). Flow rate: 1.0 mL/min 0.15% ACN in hexane, isocratic. UV detection at 251 (A) and 212 (B) nm. Peak 1: *trans*-9,*trans*-11-18:2; Peak 2: *cis*-9,*trans*-11-18:2. (Reproduced by kind permission of Journal of Chromatography A and redrawn from the original).

injector and column temperatures are necessary to minimize sample loss when the analyzed TAG contain FA with 2 or more double bonds (McMahon, 1985, Mares, et al., 1991).

Sample Loading

i. Analytical: While sample sizes utilized on the Chromspher Lipids columns vary from a few micrograms to milligrams, optimum sample sizes (dual-column Ag-HPLC) tend to be limited to 0.5 μg or less for optimum isomer separations. At these low sample loadings, Ag-HPLC (UV) data agreed well with results obtained by GC-FID for *trans*-Δ8/Δ9 through Δ12 (within 10–12% of peak area), but less so for the Δ13 and Δ14 *trans* isomers (10 – >50+% of peak area). The *cis* positional isomers equated with the GC-FID data to within 10% – 20% of peak area. (The Δ14 and Δ15 *cis* isomers could not be distinguished from baseline noise at these low sample loadings.) The 0.5–150 μg samples utilized by Adlof (Adlof, 1995; single and dual-column Ag-HPLC) may be contrasted with the 0.1–0.5 mg samples employed by other researchers (Christie et al., 1988; Nikolova-Damyanova et al., 1992; Toschi et al., 1993) using these types of columns. Significant resolution losses occurred when injected sample sizes were increased to even 0.1–0.2 milligrams. Sample sizes versus. FAME retention times (single Ag-HPLC column) have also been studied (Adlof et al., 1995).

ii. Semi-preparative: By manipulation of the percentage of ACN in the hexane, an Ag-HPLC system (dual-column) could be used to purify larger (8–10mg) samples of CLA isomers (Adlof, 2004). The fractionation of a 10 mg mixture of methyl *trans*-9,*trans*-11-18:2 and methyl *cis* -9,*trans*-11-18:2 on a dual-column Ag-HPLC system is depicted in Fig. 9.14.

A 1.0–2.5 mg sample of the same mixture could be fractionated within 35 minutes by using a solvent composition of ca. 0.3% ACN in hexane, while 5–7.5 mg samples were fractionated within 35 minutes at a solvent composition of 0.15% ACN in hexane. By adjustment of the percentage of ACN in the hexane, the separation of 20 μg–10 mg samples of conjugated FAME isomers was achieved. Purities of the isolated CLA isomers were found to be 97–99% by GC. Ag-HPLC was unable to resolve all of the 9,11 CLA isomers studied, but the syntheses can be manipulated (Adlof et al., 1997) to yield only one of the *cis/trans*, *trans/cis* isomer pair. Ag-HPLC can then be applied for the semi-preparative isolation of the desired isomer.

Ag-HPLC: Variables and Limitations.

i. Care should be taken to control solvent/column temperature(s) when

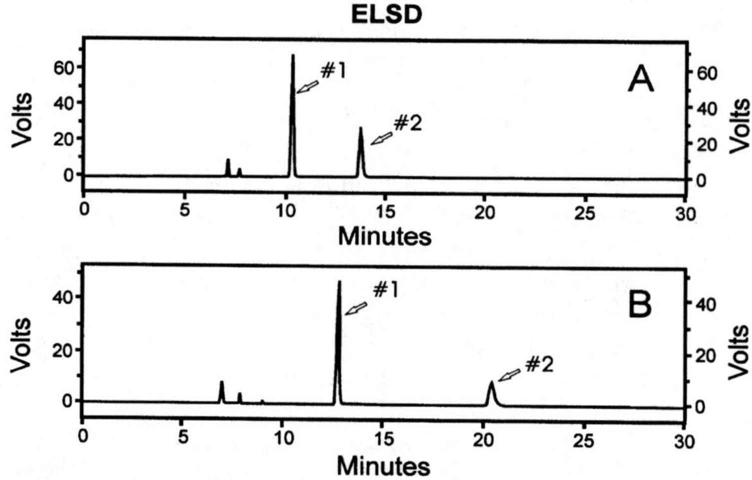

Fig. 9.15. Separation of FAME standard at 10°C (A) and 40°C (B) by dual-column Ag-HPLC; injection 2.0 μL (5mg/ mL in isooctane). Solvent: 1.0 mL/ min 1% ACN in hexane, isocratic elution. Detect1on: ELSD. Peak #1 = Methyl arachidonate (*cis*-5,*cis*-8,*cis*-11,*cis*-14-20:4) and #2 = Methyl docosahexaenoate (*cis*-4,*cis*-7,*cis*-10,*cis*-13,*cis*-16,*cis*-19-22:6).

utilizing Ag-HPLC and the isocratic ACN in hexane solvent system (Adlof and List, 2004), as evidenced by the changes in the retention times [FAME std. 18:0, oleate (*cis*-9-18:1), linoleate (*cis*-9,*cis*-12-18:2), arachidonate (*cis*-5,*cis*-8,*cis*-11,*cis*-14-20:4), and docosahexaenoate (DHA; *cis*-4,*cis*-7, *cis*-10,*cis*-13,*cis*-16, *cis*-19-22:6) illustrated in Fig. 9.15. An increase of column temperature from 10°C to 40°C resulted in an INCREASE in both TAG and FAME retention times, increases related primarily to the number of double bonds in the molecule(s). This is demonstrated by the 48% increase (10°C to 40°C) for FAME 22:6, while the retention times (10°C to 40°C) for 18:0 FAME remained essentially unchanged (Table 9.2).

Changes in column temperatures can even result in changes in elution orders, as evidenced during the analysis of the TAG OOP and PPL at 10°C, 20°C, 30°C and 40°C (Adlof, 2005; dual-column Ag-HPLC system, 0.7% ACN in hexane as solvent, ELSD). At 10°C, PPL eluted before OOP. As column temperature was increased from 10°C to 40°C, the retention of both TAG increased, but the rate of increase (retention) was greater for PPL, the TAG containing the FA with the largest number of double bonds in the 1-/3-position on the TAG glycerol backbone. This resulted in a reversal of elution orders. PPL, well-separated and eluting before OOP at 10°C, was found to merge with the OOP peak at 30°C, and to finally to elute after (as a well-defined shoulder) OOP at 40°C, a situation analogous to the temperature-dependent reversal of Me *cis*-9, *cis*-12, *cis*-15-18:3 and *cis*-11-20:1 observed by Wolff (1994) in gas chromatography (cyanoalkyl-polysiloxane capillary columns).

TABLE 9.2
Column Temperature (°C) versus Sample Retention
System: Dual-Column Ag-HPLC; 2.0 mL/min; 1.5 % ACN in hexane, ELSD Detection
Sample: 18:0, 18:1 and 18:3 TAG std

		Elution Times (min.)		
Run #	Temp (°C)	SSS(0)*	OOO(3)	LnLnLn(9)
1	10	6.7	7.5	12.6
2	20	7.0	8.1	15.9
3	30	7.5	9.4	21.6
4	40	7.7	10.3	28.2

*TAG (# double bonds)

ii. Butterfat (Fig. 9.13), has been shown (Marini, 2000) to be a mixture of some 100,000 TAG isomers containing some 400 different (FA chain length, saturation, position/number of sites of unsaturation, and/or FA location on the glycerol backbone, and illustrates the difficulties inherent in analyzing complex mixtures of TAG. If one assumed a totally random distribution of FA, there would, in theory, be more than 83 million (437^3) different TAG in buttermilk (Patton and Jensen, 1975).

Future Considerations

The application of Ag-HPLC as a stand-alone technique or one part of a series of tools for analysis of complex mixtures of *trans* isomers (as FAs, FA esters, or in TAG formulations) will continue to evolve. The development of standard Ag-HPLC methods (AOAC, 1990; AOCS, 1997) require reproducibility. To obtain reproducible results, potential sources of error should be addressed. These include a) Batch-to-batch variations in silver loading of the ChromSpher Lipids columns (Cross et al., 2000 and in private communications), b) Changes in sample retention times/ resolution due to i) size of sample injected (Adlof, 2004), ii) column usage/age, iii) and solvent(s) used/solvent composition changes over time (Yurawecz and Moorhouse, 2001), c) Lack of internal standards for column testing/quantitation, and d) Control of column temperature. Improvements in sample separations and column efficiencies require that factors like the composition and flow-rate of the mobile phase and column temperature be accurately controlled.

Some of these issues were recently addressed by Buchgraber and co-workers (Buchgraber and Ulberth, 2001; Buchgraber et al., 2004) in an intercomparison study to determine the *trans* octadecenoic acid content of soybean oil (SBO), spiked SBO, partially hydrogenated sunflower oil (PHSO) and a blend of palm oil and PHSO samples. A total of 12 laboratories participated, some utilizing Ag-TLC/GC and the others Ag-HPLC/GC. The precision and accuracy of both methods was found to be similar. AOCS Official Method Ce 1g-96 (1997) prescribes the use of Ag-HPLC, gradient elution, and detection by ELSD for separation of *trans* FA, but the precision of the method has yet to be established. Similar work must be done to determine the *trans* content in TAG formulations and, given the different effects linked to different isomers, the content of CLA isomers in foods, formulations and tissues. [See Daun and Cantrill (2003), for a detailed review of the process for developing and validating standard methods.]

Until these issues are addressed, quantitation by Ag-HPLC alone will remain a problem. Detectors have become somewhat standardized with ELSD, UV, and photodiode array detectors. Yet continual adjustment of solvent composition, the injection of standards (as in GC), or the addition of internal standards may be required to obtain reproducible separations and elution times. Addition of an internal standard (FA ester or TAG) to a given weight of *trans*-containing sample before analysis by Ag-HPLC (detection by ELSD) should allow calculation of total % *trans*, but only for samples containing only a limited number of *trans* isomers or TAG structures.

A final area for potential advancement is the application of nano technology to Ag-HPLC. Jeffrey (1991) utilized silver ion HPLC on a 3-micron silica (Nucleosil 100-3™) column impregnated with 10% (*w/w*) silver nitrate to separate positional isomers of triacylglycerols, while more recent advances in nano-HPLC pumps, columns and detectors, offer the option of 0.3 x 150mm or longer columns, Ag-impregnated, a situation similar to that encountered in gas chromatography with the advent of capillary GC versus packed-column technologies.

As noted by Nikolova-Damyanova et al. (2000), "On the other hand, there is considerable evidence that the combination of silver ion and reversed-phase separations provides information, which cannot be achieved by any of these when applied on their own." While the recent successful coupling of silver ion HPLC with mass spectrometric detection/identification presents new opportunities for lipid analysis, developments in nanotechnologies have allowed the coupling of silver-ion and reverse-phase HPLC (RP-HPLC) systems (two-dimensional HPLC). Mondello and co-workers (2004) recently utilized a combination of micro-bore Ag-HPLC and RP-HPLC with APCI-MS to completely separate and analyze rice oil TAG. The use "of a micro-bore column in the first dimension allowed the injection of a small sample volume onto the secondary column; the transfer of incompatible solvents from the first to the second dimension without deterioration in peak shapes or loss of resolution(s)."

References

Adlof, R.O. Fractionation of Egg and Soybean Phosphatidylcholines by Silver Resin Chromatography. *J. Chromatogr.* **1991,** *538*, 469–473.

Adlof, R.O. Separation of *cis* and *trans* Unsaturated Fatty Acid Methyl Esters by Silver ion High-performance Liquid Chromatography. *J. Chromatogr. A.* **1994a,** *659*, 95–99.

Adlof, R.O. The preparative separation of lipids by silver resin chromatography. In: *Separation Technology*. E.F. Vansant, Ed. Elsevier Science B.V. 1994b; pp. 777–781.

Adlof, R.O. Analysis of Triacylglycerol Positional Isomers by Silver Ion High Performance Liquid Chromatography. *J. High Resol. Chromatog.* **1995,** *18*, 105–107.

Adlof, R.O. Normal-phase Separation Effects with Lipids on a Silver Ion High-performance Liquid Chromatography Column. J. *Chromatogr. A.* **1997,** *764*, 337–340.

Adlof, R.O. Separation of Conjugated Linoleic Acid-containing Mono-, Di- and Triacylglycerols by Silver-ion High Performance Liquid Chromatography. *European Journal Lipids Sci. Technol.* **2001,** *103*, 614–617.

Adlof, R.O. Application of silver ion chromatography to the separation of conjugated linoleic acid isomers. In *Advances in Conjugated Linoleic Acid Research, Vol. 2*. Christie, W.; J. Sébédio; R. Adlof, Eds. AOCS Press: Champaign, IL, 2003; pp. 21–38.

Adlof, R.O. Separation of Conjugated Linoleic Acid Methyl Esters by Silver-ion High Performance Liquid Chromatography in Semi-preparative Mode. *J. Chromatogr. A.* **2004,** *1033*, 369–371.

Adlof, R.O. Recent advances in silver-ion HPLC utilizing acetonitrile in hexane as solvent: Temperature effects on lipid elution orders/ resolution. In *New Techniques and Applications in Lipid Analysis and Lipidomics*. Mossoba, M.; J. Kramer; W. Brenna; R. McDonald, Eds. AOCS Press: Champaign, IL, 2006.

Adlof, R.O.; W.R. Miller; E.A. Emken. Synthesis of *cis* and *trans* Methyl 8- and 13- Octadecenoate-d2 and d4 Isomers. *J. Label. Compd. Radiopharm.* **1978,** *15*, 625–636.

Adlof, R.O.; H. Rakoff; E.A. Emken. Partial Argentation Resin Chromatograph (PARC) I: Effect of Percent Silver on the Elution and Separation of Methyl Octadecadienoate Isomers. *J. Amer. Oil Chem. Soc.* **1980a,** *57*, 273–275.

Adlof, R.O.; Emken, E.A. Partial Argentation Resin Chromatography (PARC) II: Separation of Saturated and Mono-, Di-, Tri-, and Tetraenoic Fatty Esters. *J. Amer. Oil Chem. Soc.* **1980b,** *57*, 276–278.

Adlof, R.O.; E.A. Emken. Partial Argentation Resin Chromatography (PARC): III. The Effects of Sodium Ion Incorporation and Solvent on the Separation of Mixtures of Fatty Acids, of Fatty Esters, and of Triacylglycerols. *J. Amer. Oil Chem. Soc.* **1981,** *58*, 99–101.

Adlof, R.O.; L.C. Copes; E.A. Emken. Analysis of the Monoenoic Fatty Acid Distribution in Hydrogenated Vegetable Oils by Silver-ion High-performance Liquid Chromatography. *J. Amer. Oil Chem. Soc.* **1995,** *72*, 571–574.

Adlof, R.O.; E.L. Walter; E.A. Emken. Synthesis of Five Conjugated Linoleic Acid Isomers Labelled with Deuterium Atoms. Proceedings International Symposium on the Synthesis and Applications of Isotopes and Isotopically Labelled Compounds 6. 1997; pp. 387–390.

Adlof, R.O.; T. Lamm. Fractionation of *cis*- and *trans*- Oleic, Linoleic and Conjugated

Linoleic Fatty Acid Methyl Esters by Silver Ion High-performance Liquid Chromatography. *J. Chromatogr. A.* **1998,** *799*, 329–332.

Adlof, R.O.; S.M. Duval; E.A. Emken. Biosynthesis of Conjugated Linoleic Acid in Humans. *Lipids*. **2000,** *35*, 131–135.

Adlof, R.O.; L.C. Copes; E.L. Walter. Changes in Conjugated Linoleic Acid Composition Within Samples Obtained From a Single Source. *Lipids*. **2001,** *36*, 315–317.

Adlof, R.O.; G. List. Analysis of Triglyceride Isomers by Silver-ion High-performance Liquid Chromatography: Effect of Column Temperature on Retention Times. *J. Chromatogr. A.* **2004,** *1046*, 109–113.

AOAC. Isolated *trans* isomers in margarines and shortenings: Infrared spectroscopic method. AOAC Official Methods of Analysis 969 (1990).

AOCS. Sampling and analysis of commercial fats and oils. *Trans* fatty acids by silver-ion exchange HPLC. AOCS Recommended Practice Ce 1g-96 (1997).

Battaglia, R.; D. Frohlich. HPLC-Separation of *cis* and *trans* Monounsaturated Fatty Acids. *Chromatographia* **1980,** *13*, 428–431.

Buchgraber, M.; F. Ulberth; E.F. Anklam. Interlaboratory Evaluation of Injection Techniques for Triglyceride Analysis of Cocoa Butter by Capillary Gas Chromatography. *J. Chromatogr. A.* **2004,** *1036*, 197–203.

Buchgraber, M.; F. Ulberth. Determination of *trans* Octadecenoic Acids by Silver-ion Chromatography-Gas Liquid Chromatography: An Intercomparison of Methods. *J AOAC Intern.* **2001,** *84*, 1490–1498.

Christie, W.W. A Stable Silver-Loaded Column for the Separation of Lipids by High-Performance Liquid Chromatography. *J. High Res. Chromatog. Chomatogr. Comm.* **1987a,** *10*, 148–150.

Christie, W.W. *High Performance Liquid Chromatography and Lipids, A Practical Guide*. Pergamon Press: Oxford, 1987b.

Christie, W.W. Separation of Molecular Species of Triacylglycerols by High-performance Liquid Chromatography with a Silver Ion Column. *J. Chromatogr.* **1988,** *454*, 273–284.

Christie, W.W. Silver-ion High-performance Liquid Chromatography. In *New Trends In Lipid and Lipoprotein Analysis*. Sébédio, J.-L.; E.-G. Perkins, Eds. AOCS Press: Champaign, 1995; pp. 59–73.

Christie, W.W. Structural analysis of fatty acids. In *Advances in Lipid Methodology Four*, Christie, W.W., Ed. The Oily Press: Dundee, Scotland, 1997; pp. 119–169.

Christie, W.W. Molecular species of triacylglycerols, diacylglycerols derived from complex lipids and related lipids. In *Lipid Analysis, 3rd. Edn.* The Oily Press: Bridgwater, UK, 2003; pp. 302–342.

Christie, W.W.; G.H.M. Breckenridge. Separation of *cis* and *trans* Isomers of Unsaturated Fatty Acids by High-performance Liquid Chromatography in the Silver Ion Mode. *J. Chrom.* **1989,** *469*, 261–269.

Christie, W.W.; E.Y. Brechany; K. Stefanov. Silver Ion High-performance Liquid Chromatography and Gas Chromatography - Mass Spectrometry in the Analysis of Complex Fatty Acid Mixtures: Application to Marine Invertebrates. *Chem. Phys. Lip.* **1988,** *46*, 127–135.

Christie, W.W.; E.Y. Brechany; K. Stefanov; S. Popov. The Fatty Acids of the Dysidea Fragilis From the Black Sea. *Lipids*. **1992,** *27*, 640–644.

Clifton, P.M.; J.B. Keogh; M. Noakes. *trans* Fatty Acids in Adipose Tissue and the Food Supply are Associated with Myocardial Infarction. *J. Nutr.* **2004,** *134*, 874–879.

Cross, R.F.; E. Ostrowska; M. Muralitharan; F.R. Dunshea. Mixed Mode Retention and the Use of Competing Acid for the Ag-HPLC Analysis of Underivatized Conjugated Linoleic Acids. *J. High Res. Chromatog.* **2000,** *23*, 317–323.

Cruz-Hernandez, C.; Z. Dena; J. Zhou; A.R. Hill; M.P. Yurawecz; P. Delmonte; M.M. Mossoba; M.E.R. Dugan; J.K.G. Kramer. Methods for Analysis of Conjugated Linoleic Acids and *trans*-18:1 Isomers in Dairy Fats by Using a Combination of Gas Chromatography, Silver-ion Thin-layer Chromatography/Gas Chromatography, and Silver-ion Liquid Chromatography. *J. AOAC. Int.* **2004,** *87*, 545–560.

Daun, J.K.; R. Cantrill. Process for development of standard methods for the analysis of fats, oils and lipids. In *Advances in Lipid Methodology – Five.* Adlof, R., Ed. The Oily Press: Bridgwater, UK, 2003; pp. 273–299.

DeJarlais, W.J.; R.O. Adlof; E.A. Emken. Acetonitrile as Solvent in Silver Resin Chromatography. *J. Amer. Oil Chem. Soc.* **1983,** *60*, 975–978.

Delmonte, P.; J.A.G. Roach; M.M. Mossoba; G. Losi; M.P. Yurawecz. Synthesis, Isolation, and GC Analysis of All the 6, 8-to 13, 15-*cis*/*trans* Conjugated Linoleic Acid Isomers. *Lipids* **2004a,** *39*, 185–191.

Delmonte, P.; M.P. Yurawecz; M.M. Mossoba; C. Cruz-Hernandez; J.K.G. Kramer. Improved Identification of Conjugated Linoleic Acid Isomers Using Silver-ion HPLC Separations. *Journal of AOAC International.* **2004b,** *87*, 563–568.

De Ligny, C.L. The investigation of complex association by gas chromatography and related chromatographic and electrophoretic methods. In *Advances in Chromatography, Vol 14*. E. Grushka; J. Cazes; P.R. Brown; Eds. Marcel Dekker: New York, 1976; pp 284–294.

de Roos, N.M.; E.G. Schouten; M.B. Katan. *Trans* Fatty Acids, HDL-cholesterol, and Cardiovascular Disease. Effects of Dietary Changes on Vascular Reactivity. *Eur. J. Med. Res.* **2003,** *8*, 355–357.

Dionisi, F.; P.A. Dionisi; F.L.B. Golay. Influence of Milk Fat Presence on the Determination of *trans* Fatty Acids in Fats Used for Infant Formulae. *Analytical Chimica Acta* **2002,** *465*, 395–407.

Dlouhy, P.; E. Tvrzicka; B. Stankova; M. Vecka; A. Zak; Z. Straka; J. Fanta; J. Pachl; D. Kubisova; J. Rambouskova; D. Bilkova; M. Andel. Higher Content of 18:1 *trans* Fatty Acids in Subcutaneous Fat of Persons with Coronarographically Documented Atherosclerosis of the Coronary Arteries. *Ann. Nutr. Metab.* **2003,** *47*, 302–305.

Dobson, G.; W.W. Christie; B. Nikolova-Damyanova. Silver Ion Chromatography of Lipids and Fatty Acids. *J. Chromatogr. B.* **1995,** *671*, 198–222.

Dugo, P.; O. Favoino; P.Q. Tranchida; G. Dugo; L. Mondello. Off-line Coupling of Non-aqueous Reversed-phase and Silver Ion High-performance Liquid Chromatography-Mass Spectrometry For the Characterization of Rice Oil Triacylglycerol Positional Isomers. *J. Chromatogr. A.* **2004,** *1041*, 135–142.

Dyerberg, J.; D.C. Eskesen; P.W. Andersen; A. Astrup; B. Buemann; J.H. Christensen; P. Clausen; B.F. Rasmussen; E.B. Schmidt; T. Tholstrup; E. Toft; S. Toubro; S. Stender. Effects of *trans*- and n-3 Unsaturated Fatty Acids on Cardiovascular Risk Markers in Healthy Males. An 8 weeks Dietary Intervention Study. *Eur. J. Clin. Nutr.* **2004,** *58*, 1062–1070.

Elfman-Borjesson, I.; S. van den Hark; M. Harrod. Gradients of n-heptane and

Acetonitrile in Silver-ion High-performance Liquid Chromatography Analyses of *cis* and *trans* Bonds in Lipids. *J. Amer. Oil Chem. Soc.* **1997,** *74*, 1177–1180.

Emken, E.A.; C.R. Scholfield; H.J. Dutton. Chromatographic Separation of *cis* and *trans* by Argentation with a Macroreticular Exchange Resin. *J. Amer. Oil Chem. Soc.* **1964,** *41*, 388–390.

Eulitz, K.; M.P. Yurawecz; N. Sehat; J. Fritsche; J.A.G. Roach; M.M. Mossoba; J.K.G. Kramer; R.O. Adlof; Y. Ku. Preparation, Separation, and Confirmation of the Eight Geometrical *cis/trans* Conjugated Linoleic Acid Isomers 8,10 - through 11,13-18:2. *Lipids*. **1999,** *34*, 873–877.

Evershed, R.P. In *Lipid Analysis, A Practical Approach*. R.J. Hamilton; S. Hamilton, Eds. Oxford University Press: New York, 1993; pp 130–133.

Fevrier, P.; A. Binet; L. Dufosse; R. Gree; F. Yvergnaux. Separation of Glyceride Positional Isomers by Silver Ion Chromatography. *J. Chromatogr. A.* **2001,** *923*, 53–57.

Firestone, D.; A. Sheppard. Determination of *trans* fatty acids. In *Advances in Lipid Methodology - One* W.W. Christie, Ed. The Oily Press: Ayr, UK, 1992; pp 273–322.

Frenking, G.; N. Frohlich. The Nature of the Bonding in Transition-metal Compounds, *Chem. Rev.*, **2000,** *100*, 717–774.

Fritsche, J.; H. Steinhart. Analysis of *trans* fatty acids. In *New Techniques and Applications in Lipid Analysis*. McDonald, R.E.; M.M. Mossoba, Eds. AOCS Press: Champaign, IL, 1997; pp 234–255.

Fritsche, J.; S. Fritsche; M.B. Solomon; M.M. Mossoba; M.P. Yurawecz; Y. Ku. Quantitative Determination of Conjugated Linoleic Acid Isomers in Beef Fat. *Eur. J. Lipid Sci. Technol.* **2000,** *102*, 667–672.

Gavino, V.C.; G. Gavino; M.J. Leblanc; B. Tuchweber. An Isomeric Mixture of Conjugated Linoleic Acids But Not Pure *cis*-9, *trans*-11-octadecadienoic Acid Affects Body Weight Gain and Plasma Lipids in Hamsters. *J. Nutr.* **2000,** *130*, 27–29.

Gunstone F.D.; I.A. Ismail; M. Lie Ken Jie. Fatty Acids, Part 16: Thin-layer and Gas-liquid Chromatographic Properties of the *cis* and *trans* Methyl Octadecenoates and of Some Acetylenic Esters. *Chem. Phys. Lipids.* **1967,** *1*, 376–385.

Gunstone, F.D.; M. Lie Ken Jie. Fatty Acids, Part 26. Thin Layer and Gas Liquid Properties of Some *cis*-octadecenyl Compounds. *Chem. Phys. Lipids.* **1970,** *4,* 139–146.

Han, J.J.; Y. Iwasaki; T. Yamane. Use of Isopropanol as a Modifier in a Hexane-acetonitrile Based Mobile Phase For the Silver Ion HPLC Separation of Positional Isomers of Triacylglycerols Containing Long Chain Polyunsaturated Fatty Acids. *J. High Resol. Chromatog.* **1999,** *22*, 357–361.

Heath, R.R.; J.H. Tumlinson; R.E. Doolittle. Analytical and Preparative Separation of Geometrical Isomers by High Efficiency Silver Nitrate Liquid Chromatography. *J. Chrom. Sci.* **1977,** *15*, 10–13.

Jeffrey, B.S.J. Silver-complexation Liquid Chromatography For Fast, High-resolution Separations of Triacylglycerols. *Soc J. Amer. Oil Chem.* **1991,** *68*, 289–293.

Joh, Y.-G.; S.-J. Kim. Analysis of Molecular Species of Triacylglycerols From Vegetable Oils Containing Fatty Acids with Non-methylene-interrupted Double Bonds, by HPLC in the Silver-ion Mode. *J. Jpn. Oil Chem. Soc.* **1998,** *47*, 927–

936.

Juanéda, P.; O. Cordier; S. Gregoire; J.-L. Sébédio. Conjugated Linoleic Acid (CLA) Isomers in Heat-treated Vegetable Oils. *Corps Gras Lipides.* **2001,** *8*, 94–97.

Juanéda, P. Utilisation of Reversed-phase High-performance Liquid Chromatography as an Alternative to Silver-ion Chromatography for the Separation of *cis*- and *trans*-C18:1 Fatty Acid Isomers. *J. Chromatogr. A.* **2002,** *954*, 285-289.

Juanéda, P.; J.-L. Sébédio. Combined Silver-ion and Reversed-phase High-performance Liquid Chromatography for the Separation and Identification of C20 Metabolites of Conjugated Linoleic Acid Isomers in Rat Liver Lipids. *J. Chromatogr. B.* **1999,** *724*, 213–219.

Juanéda, P.; J.-L. Sébédio; W.W. Christie. Complete Separation of the Geometrical Isomers of Linolenic Acid by High Performance Liquid Chromatography with a Silver Ion Column. *J. High Resol. Chromatog.* **1994,** *17*, 321–324.

Jung, M.Y.; M.O. Jung. Identification of Conjugated Linoleic Acids in Hydrogenated Soybean Oil by Silver Ion-impregnated HPLC and Gas Chromatography-ion Impacted Mass Spectrometry of Their 4,4-dimethyloxazoline Derivatives. *J. Agric. and Food Chem.* **2002,** *50*, 6188–6193.

Kasai, P.H.; D. McLeod; T. Watanabe. Acetylene and Ethylene Complexes of Copper and Silver Atoms, Matrix Isolation ESR Study. *J Amer Oil Chem Soc* **1980,** *102*, 179–190.

Katan, M.B. *trans* fatty acids and plasma lipoproteins. Nutrition Reviews. **2000,** *58*, 188–191.

Kim, H.S.; J.H. Ryu; H. Kim; B.S. Ahn; Y.S. Kang. Reversible Olefin Complexation by Silver Ions in Dry Poly(vinyl methyl ketone) Membrane and its Application to Olefin/Paraffin Separations. *Chem. Commun.* **2000,** *14*, 1261–1262.

Kramer, J.K.G.; C. Cruz-Hernandez; Z.Y. Deng; J.Q. Zhou; G. Jahreis; M.E.R Dugan. Analysis of Conjugated Linoleic Acid and *trans* 18:1 Isomers in Synthetic and Animal Products. *Am. J. Clin. Nutr.* **2004,** *79*, 1137s–1145s.

Kramer, J.K.G.; N. Sehat; M.E.R. Dugan; M.M. Mossoba; M.P. Yurawecz; J.A.G. Roach; K. Eulitz; J.L. Aalhus; A.L. Schaefer; Y. Ku. Distributions of Conjugated Linoleic Acid (CLA) Isomers in Tissue Lipid Classes of Pigs Fed a Commercial CLA Mixture Determinated by Gas Chromatography and Silver Ion-high-performance Liquid Chromatography. *Lipids.* **1998,** *33*, 549–558.

Laakso, P. Analysis of Triacylglycerols - Approaching the Molecular Composition of Natural Mixtures. *Food Rev. Int.* **1996,** *12*, 199–250.

Laakso, P.H.; K.V.V. Nurmela; D.R. Homer. Composition of the Triacylglycerols of Butterfat and its Fractions Obtained by an Industrial Melt Crystallization Process. *J. Agric. Food Chem.* **1992,** *40*, 2472–2482.

Laakso, P.H.; H. Kallio. Triacylglycerols of Winter Butterfat Containing Configurational Isomers of Monoenoic Fatty Acyl Residues. II. Saturated Dimonoenoic Triacylglycerols. *J. Amer. Oil Chem. Soc.* **1993,** *70*, 1173–1176.

Lanser, A.C.; E.A. Emken. Purification of Methyl Arachidonate Using Silver Resin Chromatography. *J. Chromatogr.* **1983,** *256*, 460–464.

Lamaitre, R.N.; I.B. King; T.E. Raghunathan; R.M. Pearce; S. Weinmann; R.H. Knopp; M K. Copass; L.A. Cobb; D.S. Siscovick. Cell Membrane *trans*-fatty Acids and the Risk of Primary Cardiac Arrest. *Circulation.* **2002,** *105*, 697–701.

Ledoux, M.; L. Laloux; R.L. Wolff. Analytical Methods for Determination of *trans*-C18

Fatty Acid Isomers in Milk Fat. A Review. *Analusis*. **2000,** *28*, 402–412.

Louter, A.J.H.; H.G. Janssen; H. Steenbergen; E. Floter. Comprehensive LCxGC: A Valuable Tool for the Detailed Characterization of Edible Oils and Fats. *Lipid Technol.* **2003,** *5*, 64–67.

Macher, M.-B.; A. Holmquist. Triacylglycerol Analysis of Partially Hydrogenated Vegetable Oils by Silver Ion HPLC. *J. Sep. Sci.* **2001,** *24*, 179–185.

Mares, P.; T. Rezanka; M. Novak. Analysis of Human Blood Plasma Triacylglycerols Using Capillary Gas Chromatography, Silver Ion Thin-layer Chromatographic Fractionation and Desorption Chemical Ionization Mass Spectrometry. *J. Chromatogr.* **1991,** *568*, 1–10.

Marini, D. HPLC of lipids. In *Food Analysis by HPLC, 2nd Edn*. L.M.L. Nollet, Ed. Marcel Dekker Inc.: New York, 2000; pp 160–249.

McMahon, D.H. A Collaborative Study to Evaluate Quantitation Utilizing Different Injection Modes for Capillary GC. *J. Chromatogr. Sci.* **1985,** *23*, 137–143.

Mikes, F.; V. Schurig; E. Gil-Ave. Complex-forming Stationary Phases in High-speed Liquid Chromatography. *J. Chromatogr. A.* **1973,** *83*, 91–97.

Molkentin, J.; D. Precht. Content of Individual *cis/trans* Isomers of 16:1, 18:1 and 18:2 Fatty Acids in the Reference Milk Fat CRM164. *Kieler Milchwirtschafliche Forschungsberichte* **2004,** *56*, 53–63.

Momchilova, S.; B. Nikolova-Damyanova; W.W. Christie. Silver Ion High-performance Liquid Chromatography of Isomeric *cis*- and *trans*-Octadecenoic Acids. Effect of the Ester Moiety and Mobile Phase Composition. *J. Chromatogr. A.* **1998,** *793*, 275–282.

Momchilova, S.; B. Nikolova-Damyanova. Silver-ion HPLC of p-Methoxyphenacyl Derivatives of Unsaturated Fatty Acids. I. Mobile Phase Effects. *J. Liq. Chrom. Rel. Technol.* **2000a,** *23*, 2303–2316.

Momchilova, S.; B. Nikolova-Damyanova. Silver-ion HPLC of p-Methoxyphenacyl Derivatives of Unsaturated Fatty Acids. II. Chain Length versus. Double Bond Position. *J. Liq. Chrom. and Rel. Technol.* **2000b,** *23*, 2317–2325.

Momchilova, S.; B. Nikolova-Damyanova. Silver Ion HPLC of p-Methoxyphenacyl Derivatives of Unsaturated Fatty Acids. III. Mobile Phase Effects on *trans* 6-,9- and 11-18:1. *J. Liq. Chrom. and Rel. Technol.* **2002a,** *25*, 615–625.

Momchilova, S.; B. Nikolova-Damyanova. Silver Ion High Performance Liquid Chromatographic Separation of Naturally Occurring Positionally Isomeric Monoenoic Fatty Acids as p-Methoxyphenacyl Esters: The Mobile Phase Effects. *Compt. Rand. Acad. Bul. Sci.* **2002b,** *55*, 47–50.

Mondello, L.; A. Caslli; P.Q. Tranchida; R. Costa; B. Chiofalo; P. Dugo; G. Dugo. Evaluation of Fast Gas Chromatography and Gas Chromatography-Mass Spectrometry in the Analysis of Lipids. *J. Chromatogr. A* **2004,** *1035*, 237–247.

Mossoba, M.M.; R.E. McDonald; D.J. Armstrong; S.W. Page. Hydrogenation of Soybean Oil: A TLC and GC/matrix Isolation/Fourier Transform Infrared Study. *J. Agric. Food Chem.* **1991,** *39*, 695–699.

Nikolova-Damyanova, B. Lipid analysis by silver ion chromatography. In *Advances in Lipid Methodology –Five*. Adlof, R.O., Ed., PJ Barnes,The Oily Press: Ayr, UK, 2003; pp 43–123.

Nikolova-Damyanova, B.; W.W. Christie; B. Herslof. The Structure of the Triacylglycerols of Meadowfoam Oil. *J. Amer. Oil Chem. Soc.* **1990,** *67*, 503–507.

Nikolova-Damyanova, B.; W.W. Christie; B. Herslof. Silver Ion High-performance Liquid Chromatography of Derivates of Isomeric Fatty Acids. *J. Chromatogr.* **1992,** *609*, 133–140.

Nikolova-Damyanova, B.; S. Momchilova; W.W. Christie. Silver-ion High-performance Liquid Chromatographic Separation of Conjugated Linoleic Acid Isomers and Other Fatty Acids, After Conversion to Methoxyphenacyl Derivatives. *J. High Resol. Chromatog.* **2000,** *23*, 348–352.

Nikolova-Damyanova, B.; S. Momchilova. Silver ion HPLC for the Analysis of Positionally Isomeric Fatty Acids. *J. Liq. Chromatog. and Rel. Technol.* **2002,** *25*, 1947–1965.

Nonadek, L. Complexation in chromatography. In *Complexation Chromatography.* Cagniant, D., Ed. Marcel Dekker: New York, 1991; pp 1–31.

Ostrowska, E.; F.R. Dunshea; M. Muralitharan; R.F. Cross. Comparison of Silver-ion High-performance Liquid Chromatographic Quantification of Free and Methylated Conjugated Linoleic Acids. *Lipids*. **2000,** *35*, 1147–1153.

Ozawa, I.; M. Kim; K. Saito; K. Sugita; T. Baba; S. Moriyama; T. Sugo. Purification of Docosahexaenoic Acid Ethyl Ester Using a Silver-ion -immobilized Porous Hollow-fiber Membrane Module. *Biotechnol. Prog.* **2001,** *17*, 893–896.

Park, Y.; J.M. Storkson; K.J. Albright; W. Liu; M.W. Pariza. Evidence That the *trans*-10,*cis*-12 Isomer of Conjugated Linoleic Acid Induces Body Composition Changes in Mice. *Lipids*. **1999,** *34*, 235–241.

Patton, S.; R.G. Jensen Lipid Metabolism and Membrane Functions of the Mammary Gland. *Prog. Chem. Fats Other Lipids*. **1975,** *14*, 163–277.

Phillips, K.M.; R.D. Lugogo; R.F. Harris; M.-T. Tarrago-Trani; J.A. Bailey; K.K. Steward. Direct Determination of *trans*-Octadecenoic Acid in Diet Composites Using a Combination of Silver Ion-high Performance Liquid Chromatography and Capillary Gas Chromatography. *J. Food Lipids.* **1997,** *4*, 173–188.

Powell, W.S. Separation of Icosenoic Acids, Monoxydroxyicoesenoic Acids and Prostaglandins by High-performance Liquid Chromatography on a Silver Loaded Cation-exchange Column. *Anal. Biochem.* **1981,** *115*, 267–277.

Powell, W.S. Argentation-high-pressure Liquid Chromatography of Prostaglandins and Monohydroxyeicosenoic Acids. *Methods in Enzymology.* **1982,** *86*, 530–543.

Precht, D.; J. Molkentin. Overestimation of Linoleic Acid and *trans*-C18:2 Isomers in Milk Fats with Emphasis on *trans* Delta 9, *trans* Delta 12-octadecadienoic Acid. *Milchwiss.-Milk Sci. Int.* **2003,** *58*, 30–34.

Ratnayake, W.M.N. Analysis of *trans* fatty acids. In *trans Fatty Acids in Human Nutrition.* Sébédio, J.-L.; W.W. Christie, Ed. Oily Press: Dundee, UK, 1998; pp 115–161.

Ratnayake, W.M.N. Analysis of Dietary *trans* Fatty Acids. *J. Oleo Sci.* **2001,** *50*, 73–86.

Ratnayake, W.M.N. Reliable Methods for the Determination of *trans* Fatty Acids and Conjugated Linoleic Acid Isomers: an Overview. *J. AOAC. Int.* **2004a,** *87*, 520–522.

Ratnayake, W.M.N. Overview of Methods for the Determination of *trans* Fatty Acids by Gas Chromatography, Silver-ion Thin-layer Chromatography, Silver-ion Liquid Chromatography, and Gas Chromatography/Mass Spectrometry. *J. AOAC. Int.* **2004b,** *87*, 523–539.

Ratnayake, W.M.N.; G. Pelletier. Positional and Geometrical Isomers of Linoleic Acid in Partially Hydrogenated Oils. *J. Amer. Oil Chem. Soc.* **1992,** *69*, 95–105.

Sanders, T.A.B.; F.R. Oakley; D. Crook; J.A. Cooper; G.J. Miller. High Intakes of *trans* Monounsaturated Fatty Acids Taken for 2 Weeks do not Influence Procoagulant and Fibrinolytic Risk Markers for CHD in Young Healthy Men. *Br. J. Nutr.* **2003,** *89*, 767-776.

Sehat, N.; R. Rickert; M.M. Mossoba; J.K.G. Kramer; M.P. Yurawecz; J.A.G. Roach; R.O. Adlof; K.M. Morehouse; J. Fritsche; K.D. Eulitz; H. Steinhart; Y. Ku. Improved Separation of Conjugated Fatty Acid Methyl Esters by Silver Ion-high-performance Liquid Chromatography. *Lipids.* **1999,** *34*, 407–413.

Slover, H.T.; R.H. Thompson, Jr.; C.S. Davis; G.V. Merola. Lipids in Margarines and Margarine-like Foods. *J. Amer. Oil Chem. Soc.* **1985,** *62*, 775–786.

Smedman, A.; B. Vessby; S. Basu. Isomer-specific Effects of Conjugated Linoleic Acid on Lipid Peroxidation in Humans: Regulation by alpha-tocopherol and Cyclo-oxygenase-2 Inhibitor. *Clin. Sci.* **2004,** *106*, 67–73.

Smith, K.W.; J.M. Perkins; B.S.J. Jeffrey; D.L. Phillips. Separation of Molecular Species of *cis*- and *trans*- triacylglycerols in *trans*-hardened Confectionery Fats by Silver-ion High-performance Liquid Chromatography. *J. Amer. Oil Chem. Soc.* **1994,** *71*, 1219-1222.

Steinhart, H.; K. Winkler; R. Rickert. *trans*- and Conjugated Fatty Acids in Food-contents and Analytical Aspects. *Corps Gras Lipides.* **2001,** *8*, 29–32.

Steinhart, H.; R. Rickert; K. Winkler. *trans* FATTY acids (TFA) in Food of Plant Origin: Analysis, Occurrence, Intake, and Clinical Relevance. *Polish J. Food Nutr. Sci.* **2002,** *11*, 82–85.

Toschi, T.G.; P. Capella; C. Holt; W.W. Christie. A Comparison of Silver Ion HPLC Plus GC With Fourier-Transform IR Spectroscopy for the Determination of *trans* Double Bonds in Unsaturated Fatty Acids. *J. Sci. Food Agric.* **1993,** *61*, 261–266.

Ulberth, F.; M. Henninger. Simplified Method for the Determination of *trans* Monoenes in Edible Fats by TLC-GLC. *J. Amer. Oil Chem. Soc.* **1992,** *69*, 829–831.

Ulberth, F.; M. Buchgraber. Analytical Platforms to Assess the Authenticity of Cocoa Butter. *Eur. J. Lipid Sci. Technol.* **2003,** *105*, 32–42.

Wolff, R.L. Analysis of alpha-linolenic Acid Geometrical Isomers in Deodorized Oils by Capillary GC on Cyanoalkyl Polysiloxane Stationary Phases: a Note of Caution. *J. Amer. Oil Chem. Soc.* **1994,** *71*, 907–909.

Wood, E.J. *Trans* Fats: the Wrong Isomers. Retinoids and Lipid-Soluble Vitamins in Clinical Practice. **2004,** *20*, 54–57.

Yurawecz, M.P.; K. Morehouse. Silver-ion HPLC of Conjugated Linoleic Acid Isomers. *Eur. J. Lipid Sci. Technol.* **2001,** *103*, 594–632.

Index

A

B

C

D

E

F

G

H

I

L

M

N

O

P

R

S